VOLUME SIX HUNDRED AND TWENTY FIVE

METHODS IN ENZYMOLOGY

DNA Sensors and Inflammasomes

METHODS IN ENZYMOLOGY

VOLUME SIX HUNDRED AND TWENTY FIVE

METHODS IN ENZYMOLOGY

DNA Sensors and Inflammasomes

Edited by

JUNGSAN SOHN
Department of Biophysics and Biophysical Chemistry, Johns Hopkins University School of Medicine, Baltimore, MD, United States

ELSEVIER

ACADEMIC PRESS
An imprint of Elsevier

Academic Press is an imprint of Elsevier
50 Hampshire Street, 5th Floor, Cambridge, MA 02139, United States
525 B Street, Suite 1650, San Diego, CA 92101, United States
The Boulevard, Langford Lane, Kidlington, Oxford OX5 1GB, United Kingdom
125 London Wall, London, EC2Y 5AS, United Kingdom

First edition 2019

ISBN: 978-0-12-818359-5
ISSN: 0076-6879

For information on all Academic Press publications
visit our website at https://www.elsevier.com/books-and-journals

Publisher: Zoe Kruze
Acquisition Editor: Ashlie M. Jackman
Editorial Project Manager: Shellie Bryant
Production Project Manager: James Selvam
Cover Designer: Greg Harris

Typeset by SPi Global, India

Contents

Contributors

Brendan Antiochos
Johns Hopkins Division of Rheumatology, Baltimore, MD, United States

Petr Broz
Department of Biochemistry, University of Lausanne, Lausanne, Switzerland

Michael A. Chattergoon
Johns Hopkins University School of Medicine, Baltimore, MD, United States

Yashao Chen
Key Laboratory of Applied Surface and Colloid Chemistry, Ministry of Education, School of Chemistry and Chemical Engineering, Shaanxi Normal University, Xi'an, China

Ileana M. Cristea
Department of Molecular Biology, Princeton University, Princeton, NJ, United States

Eva de Alba
Department of Bioengineering, School of Engineering, University of California, Merced, CA, United States

Pedro Diaz-Parga
Department of Bioengineering, School of Engineering; Quantitative Systems Biology Graduate Program, University of California, Merced, CA, United States

Saskia F. Erttmann
Department of Molecular Biology, Umeå University, Umeå, Sweden

Nelson O. Gekara
The Laboratory for Molecular Infection Medicine Sweden (MIMS), Umeå Centre for Microbial Research (UCMR), Umeå University, Umeå; Department of Molecular Biosciences, The Wenner-Gren Institute, Stockholm University, Stockholm, Sweden

Patricia Grob
Howard Hughes Medical Institute, UC Berkeley, Berkeley, CA, United States

Justin Hall
Worldwide Medicinal Chemistry, Pfizer, Groton, CT, United States

Nicole Haloupek
Department of Molecular and Cell Biology, UC Berkeley, Berkeley, CA, United States

Min Hao
Key Laboratory of Applied Surface and Colloid Chemistry, Ministry of Education, School of Chemistry and Chemical Engineering, Shaanxi Normal University, Xi'an, China

Jörg S. Hartig
Department of Chemistry; Konstanz Research School Chemical Biology (KoRS-CB), University of Konstanz, Konstanz, Germany

Wayne O. Hemphill
Department of Biochemistry, Wake Forest School of Medicine, Winston-Salem, NC, United States

Sebastian Hiller
Biozentrum, University of Basel, Basel, Switzerland

Richard Hooy
Department of Biophysics and Biophysical Chemistry, John Hopkins University School of Medicine, Baltimore, MD, United States

Timothy R. Howard
Department of Molecular Biology, Princeton University, Princeton, NJ, United States

Hui Jiang
The Laboratory for Molecular Infection Medicine Sweden (MIMS), Umeå Centre for Microbial Research (UCMR), Umeå University, Umeå, Sweden

Philip J. Kranzusch
Department of Microbiology, Harvard Medical School; Department of Cancer Immunology and Virology, Dana-Farber Cancer Institute; Parker Institute for Cancer Immunotherapy at Dana-Farber Cancer Institute, Boston, MA, United States

Alexey Krasnoslobodtsev
Nanoimaging Core Facility at the University of Nebraska Medical Center; Department of Physics, University of Nebraska Omaha, Omaha, NE, United States

Karolin Luger
Department of Biochemistry; Howard Hughes Medical Institute, University of Colorado Boulder, Boulder, CO, United States

Alexander Lushnikov
Nanoimaging Core Facility at the University of Nebraska Medical Center, Omaha, NE, United States

Guido Massaccesi
Johns Hopkins University School of Medicine, Baltimore, MD, United States

Mariusz Matyszewski
Department of Biophysics and Biophysical Chemistry, Johns Hopkins University School of Medicine, Baltimore, MD, United States

Zachary Mazanek
Department of Biophysics and Biophysical Chemistry, Johns Hopkins University School of Medicine, Baltimore, MD, United States

Adam Mazur
Biozentrum, University of Basel, Basel, Switzerland

Ramya Nandakumar
Department of Biomedicine; Aarhus Research Center for Innate Immunology, University of Aarhus, Aarhus, Denmark

Eva Nogales
Department of Molecular and Cell Biology; Howard Hughes Medical Institute, UC Berkeley; Molecular Biophysics and Integrated Bioimaging Division, Lawrence Berkeley National Lab, Berkeley, CA, United States

Søren R. Paludan
Department of Biomedicine; Aarhus Research Center for Innate Immunology, University of Aarhus, Aarhus, Denmark

Swarupa Panda
The Laboratory for Molecular Infection Medicine Sweden (MIMS), Umeå Centre for Microbial Research (UCMR), Umeå University, Umeå, Sweden

Fred W. Perrino
Department of Biochemistry, Wake Forest School of Medicine, Winston-Salem, NC, United States

Qianqian Qi
Key Laboratory of Applied Surface and Colloid Chemistry, Ministry of Education, School of Chemistry and Chemical Engineering, Shaanxi Normal University, Xi'an, China

Erik C. Ralph
Worldwide Medicinal Chemistry, Pfizer, Groton, CT, United States

Jianbin Ruan
Program in Cellular and Molecular Medicine, Boston Children's Hospital; Department of Biological Chemistry and Molecular Pharmacology, Harvard Medical School, Boston, MA, United States

Johannes Rudolph
Department of Biochemistry, University of Colorado Boulder, Boulder, CO, United States

Jungsan Sohn
Department of Biophysics and Biophysical Chemistry, John Hopkins University School of Medicine, Baltimore, MD, United States

Bokai Song
Department of Molecular Biology, Princeton University, Princeton, NJ, United States

Patrycja Swacha
The Laboratory for Molecular Infection Medicine Sweden (MIMS), Umeå University, Umeå, Sweden

Jeannette Tenthorey
Department of Molecular and Cell Biology, UC Berkeley, Berkeley, CA, United States

Russell E. Vance
Department of Molecular and Cell Biology; Howard Hughes Medical Institute, UC Berkeley, Berkeley, CA, United States

Changhao Wang
Key Laboratory of Applied Surface and Colloid Chemistry, Ministry of Education, School of Chemistry and Chemical Engineering, Shaanxi Normal University, Xi'an, China

Aaron T. Whiteley
Department of Microbiology, Harvard Medical School; Department of Cancer Immunology and Virology, Dana-Farber Cancer Institute, Boston, MA, United States

Samuel Joseph Windross
Department of Biomedicine; Aarhus Research Center for Innate Immunology, University of Aarhus, Aarhus, Denmark

Hao Wu
Program in Cellular and Molecular Medicine, Boston Children's Hospital; Department of Biological Chemistry and Molecular Pharmacology, Harvard Medical School, Boston, MA, United States

Shiyu Xia
Program in Cellular and Molecular Medicine, Boston Children's Hospital; Department of Biological Chemistry and Molecular Pharmacology, Harvard Medical School, Boston, MA, United States

Wen Zhou
Department of Microbiology, Harvard Medical School; Department of Cancer Immunology and Virology, Dana-Farber Cancer Institute, Boston, MA, United States

Preface

The innate immune system plays a central role in defense against various maladies, ranging from pathogen invasion to cancer development. In this volume of *Methods in Enzymology*, we present cutting-edge techniques that can delineate how two key players of innate immunity operate at the molecular level, *DNA sensors and inflammasomes.*

We first open the volume with a minireview on the cGAS-STING signaling axis, which is the central pathway that initiates type-1 interferon responses against cytoplasmic DNA. We then delve into biochemical methods for tracking the signaling activity of the cGAS-STING pathway in vitro, which include protein purification and crystallography (Chapter 2), second messenger synthesis (Chapter 3), and enzymatic assays (Chapters 4 and 5). Next, we describe biochemical methods for monitoring the signaling activities of inflammasomes, which are filamentous signaling platforms that trigger interleukin and cell-death responses against various maladies such as cytoplasmic DNA and bacterial toxins (Chapters 6 and 7). We then describe methods for tracking biochemical activities of sensors implicated in degrading cytoplasmic DNA (Chapter 8) and repairing endogenous DNA damage (Chapter 9).

The next five chapters discuss methods for visualizing the structures of DNA sensors and inflammasomes at various resolutions: atomic force microscopy (Chapter 10), negative-stain electron microscopy (nsEM; Chapter 11), cryo-EM (Chapter 12), an integral approach combining solid-state nuclear magnetic resonance (NMR) and cryo-EM (Chapter 13), and solution NMR (Chapter 14). The final six chapters then describe *in cellulo* methods for tracking DNA sensors and inflammasomes: visualizing DNA·sensor complexes via confocal microscopy (Chapter 15), tracking type-1 interferon responses (Chapter 16), monitoring inflammasome formation (Chapter 17), tracking DNA damage (Chapter 18), deciphering protein–protein interaction networks (i.e., proteomics; Chapter 19), and finally monitoring ubiquitination (Chapter 20).

I hope this compendium will be of interest to a diverse group of researchers who study DNA sensors and inflammasomes. I also hope it will facilitate cross-disciplinary approaches in the field.

JUNGSAN SOHN
Department of Biophysics and Biophysical Chemistry,
Johns Hopkins University School of Medicine,
Baltimore, MD, United States

CHAPTER ONE

Intercellular communication in the innate immune system through the cGAS-STING pathway

Ramya Nandakumar[a,b], Samuel Joseph Windross[a,b], Søren R. Paludan[a,b,*]
[a]Department of Biomedicine, University of Aarhus, Aarhus, Denmark
[b]Aarhus Research Center for Innate Immunology, University of Aarhus, Aarhus, Denmark
*Corresponding author: e-mail address: srp@biomed.au.dk

Contents

Abstract

Type I interferons (IFNα/β) are pivotal to anti-viral defense. Although an absolute requirement for viral clearance, the presence of IFN in high doses sustained over time has been implicated to be the underlying cause of clinical outcomes such as interferonopathies. Therefore, tight regulation of cellular communication processes is required for tissue homeostasis. This review covers the cellular signaling pathways initiated by DNA sensing through cGAS in the cellular cytosol, with particular focus on the subsequent establishment of immune responses in bystander cells.

1. Introduction

The innate immune system exerts defense against infections, but an unregulated immune response resulting in chronic type I interferon (IFN) production leads to inflammatory diseases (Crow & Manel, 2015; Rice et al., 2013).

Methods in Enzymology, Volume 625
ISSN 0076-6879
https://doi.org/10.1016/bs.mie.2019.07.007

Under both physiological and pathological conditions, cellular communication and cooperation is an absolute prerequisite in the maintenance and restoration of homeostasis. Communication between cells is mediated not only through direct cell-to-cell contact but also through secretion and export of small molecules, proteins and vesicles to the extracellular milieu.

The most widely acknowledged mode of cellular cross talk, in response to cellular insult or injury, is through secreted cytokines and chemokines. Of these, the most clinically relevant in the context of viral infections, are the secreted IFN family of cytokines and the resulting activation of interferon stimulated genes (ISGs), which orchestrate anti-viral and anti-proliferative activities. Of the three classes of interferon, type I (IFN α/β) and types III (IFN λ) IFNs are the most relevant in the context of immediate innate immune activation and the subsequent induction of the adaptive arm. Type II interferons (IFN-γ), which act primarily to drive undifferentiated Th0 CD4+ cells into Th1 differentiated cells. The effect of IFN-γ is mainly associated with the adaptive arm of the immune response (da Silva, Fonseca, Alvarez, & D'Imperio Lima, 2015).

Pathogen associated molecular patterns (PAMPs) or cellular danger associated molecular patterns (DAMPs) recognized by distinct pattern recognition receptors (PRRs) lead to the induction of IFN. Secreted IFN can signal through autocrine and paracrine ways to continue and propagate an immune response mediated by induction of IFN-stimulated genes (ISGs) in infected as well as neighboring cells. Although the field of IFN signaling has been one of intense research, the list of PRRs and its effectors is still growing.

In this article, we review the effectors of intercellular communication in innate immunity, and discuss how recent advances in this field have led to new insights into how signals are propagated in a multicellular environment in response to immune recognition of infection.

2. cGAS-STING signaling

PRRs are localized in different cellular compartments, and recognize a broad range of molecular signatures leading to host defense and inflammation. For instance, Toll-like receptors localize to the plasma membrane and endosomes, and cyclic-GMP-AMP (cGAMP) synthase (cGAS) localizes to the cytoplasm (Alexopoulou, Holt, Medzhitov, & Flavell, 2001; Gurtler & Bowie, 2013; Paludan & Bowie, 2013; Sun, Wu, Du, Chen, & Chen, 2013). Therefore, PAMP compartmentalization is a determinant for immune activation. Foreign DNA in the cytoplasm is reminiscent of a virus, bacteria, or parasite challenge. While the detection and subsequent immune response might result in pathogen restriction, the detection of host DNA in the cytoplasm can trigger autoimmune conditions such as Sjøgren's syndrome and Aicardi-Goutières syndrome in humans (Crow & Manel, 2015).

Cytoplasmic DNA is detected by cGAS, acknowledged now as the universal sensor of cytosolic DNA, although other sensors and accessory molecules are also involved in full immunological response to DNA (Almine et al., 2017; Civril et al., 2013; Jonsson et al., 2017; Sun et al., 2013). Detection of dsDNA in the cytosol by cGAS, results in the structural reorganization of the catalytic pocket allowing for the synthesis of second messenger 2′3′-cyclic guanosine-adenosine mono phosphate (2′3′-cGAMP) from guanosine-5′-triphosphate and adenosine-5′-triphosphate (Ablasser et al., 2013a; Gao et al., 2013). cGAMP binds to the ER resident protein STING to activate recruitment of kinase TBK1 to IRF3 phosphorylating it. This results in the subsequent nuclear translocation of IRF3 resulting in the transcription of immune and inflammatory genes. Structural elucidation of co-crystallized cGAS-DNA complex revealed electrostatic interactions and hydrogen bonds between cGAS and the sugar phosphate backbone of B-DNA in a relatively sequence—independent manner (Civril et al., 2013; Gao et al., 2013; Li et al., 2013).

The activation of cGAS and the ensuing IFN response, although DNA sequence-independent displays length-dependency (Andreeva et al., 2017; Luecke et al., 2017). This observation is validated by the increased amount of cGAMP produced in a DNA-length dependent manner, plausibly due to longer DNA being relatively more efficient at sequestrating DNA: cGAS complexes within confined phase separated liquid droplets (Du & Chen, 2018).

Binding of cGAMP to STING, allows for a conformational rearrangement that dissociates STING from STIM1, an ER calcium sensor protein, with a newly ascribed role as the exclusive STING-ER retention factor (Srikanth et al., 2019). Binding of STING to STIM1 in resting state maintains cellular equilibrium, as studies using cells from SAVI patients; a disease characterized by chronic STING activation and IFN production, was shown to disrupt the STIM1-STING interaction leading to uncontrolled translocation of STING from ER to ER-golgi intermediate compartment (ERGIC).

Translocation of STING to the golgi membrane enables the palmitoylation of STING, wherein palmitic acid moieties attach to cysteine residues (C88/91) (Mukai et al., 2016). STING then recruits downstream proteins TBK1 and IRF3, which binds STING and phosphorylates Interferon regulatory factor (IRF3). Nuclear translocation of IRF3 leads to the transcription of type I IFN genes as well as inflammatory cytokines.

2.1 Inter-cellular communication

A key purpose of intercellular communication in the immune system is to protect uninfected cells and tissues from infection and damage. While there is strong evidence for the induction of immune activation in infected cells, knowledge on the mechanisms through which anti-viral programs are disseminated to bystander cells has emerged in greater detail in recent years.

2.2 Type 1 IFN

The IFN receptor ligation and subsequent JAK-STAT activation results in the transcription of several hundred ISGs, inducing an antiviral state in the cell. Although all three families of IFNs stimulate the induction of an overlapping set of ISGs, their transcription has been observed to be cell-type and virus specific.

ISGs mainly act through the restriction of viral replication, transcription and virion formation and release. The mode of inhibition for ISG families such as PKR (protein kinase R), Mx (Myxovirus resistance) and Viperin (Virus inhibitory protein, endoplasmic reticulum-associated, interferon-inducible) are well documented (Schneider, Chevillotte, & Rice, 2014). However, in depth mechanisms for ISG inhibition of both well-studied and emerging viruses continue to be described.

The Mx2 gene, which encodes the restriction factor MxB, is part of the family of dynamin like large GTPases. However, unlike the closely related MxA, it has recently been linked to the restriction of HSV1/HSV2 through inhibition of genome uncoating at the nuclear envelope. Subsequent knockdown of MxB demonstrated a 10-fold increase in viral titer in cells pretreated with IFN (Crameri et al., 2018). Recent studies on ZIKA virus have demonstrated the ability of the NS5A protein to degrade STAT2 and perturb IFN production. IFN production and subsequent transcription of ISGs has been shown to restrict ZIKA virus replication (Bowen et al., 2017). Viperin, a known restriction factor against other flaviviridae family members, is observed at elevated levels in ZIKA infected cells (Van der Hoek et al., 2017). Cells expressing Viperin conferred lower amounts of ZIKA RNA, suggesting possible association with viral replication complexes as a mode of restriction. Subsequent mutational analysis identified the C terminal domain as critical for Viperin mediated restriction of ZIKA virus (Van der Hoek et al., 2017).

3. Immuno-signaling to neighboring cells through gap junctions

A more recently identified mode of direct cellular signaling, is the transfer of the second messenger cGAMP across plasma membranes through gap junctions into adjacent cells (Fig. 1). Structurally similar to bacterially derived *c*-di-AMP and *c*-di-GMP, cGAMP diffuses through

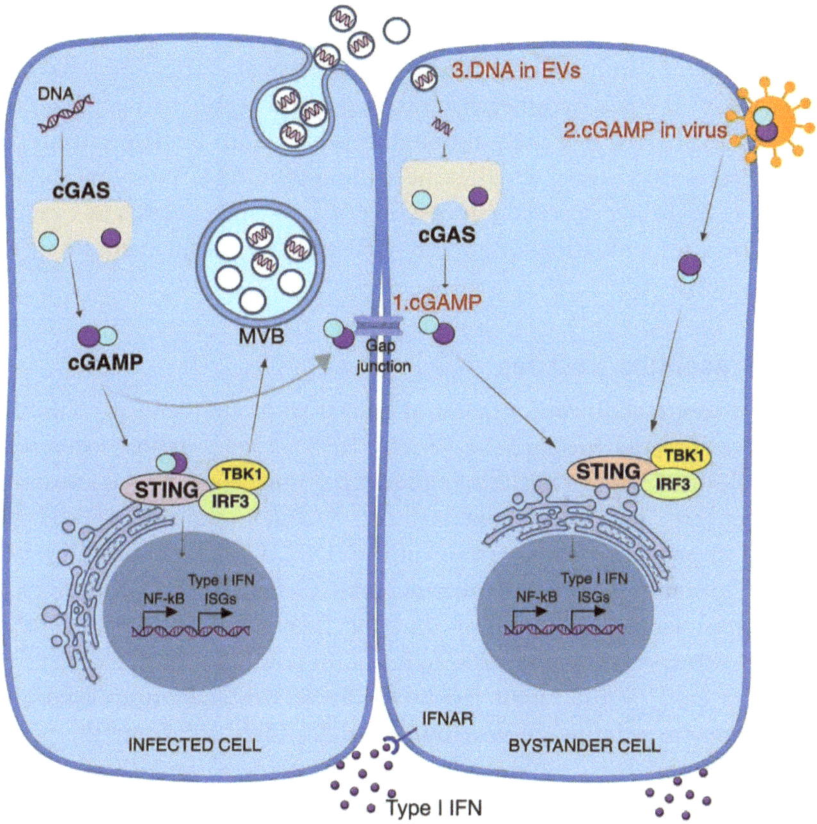

Fig. 1 Mechanisms of intracellular communication in the STING pathway. Cytosolic DNA in infected or DNA-stressed cells is recognized by cGAS, which results in the synthesis of second messenger cGAMP. (1) cGAMP activates the ER-associated protein STING in the cell, but the cyclic-dinucleotide can also traverse gap junctions to activate STING in neighboring cells. (2) Distant cells can also be conferred protection by viral particles budding out of infected cells that can package cGAMP, which is released into bystander cells during infection. (3) Additionally, pathogenic DNA can be packaged into EVs which when fused with bystander cells, activates cGAS to induce an IFN response in the cell.

gap junctions to neighboring cells and activates STING signaling (Ablasser et al., 2013b; Bridgeman et al., 2015; Gentili et al., 2015). This transfer and the subsequent STING-mediated response, leads to the establishment of an anti-viral state in bystander cells preparing it for a potential pathogen challenge.

3.1 cGAMP in virus particles

An additional mode of intercellular communication in the early immune response to infections is through packaging of cGAMP into virus particles, thus leading to transfer of danger signals to the target cells along with the infectious agent (Bridgeman et al., 2015; Gentili et al., 2015). This has been reported for HIV and cytomegalovirus, and could represent a general principle in the antiviral response, although this has yet to be demonstrated. It will be interesting to learn whether the loading of cGAMP into viral particles is an active or passive process. Hitchhiking of anti-viral effectors in viral particles allows for induction of an anti-viral state in distant tissues and also between hosts.

3.2 Extracellular vesicles

The most recent addition to the list of mediators of intercellular communication, are secreted extracellular vesicles (EVs). Of substantial interest are the double-membrane vesicles of endocytic origin referred to as exosomes. Exosomes are nano-sized extracellular vesicles, typically 50–150 nm enriched in proteins such as tetraspanins (CD81, CD63), ESCRT- related family of proteins, heat-shock proteins (HSP70 and HSP90), proteins of the cytoskeleton (actin and tubulin), and the Rab family of small GTPases (Thery, Ostrowski, & Segura, 2009). Exosomes distinguish themselves from other EVs (such as microvesicles) by virtue of size and origin (exosomes have endocytic origins whereas microvesicles bleb out from the plasma membrane), and have different subsets of protein markers (Kowal et al., 2016; Raposo & Stoorvogel, 2013; Thery et al., 2009; Yuana, Sturk, & Nieuwland, 2013). However, there is considerable overlap of characteristics between the different EVs isolated using the current methods of isolation.

First recognized in reticulocytes to be involved in elimination of unwanted cellular content (Harding, Heuser, & Stahl, 1984; Pan & Johnstone, 1983) exosomes have since been shown to facilitate cell-to-cell communication. Once released from infected cells, exosomes can activate bystander cells by aiding in receptor-ligand interactions between vesicular and receptor cell surfaces, such as the interaction of MIIC of B-cells derived exosomes with T-cells resulting in their activation (Raposo et al., 1996). Additionally, vesicles from infected cells, internalized by bystander cells resulting in the cytosolic release of vesicular mRNA, DNA and miRNA can alter the immune-regulatory protein landscape in these cells (Dreux et al., 2012; Mittelbrunn et al., 2011; Nandakumar et al., 2019).

4. Exosomes and immune responses to virus infections

Given that viruses entering cells via endocytosis can converge at the endosome where intra-luminal vesicles are formed, it is conceivable that viruses evolved to hijack this communication system. Indeed, exosomes have been found to shuttle infectious Hepatitis-A virus particles, that are capable of establishing productive infection while being resistant to neutralizing antibodies (Feng et al., 2013; Ramakrishnaiah et al., 2013).

Co-culture experiments with Hepatitis-C virus infected hepatoma cells and pDCs revealed the presence of a mechanism that induced transfer of viral RNA to pDCs resulting in secretion of high levels of IFN (Dreux et al., 2012; Pan & Johnstone, 1983). The IFN-inducing ability of pDCs directly correlated with the amount of HCV RNA in the hepatoma cells.

Packaging of viral RNA into vesicles from the cytosol of infected cells into exosomes and their delivery into pDCs are a way of "presenting" viral genome to initiate an interferon response. Although it is tempting to speculate that this could be a general strategy employed by the host for anti-viral defense, not all RNA virus infected cells elicit this response.

Not surprisingly, viruses seem to hijack the communicative property of exosomes to further infection. Studies have described exosome mediated transfer of Hepatitis-C virus to uninfected cells where it could establish persistent infection (Ramakrishnaiah et al., 2013). This "Trojan horse" mechanism enables the virus to travel undetected in the blood stream and infect distant cells. In line with this finding, plasma from HCV-infected patients has been shown to contain viral RNA (Masciopinto et al., 2004). Additionally, it has also been hypothesized that the transfer of anti-viral effectors and viral nucleic acids may help in silencing viral genes aiding establishment of latency (Kalamvoki, Du, & Roizman, 2014).

5. Exosomes and immune responses to bacterial infections

Bacteria can also subvert or avert the exosome pathway to alter host defense to facilitate bacterial dissemination. THP1 cells stimulated with exosomes derived from *M. tuberculosis* or *S. typhimurium* infected THP1 cells induce a robust pro-inflammatory TNF-α response (Bhatnagar, Shinagawa, Castellino, & Schorey, 2007; Singh, LeMaire, Tan, Zeng, & Schorey, 2011). Additionally, mice administered with EVs from *M. tuberculosis*-infected macrophages, showed increased recruitment of neutrophils and macrophages in their lungs (Bhatnagar et al., 2007).

The more recent discovery of pathogenic DNA hitchhiking within extracellular vesicles released from *Listeria* infected cells acts as an emerging mode of cellular communication (Nandakumar et al., 2019). This work revealed that secretion of pathogenic DNA present in the cellular cytosol could encapsulate in vesicles within multivesicular bodies (MVB) to be secreted into extracellular space. These vesicles, internalized by bystander cells could activate a cGAS-mediated IFN response (Fig. 1).

DNA sequestration into vesicles was dependent on STING and phosphorylation by TBK1. Phosphorylation by TBK1 of protein MVB12b, a component of the ESCRT machinery was a prerequisite for such a packaging and transfer phenomenon and phosphorylation of Serine residue at position S222 was critical for this function. How exactly DNA from the cytosol is recognized, cleaved and is packaged into vesicles is still elusive. Importantly, in vivo experiments demonstrated that the bacteria benefitted from activation of the exosome-mediated amplification of the immune response. This was at least partly due to exosome-mediated elevation of STING-dependent T cell apoptosis, thus hampering anti-bacterial defense. Since a key response evoked in cells receiving DNA-containing exosomes was expression of type I IFN, it will be interesting to learn how this phenomenon impacts control of virus infections.

6. Concluding remarks

The ability of cells to communicate is essential for rapid spread of danger and defense signals following detection of pathogenic microorganisms. In this review, we have provided an overview of how the cGAS-STING system activates danger signaling on both the infected and bystander cells. The IFN system is adept at recognizing pathogen attack and is capable of inducing a suitable anti-viral response. However, many pathogens have co-evolved with the host to evade and successfully escape host defense. The release of vesicles loaded with cellular/pathogenic cargo to be exported into distant uninfected cells independent of cellular contact enables establishment of anti-viral pathways, which precede a possible infection.

Conversely, although viruses could deplete the IFN pathway in the infected cell, the neighboring cells could be instrumental in restoring anti-viral defense in the infected cell via anti-viral molecules packaged in exosomes.

The loading of ligands and receptors along with cytosolic DNA, RNA and proteins in a protected environment with the added ability to travel to distant sites and deliver these signals appears to be an attempt at furthering their cellular function.

References

Ablasser, A., et al. (2013a). cGAS produces a 2'-5'-linked cyclic dinucleotide second messenger that activates STING. *Nature, 498*, 380–384. https://doi.org/10.1038/nature12306.

Ablasser, A., et al. (2013b). Cell intrinsic immunity spreads to bystander cells via the intercellular transfer of cGAMP. *Nature, 503*, 530–534. https://doi.org/10.1038/nature12640.

Alexopoulou, L., Holt, A. C., Medzhitov, R., & Flavell, R. A. (2001). Recognition of double-stranded RNA and activation of NF-kappaB by Toll-like receptor 3. *Nature, 413*, 732–738. https://doi.org/10.1038/35099560.

Almine, J. F., et al. (2017). IFI16 and cGAS cooperate in the activation of STING during DNA sensing in human keratinocytes. *Nature Communications, 8*, 14392. https://doi.org/10.1038/ncomms14392.

Andreeva, L., et al. (2017). cGAS senses long and HMGB/TFAM-bound U-turn DNA by forming protein-DNA ladders. *Nature, 549*, 394–398. https://doi.org/10.1038/nature23890.

Bhatnagar, S., Shinagawa, K., Castellino, F. J., & Schorey, J. S. (2007). Exosomes released from macrophages infected with intracellular pathogens stimulate a proinflammatory response in vitro and in vivo. *Blood, 110*, 3234–3244. https://doi.org/10.1182/blood-2007-03-079152.

Bowen, J. R., et al. (2017). Zika virus antagonizes type I interferon responses during infection of human dendritic cells. *PLoS Pathogens, 13*, e1006164. https://doi.org/10.1371/journal.ppat.1006164.

Bridgeman, A., et al. (2015). Viruses transfer the antiviral second messenger cGAMP between cells. *Science, 349*, 1228–1232. https://doi.org/10.1126/science.aab3632.

Civril, F., et al. (2013). Structural mechanism of cytosolic DNA sensing by cGAS. *Nature, 498*, 332–337. https://doi.org/10.1038/nature12305.

Crameri, M., et al. (2018). MxB is an interferon-induced restriction factor of human herpesviruses. *Nature Communications, 9*, 1980. https://doi.org/10.1038/s41467-018-04379-2.

Crow, Y. J., & Manel, N. (2015). Aicardi-Goutieres syndrome and the type I interferonopathies. *Nature Reviews. Immunology, 15*, 429–440. https://doi.org/10.1038/nri3850.

da Silva, H. B., Fonseca, R., Alvarez, J. M., & D'Imperio Lima, M. R. (2015). IFN-gamma priming effects on the maintenance of effector memory CD4(+) T cells and on phagocyte function: Evidences from infectious diseases. *Journal of Immunology Research, 2015*, 202816. https://doi.org/10.1155/2015/202816.

Dreux, M., et al. (2012). Short-range exosomal transfer of viral RNA from infected cells to plasmacytoid dendritic cells triggers innate immunity. *Cell Host & Microbe, 12*, 558–570. https://doi.org/10.1016/j.chom.2012.08.010.

Du, M., & Chen, Z. J. (2018). DNA-induced liquid phase condensation of cGAS activates innate immune signaling. *Science, 361*, 704–709. https://doi.org/10.1126/science.aat1022.

Feng, Z., et al. (2013). A pathogenic picornavirus acquires an envelope by hijacking cellular membranes. *Nature, 496*, 367–371. https://doi.org/10.1038/nature12029.

Gao, P., et al. (2013). Cyclic [G(2',5')pA(3',5')p] is the metazoan second messenger produced by DNA-activated cyclic GMP-AMP synthase. *Cell, 153*, 1094–1107. https://doi.org/10.1016/j.cell.2013.04.046.

Gentili, M., et al. (2015). Transmission of innate immune signaling by packaging of cGAMP in viral particles. *Science, 349*, 1232–1236. https://doi.org/10.1126/science.aab3628.

Gurtler, C., & Bowie, A. G. (2013). Innate immune detection of microbial nucleic acids. *Trends in Microbiology, 21*, 413–420. https://doi.org/10.1016/j.tim.2013.04.004.

Harding, C., Heuser, J., & Stahl, P. (1984). Endocytosis and intracellular processing of transferrin and colloidal gold-transferrin in rat reticulocytes: Demonstration of a pathway for receptor shedding. *European Journal of Cell Biology, 35*, 256–263.

Jonsson, K. L., et al. (2017). IFI16 is required for DNA sensing in human macrophages by promoting production and function of cGAMP. *Nature Communications, 8*, 14391. https://doi.org/10.1038/ncomms14391.

Kalamvoki, M., Du, T., & Roizman, B. (2014). Cells infected with herpes simplex virus 1 export to uninfected cells exosomes containing STING, viral mRNAs, and microRNAs. *Proceedings of the National Academy of Sciences of the United States of America, 111*, E4991–E4996. https://doi.org/10.1073/pnas.1419338111.

Kowal, J., et al. (2016). Proteomic comparison defines novel markers to characterize heterogeneous populations of extracellular vesicle subtypes. *Proceedings of the National Academy of Sciences of the United States of America, 113*, E968–E977. https://doi.org/10.1073/pnas.1521230113.

Li, X., et al. (2013). Cyclic GMP-AMP synthase is activated by double-stranded DNA-induced oligomerization. *Immunity, 39*, 1019–1031. https://doi.org/10.1016/j.immuni.2013.10.019.

Luecke, S., et al. (2017). cGAS is activated by DNA in a length-dependent manner. *EMBO Reports, 18*, 1707–1715. https://doi.org/10.15252/embr.201744017.

Masciopinto, F., et al. (2004). Association of hepatitis C virus envelope proteins with exosomes. *European Journal of Immunology, 34*, 2834–2842. https://doi.org/10.1002/eji.200424887.

Mittelbrunn, M., et al. (2011). Unidirectional transfer of microRNA-loaded exosomes from T cells to antigen-presenting cells. *Nature Communications, 2*, 282. https://doi.org/10.1038/ncomms1285.

Mukai, K., et al. (2016). Activation of STING requires palmitoylation at the Golgi. *Nature Communications, 7*, 11932. https://doi.org/10.1038/ncomms11932.

Nandakumar, R., et al. (2019). Intracellular bacteria engage a STING-TBK1-MVB12b pathway to enable paracrine cGAS-STING signalling. *Nature Microbiology, 4*, 701–713. https://doi.org/10.1038/s41564-019-0367-z.

Paludan, S. R., & Bowie, A. G. (2013). Immune sensing of DNA. *Immunity, 38*, 870–880. https://doi.org/10.1016/j.immuni.2013.05.004.

Pan, B. T., & Johnstone, R. M. (1983). Fate of the transferrin receptor during maturation of sheep reticulocytes in vitro: Selective externalization of the receptor. *Cell, 33*, 967–978.

Ramakrishnaiah, V., et al. (2013). Exosome-mediated transmission of hepatitis C virus between human hepatoma Huh7.5 cells. *Proceedings of the National Academy of Sciences of the United States of America, 110*, 13109–13113. https://doi.org/10.1073/pnas.1221899110.

Raposo, G., & Stoorvogel, W. (2013). Extracellular vesicles: Exosomes, microvesicles, and friends. *The Journal of Cell Biology, 200*, 373–383. https://doi.org/10.1083/jcb.201211138.

Raposo, G., et al. (1996). B lymphocytes secrete antigen-presenting vesicles. *The Journal of Experimental Medicine, 183*, 1161–1172.

Rice, G. I., et al. (2013). Assessment of interferon-related biomarkers in Aicardi-Goutieres syndrome associated with mutations in TREX1, RNASEH2A, RNASEH2B, RNASEH2C, SAMHD1, and ADAR: A case-control study. *Lancet Neurology*, *12*, 1159–1169. https://doi.org/10.1016/S1474-4422(13)70258-8.

Schneider, W. M., Chevillotte, M. D., & Rice, C. M. (2014). Interferon-stimulated genes: A complex web of host defenses. *Annual Review of Immunology*, *32*, 513–545. https://doi.org/10.1146/annurev-immunol-032713-120231.

Singh, P. P., LeMaire, C., Tan, J. C., Zeng, E., & Schorey, J. S. (2011). Exosomes released from *M. tuberculosis* infected cells can suppress IFN-gamma mediated activation of naive macrophages. *PLoS One*, *6*, e18564. https://doi.org/10.1371/journal.pone.0018564.

Srikanth, S., et al. (2019). The Ca(2+) sensor STIM1 regulates the type I interferon response by retaining the signaling adaptor STING at the endoplasmic reticulum. *Nature Immunology*, *20*, 152–162. https://doi.org/10.1038/s41590-018-0287-8.

Sun, L., Wu, J., Du, F., Chen, X., & Chen, Z. J. (2013). Cyclic GMP-AMP synthase is a cytosolic DNA sensor that activates the type I interferon pathway. *Science*, *339*, 786–791. https://doi.org/10.1126/science.1232458.

Thery, C., Ostrowski, M., & Segura, E. (2009). Membrane vesicles as conveyors of immune responses. *Nature Reviews. Immunology*, *9*, 581–593. https://doi.org/10.1038/nri2567.

Van der Hoek, K. H., et al. (2017). Viperin is an important host restriction factor in control of Zika virus infection. *Scientific Reports*, *7*, 4475. https://doi.org/10.1038/s41598-017-04138-1.

Yuana, Y., Sturk, A., & Nieuwland, R. (2013). Extracellular vesicles in physiological and pathological conditions. *Blood Reviews*, *27*, 31–39. https://doi.org/10.1016/j.blre.2012.12.002.

CHAPTER TWO

Analysis of human cGAS activity and structure

Wen Zhou[a,b], Aaron T. Whiteley[a,b], Philip J. Kranzusch[a,b,c,*]
[a]Department of Microbiology, Harvard Medical School, Boston, MA, United States
[b]Department of Cancer Immunology and Virology, Dana-Farber Cancer Institute, Boston, MA, United States
[c]Parker Institute for Cancer Immunotherapy at Dana-Farber Cancer Institute, Boston, MA, United States
*Corresponding author: e-mail address: philip_kranzusch@dfci.harvard.edu

Contents

Abstract

Cyclic GMP–AMP synthase (cGAS) is an innate immune system enzyme responsible for recognition of double-stranded DNA aberrantly localized in the cell cytosol. cGAS binds DNA and is activated to catalyze production of the nucleotide second messenger 2′–5′/3′–5′ cyclic GMP–AMP (2′3′ cGAMP). In spite of a major role for cGAS in the cellular immune response, a complete understanding of cGAS biology has been limited by a lack of genetic tools to rapidly screen cGAS activity, instability of human cGAS–DNA interactions *in vitro*, and a previous absence of structural information for the human cGAS–DNA complex. Here we detail procedures to map the molecular determinants of cGAS activation and describe methods developed to prepare human cGAS–DNA crystals for structural analysis. Together with earlier systems established to study mammalian homologs of cGAS, these innovations provide a foundation to understand and therapeutically target human cGAS biology.

Methods in Enzymology, Volume 625
ISSN 0076-6879
https://doi.org/10.1016/bs.mie.2019.04.012

1. Introduction

Mislocalization of DNA into the cell cytosol activates the human innate immune system. The enzyme cyclic GMP–AMP synthase (cGAS) recognizes cytosolic DNA and cGAS activity is critical for the cellular response to pathogen replication, stress, and aberrant DNA replication in cancer (Chen, Sun, & Chen, 2016; Khoo & Chen, 2018; Sun, Wu, Du, Chen, & Chen, 2013). cGAS binds directly to double-stranded DNA and undergoes a conformational change that activates the enzyme to catalyze synthesis of the nucleotide second messenger 2′–5′/3′–5′ cyclic GMP–AMP (2′3′ cGAMP) (Ablasser et al., 2013; Civril et al., 2013; Diner et al., 2013; Gao, Ascano, Wu, et al., 2013; Kranzusch, Lee, Berger, & Doudna, 2013; Li et al., 2013; Sun et al., 2013; Zhang et al., 2013). 2′3′ cGAMP diffuses throughout the cell, binds to the receptor Stimulator of Interferon Genes (STING), and initiates a signaling cascade that induces downstream type I interferon and NF-κB signaling (Burdette et al., 2011; Gao, Ascano, Zillinger, et al., 2013; Ishikawa & Barber, 2008; Jin et al., 2008; Sun et al., 2009, 2013; Zhang et al., 2013; Zhong et al., 2008).

Following the initial discovery of cGAS by Sun et al. (2013), numerous techniques have been developed to investigate regulation of cytosolic DNA recognition. Initial structural and biochemical studies revealed the basic mechanism of enzyme activation and 2′3′ cGAMP synthesis, but relied primarily on mouse cGAS and other mammalian cGAS homologs that exhibit increased activity and *in vitro* stability (Civril et al., 2013; Gao, Ascano, Wu, et al., 2013; Kato et al., 2013; Kranzusch et al., 2013; Li et al., 2013; Zhang et al., 2014). These pioneering studies are essential to our understanding of cGAS biology, but key limitations of these approaches have prevented fully understanding human cGAS function.

The *CGAS* gene has undergone strong positive-selection and is one of the most divergent genes shared between the human and mouse genomes (George et al., 2011; Hancks, Hartley, Hagan, Clark, & Elde, 2015). The high level of species-specific variation in human cGAS directly impacts enzyme function and intrinsic regulation, and human cGAS exhibits a markedly reduced rate of 2′3′ cGAMP synthesis compared to mouse cGAS (Zhou et al., 2018). Although promising small molecule inhibitors of cGAS have recently been identified (Hall et al., 2017; Vincent et al., 2017), the reduced catalytic efficiency of human cGAS has limited development of screens designed to directly identify potent human cGAS agonist and antagonists.

An additional species-specific phenomenon recognized by many groups in the field is that human cGAS aggregates *in vitro* in the presence of DNA. Unlike mouse or other mammalian cGAS homologs, human cGAS does not appear to form a stable, minimal protein–DNA complex. This challenge has limited the application of common *in vitro* techniques to monitor human cGAS interaction with activating DNA and prevented determination of high-resolution structures of human cGAS in an activated state.

Here we present detailed protocols that overcome challenges specific to human cGAS and allow determination of the structure of the human cGAS–DNA complex (Zhou et al., 2018). Through a combination of bacterial genetics and biochemical assembly of stable human cGAS–DNA complexes *in vitro*, our approach mapped a key species-specific determinant of cGAS regulation and provided a new template to explain the structural basis of cGAS small molecule inhibitor specificity. We have used these experimental advances to dissect the mechanism of human cGAS activation and explain how species-specific cGAS variations impact DNA recognition in human cells. Importantly, we anticipate that adaptation of these techniques will enable continued advances in understanding the biology and therapeutic potential of cGAS-STING signaling.

2. A genetic assay for screening cGAS activity using bacteria

In this section, we discuss the application of a bacterial assay that indirectly measures cGAS activity. We developed this assay to enable rapid analysis of human cGAS variants with the specific goal of defining the genetic basis for why human cGAS is not activated by short (<17 bp) DNA and cannot form a stable protein–DNA complex *in vitro* suitable for X-ray crystallographic structural analysis. Conventional techniques have allowed for successful determination of structures of mouse and porcine cGAS in complex with DNA (Civril et al., 2013; Gao, Ascano, Wu, et al., 2013; Li et al., 2013; Zhang et al., 2014), but these structures cannot explain how species-specific variation impacts regulation of cGAS function and interactions with small molecule therapeutics.

We coupled cGAS activity with a bacterial phenotype by expressing human cGAS variants in the bacterium *Vibrio cholerae* and monitoring bacterial chemotaxis. *V. cholerae* is a model Gram negative pathogen that is highly amenable to genetic manipulation and uses 3′–5′/3′–5′ cGAMP

(3′3′ cGAMP) to modulate chemotaxis and virulence (Davies, Bogard, Young, & Mekalanos, 2012). We confirmed that cGAS-derived 2′3′ cGAMP can also modulate chemotaxis in *V. cholerae*, and defined conditions where *V. cholerae* chemotaxis can be used as a quantitative read-out for cGAS activity *in vivo* (Zhou et al., 2018). There are 116 amino-acid differences between the enzymatic domains of human and mouse cGAS. By combining the *V. cholerae* bacterial assay with a phylogenetic chimera approach, we identified two key residues K187 and L195 in human cGAS that control enzyme regulation and, when mutated, enabled crystallization and structural analysis of the human cGAS–DNA complex.

2.1 Assay setup and considerations

The cGAS activity assay in *V. cholerae* takes advantage of endogenous 3′3′ cGAMP signaling and shared functionality between human and bacterial cyclic dinucleotide signaling. 3′3′ cGAMP in *V. cholerae* is synthesized by the enzyme dinucleotide cyclase in *Vibrio* (DncV), a structural homolog of cGAS whose expression is required for virulence (Davies et al., 2012; Kranzusch et al., 2014). During infection, expression of the *Vibrio dncV* gene increases and it was previously shown in culture that over-expression of *dncV* inhibits bacterial chemotaxis in soft-agar assays (Davies et al., 2012). We have demonstrated that overexpression of *CGAS* in *V. cholerae* also decreases bacterial chemotaxis, and chemotaxis can be used as a continuous and quantitative readout. cGAS-dependent chemotaxis repression correlates with efficiency of 2′3′ cGAMP production *in vitro* and we have used this assay to track the increased 2′3′ cGAMP synthesis activity of mouse vs human *CGAS* alleles. *dncV* is encoded in the *Vibrio* genome adjacent to a 3′3′ cGAMP receptor encoding gene called cGAMP-activated phospholipase in *Vibrio* (*capV*) (Severin et al., 2018). Although *dncV* and *capV* are conserved in all 7th pandemic *Vibrio* strains including the bacterial strain used for our cGAS-assays, it remains unclear if activation of *capV* is solely responsible for *dncV*-dependent or *CGAS*-dependent phenotypes.

The following protocol describes in detail the *Vibrio* genetic assay for screening the production of 2′3′ cGAMP by ectopic *CGAS* expression. We selected a *dncV*-deficient host strain of *V. cholerae* (strain C6706 *dncV::Tn*) to remove effects of endogenous 3′3′ cGAMP, though these assays are also expected to work with all 7th pandemic, motile *V. cholerae* isolates. Plasmids encoding arabinose inducible *CGAS* alleles are introduced into *V. cholerae* through conjugation and monitored for chemotaxis by inoculating bacteria into a soft-agar, chemotaxis medium plate. Chemotaxis is quantified after 24 h of growth and all phenotypes can be validated through biochemical assays of 2′3′ cGAMP synthesis with recombinant protein. Sufficient expression of cGAS protein is required in this assay, for this reason we used an N-terminal maltose binding protein (MBP) tag and *CGAS* genes codon-optimized for bacterial expression. 2′3′ cGAMP concentrations were not measured *in vivo* and it is unclear how this molecule might be metabolized by *V. cholerae*; however, all cGAS-dependent phenotypes require the enzyme active-site residues. We will not attempt to discuss the detail about structure-guided chimera design, western blotting, or cloning techniques for plasmid construction here. Rather, we will concentrate on the protocol for *V. cholerae* chemotaxis assays to monitor cGAS activity and relevant considerations.

2.2 Protocol

1. *Vector construction.* All genes (*dncV*, *CGAS* alleles, and controls) are overexpressed from the plasmid pBAD24T, a conjugation proficient version of pBAD24 that harbors an arabinose inducible promoter and strong ribosomal binding site (Guzman, Belin, Carson, & Beckwith, 1995). We recommend using codon-optimized variants of both mouse *CGAS* and human *CGAS* for bacterial expression to ensure efficient expression in *V. cholerae* and minimize RNA differences unrelated to cGAS function. Vectors are constructed by conventional cloning techniques and synthases were expressed fused to an N-terminal MBP-tag separated by a flexible TEV site and glycine-serine linker. Negative control vectors expressed MBP alone and catalytically inactive *CGAS*, the positive control vector expressed *dncV* or mouse *CGAS* (Fig. 1A).
2. *Vibrio cholerae strain construction.* Bacteria were cultivated at 37 °C in LB medium (1% tryptone, 0.5% yeast extract, 0.5% NaCl w/v). Where appropriate antibiotics and supplemented nutrients are used at the following concentrations: streptomycin, 100 μg mL^{-1}; carbenicillin, 100 μg mL^{-1}; and diaminopimelic acid, 300 μg mL^{-1}. Here we used a streptomycin resistant isolate of C6706 *V. cholerae* disrupted for *dncV* (*dncV::Tn*) (Davies et al., 2012). Plasmids were introduced into an isogenic host *V. cholerae* by conjugation using MFDλpir *E. coli* as the conjugation donor of each plasmid and diaminopimelic acid auxotrophy for counter selection (Ferrieres et al., 2010). *V. cholerae* strains were selected for using carbenicillin in the absence of diaminopimelic acid,

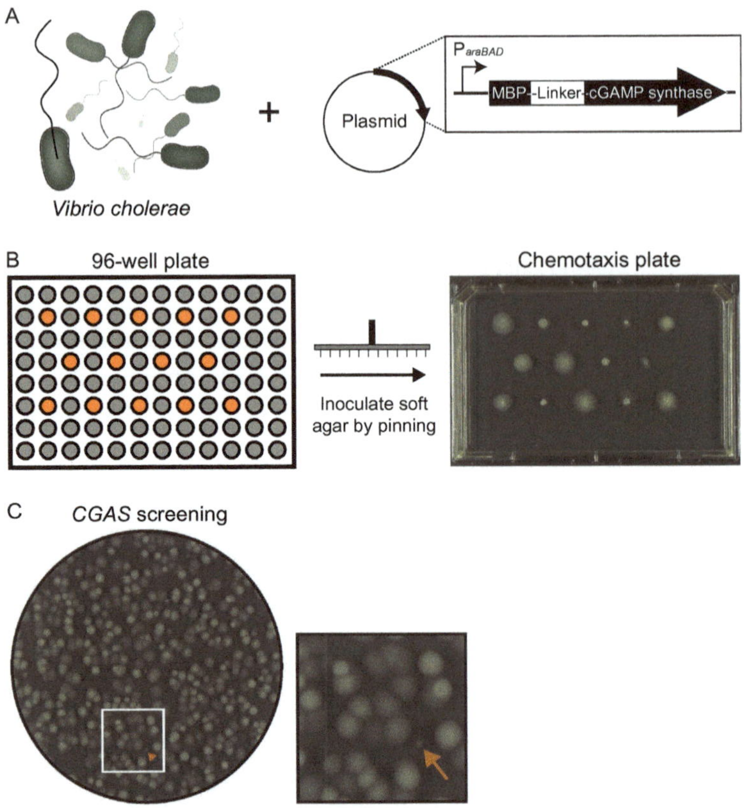

Fig. 1 A bacterial phenotypic assay for rapid screening of cGAS activity. (A) An isogenic *V. cholerae* host strain is used to interrogate different cGAMP synthases expressed from a plasmid. Strains are constructed by conjugation. Right, a schematic of the pBAD24T plasmid used to overexpress *CGAS* and relevant control genes. Genes are expressed from an arabinose inducible promoter (P_{araBAD}) and cGAMP synthases are expressed with an N-terminal maltose binding protein (MBP) fusion connected with a TEV site and glycine-serine linker. (B) Bacteria are arrayed in a 96-well plate for analysis of chemotaxis, separating each culture by at least one well and avoiding edges. Chemotaxis medium is inoculated using a pinning apparatus that has been dipped into the arrayed 96-well plate and bacteria are allowed to grow and chemotax for up to 24 h. (C) The *V. cholerae* cGAS activity assay is suitable for screening large libraries of *CGAS* alleles. In this mock screen, a mixture of bacteria expressing different *CGAS* alleles were inoculated into molten chemotaxis medium and poured into a Petri dish prior to overnight incubation at 30 °C. The majority of inoculated bacteria express an inactive mouse *CGAS* while a minority express wild type, active mouse *CGAS*. The assay can clearly differentiate *CGAS* alleles based on activity.

colony-purified, and stored at −80 °C in LB supplemented with 30% (v/v) glycerol. An important consideration is adequate cGAS expression in *V. cholerae*, which can be verified by western blot of bacteria grown under inducing conditions as previously described (Zhou et al., 2018).

3. *Chemotaxis plate preparation.* Formulate chemotaxis medium (1% tryptone, 0.5% NaCl, 0.3% agar w/v final concentration) at 2 × concentration and autoclave to sterilize and solubilize agar. 2 × chemotaxis medium is stable at room temperature for up to 2 months. Prepare chemotaxis plates 2–24 h prior to inoculation of bacteria. Melt 2 × chemotaxis medium and cool to 50–60 °C in a water bath, supplement medium with arabinose at 0.2% w/v (final concentration) and carbenicillin, and dilute to 1 × with > 37 °C sterile water. Many sterile plates or dishes can be used for chemotaxis analysis, we recommend using deep well plates (OmniTray w/ Lid, Thermo Fisher) compatible with bacterial pinning apparatuses (V&P Scientific). Pour 40 mL of molten chemotaxis medium per plate. Plates should be allowed to cool and dry for >2 h; however, a consideration in this assay is inherent plate-to-plate variability in dryness. We recommend always including appropriate controls for each assay and normalizing chemotaxis to MBP-only and *dncV* expressing strains.

4. *V. cholerae* chemotaxis assay. *V. cholerae* chemotaxis assays are adapted from a previously established protocol (Davies et al., 2012), and performed as illustrated in Fig. 1A and B. Inoculate one 14 mL culture tube containing 1–3 mL of LB medium supplemented with carbenicillin either with an overnight colony of *V. cholerae* or directly from a frozen glycerol stock. Note, arabinose should not be included in this media in order to prevent early protein induction. Cultivate overnight (16–30 h) at 37 °C with shaking at 220–300 RPM. Bacteria can be introduced into chemotaxis medium multiple ways. A key point of this assay is to measure bacterial chemotaxis through the soft agar and not colony formation on the surface or underneath medium agar. For this reason, bacteria are ideally inoculated on top of chemotaxis medium or by gently piercing the agar-surface without fully reaching the bottom of the dish. We recommend using a bacterial pinning apparatus to enhance reproducibility and uniformity. Stationary phase *V. cholerae* cultures are diluted 10-fold with fresh LB medium, arrayed in a 96-well plate then applied to the surface of a chemotaxis plate *via* pinning with a 96-pin, floating pin replicator (VP 408FH, V&P Scientific) according to the manufacturer's instructions. Space between inoculated strains is important for

unimpeded chemotaxis, thus strains are arrayed in every other well of a 96-well plate, avoiding edge wells. The floating pin replicator apparatus is dipped into the 96-well plate, removed, and applied to the surface of the chemotaxis medium plate for inoculation. Alternatively, bacteria can be introduced into the chemotaxis medium using a sterile pipette tip and manual inoculation of 1 μL individual spots on top of the soft agar. Incubate the chemotaxis plates at 30 °C for 16–24 h.

5. *Chemotaxis quantification.* Chemotaxis should be quantified by imaging plates at 16–24 h post inoculation and digitally measuring the area of motile bacteria. Each image should be analyzed using ImageJ software or equivalent, the area of bacterial growth can be traced using an oval, and area measured as number for pixels. Chemotaxis area is compared and reported by normalizing each strain to MBP-alone vector to decrease plate-to-plate variation. For each cGAMP synthase or control plasmid, divide bacterial growth area in pixels by area of MBP alone vector for each experiment and on each plate for quantification of "chemotaxis as a % of MBP." Chemotaxis is repressed by 3′3′ cGAMP and 2′3′ cGAMP, thus a decrease in bacterial chemotaxis represents increased nucleotide second messenger synthesis. To represent increased cGAS or DncV activity as a positive value related to chemotaxis, transform values into "chemotaxis repression" by calculating (chemotaxis as a % of MBP)$^{-1}$, then normalizing this value to *dncV*. Chemotaxis repression is determined and used to simplify a direct relationship between cGAMP levels and plotted values.

2.3 Future applications

The *V. cholerae* chemotaxis assay presented here enables a rapid phenotypic read-out of cGAS-dependent 2′3′ cGAMP synthesis and a unique approach to understanding mammalian signaling in a prokaryotic system. Our initial studies focused on mapping intrinsic differences of human and mouse cGAS and led to the successful identification of a human cGAS K187N/L195R variant with several novel applications (Zhou et al., 2018). We used this allele to determine the structure of human cGAS in complex with its activating ligand DNA. In addition, this allele of human cGAS and the *V. cholerae* cGAS activity assay also enable rapid analysis of cancer-associated cGAS mutations, screening of small molecule inhibitors of cGAS in bacteria, and continued analysis of metazoan cGAS homologs. The *V. cholerae* assay can also be expanded to include a second plasmid co-expressing genes that are expected to modulate cGAS activity. For example, one could express antagonists to disrupt cGAS-dependent phenotypes, such as recently identified viral inhibitors of cGAS (Biolatti et al., 2018; Huang et al., 2018; Wu et al., 2015; Zhang et al., 2016, 2018) and 2′3′ cGAMP (Eaglesham, Pan, Kupper, & Kranzusch, 2019). This approach is also amenable to unbiased screening of cDNA libraries derived from pathogens or cells for cGAS modulating proteins through screening colony size on soft-agar (Fig. 1C). We anticipate that orthogonal phenotypes in *E. coli* and other commonly studied bacteria will continue to provide additional bacterial genetic approaches to dissect cGAS function.

3. Crystallization and structural determination of the human cGAS–DNA complex

High-resolution structural information is critical to define the mechanism of enzyme function and regulation, and to enable rational development of small molecule therapeutics. A major challenge in the cGAS field has been instability of human cGAS–DNA interactions *in vitro* and an inability to assemble a minimal complex suitable for crystallization and X-ray crystallography-based structure determination. Our bacterial genetics approach comparing human and mouse cGAS identified two key substitutions in human cGAS (K187N and L195R) that stabilize interactions between cGAS and DNA, increase enzymatic activity, and enable crystallization of human cGAS in an activated state competent for 2′3′ cGAMP synthesis. Human cGAS K187 and L195 are located on the back half of the enzyme along a long helix designated the "spine" that forms part of the DNA binding interface (Civril et al., 2013; Zhou et al., 2018). Importantly, these mutations are located distantly away from the core active site and allow high-resolution structure determination of the enzyme contacts with nucleotide ligands and small molecules capable of targeting human cGAS activity. In combination with previously determined structures of human cGAS in the *apo* conformation (Hall et al., 2017; Kato et al., 2013; Kranzusch et al., 2013; Li et al., 2013; Zhang et al., 2014), these experimental approaches allow investigation of co-factors and small molecule compounds that target either the active or inactive cGAS conformations. Here we will outline procedures that overcome the challenges to crystallization and structure determination of the human cGAS–DNA complex, including design and preparation of human cGAS protein and DNA ligands, growth and harvest of crystals, and collection of X-ray diffraction data.

3.1 Preparation of human cGAS protein and DNA

The human cGAS protein is composed of a disordered N-terminus (amino acids 1–156) extended from a structured C-terminal domain (amino acids 157–522) that harbors both the DNA-binding interface and the nucleotidyltransferase domain (Civril et al., 2013; Gao, Ascano, Wu, et al., 2013; Kato et al., 2013; Kranzusch et al., 2013; Li et al., 2013; Zhang et al., 2014) (Fig. 2A). The cGAS N-terminus is highly charged (predicted isoelectric point of ~11) and is critical to mediate cGAS phase separation and enhanced enzyme activity (Du & Chen, 2018; Sun et al., 2013; Tao et al., 2017). Although the N-terminus is important for cGAS regulation, the nucleotidyltransferase domain alone is sufficient for DNA recognition and 2′3′ cGAMP synthesis *in vitro* indicating that the C-terminus of cGAS is the minimal functional domain required for cytosolic DNA recognition (Kranzusch et al., 2013; Sun et al., 2013). All crystal structures of cGAS to date use truncated versions of the enzyme that remove the unstructured N-terminus to facilitate crystallization. The protocol reported here relies on crystallization of human cGAS 157–522 with K187N and L195R substitutions that stabilize cGAS–DNA interactions (Zhou et al., 2018). Crystallization of this fragment is ideal for structural analysis of the active site and candidate small molecule therapeutics. Additionally, use of the human cGAS K187N/L195R variant may also facilitate future efforts focused on including sections of the N-terminus to define at a structural level how this region of cGAS influences DNA interactions and higher-order complex formation.

Similar to minimization of the functional cGAS domain used for crystallography, the length and sequence of DNA used during crystallization is an important optimization step. For general considerations, we recommend parameters used for protein–nucleic acid crystallization as previously reviewed (Hollis, 2007). cGAS recognizes double-stranded DNA in a sequence-independent manner (Civril et al., 2013; Gao, Ascano, Wu, et al., 2013; Kato et al., 2013; Kranzusch et al., 2013; Li et al., 2013; Sun et al., 2013; Zhang et al., 2014), and the major parameter to optimize for cGAS crystallography is therefore the length of double-stranded DNA and nature of the 5′ and 3′ termini. Structures of the mouse cGAS–DNA complex revealed that 1 copy of mouse cGAS occupies ~13 base pairs of double-stranded DNA (Gao, Ascano, Wu, et al., 2013), and that DNA substrates 16–17 bp length are ideal for allowing DNA–DNA crystal contacts between adjacent unit cells of the crystal. In order to facilitate direct comparison of the human cGAS–DNA complex with previous mouse cGAS–DNA structures, we utilized a double-stranded DNA ligand

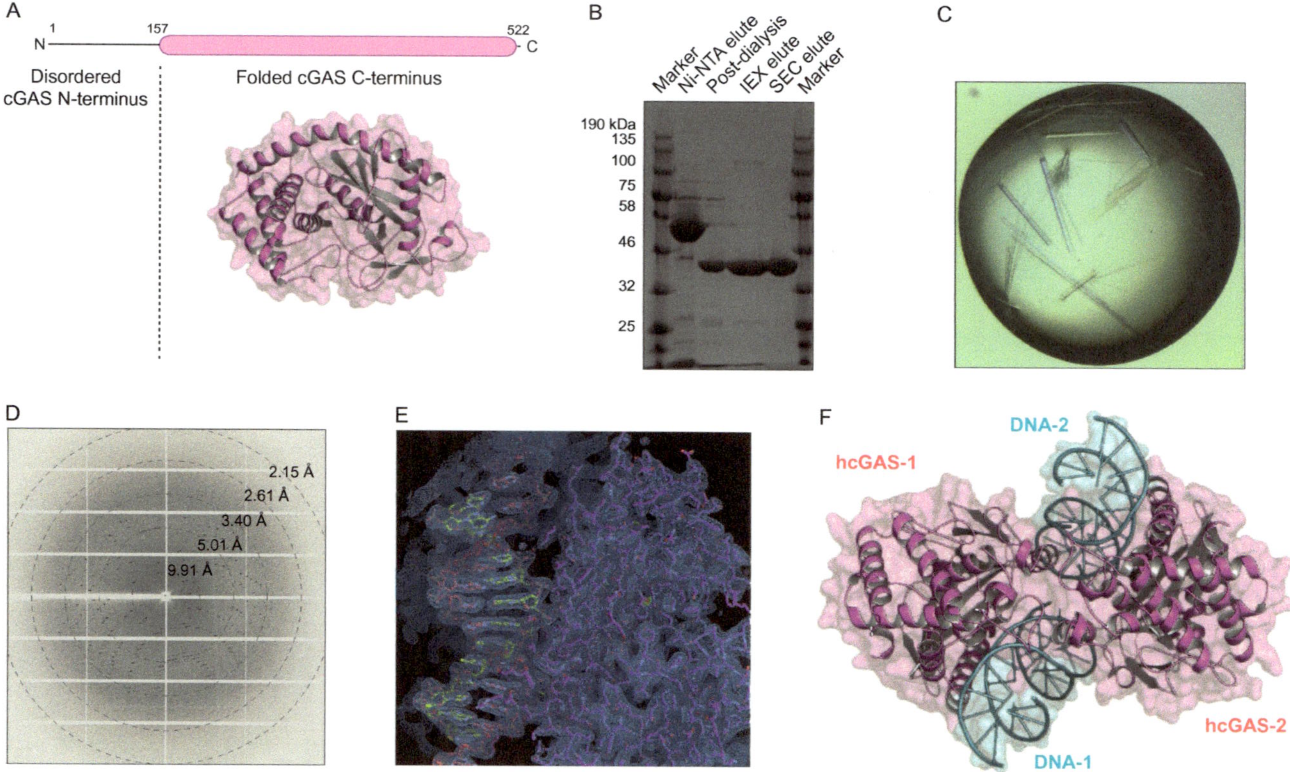

Fig. 2 Crystallization and structural determination of the human cGAS–DNA complex. (A) Schematic of human cGAS. The human cGAS N-terminus is a highly charged, disordered sequence, and no structural information is available. The human cGAS C-terminus contains the nucleotidyltransferase domain and is sufficient for DNA recognition and 2′3′ cGAMP synthesis (PDB: 4KM5). (B) SDS-PAGE gel analysis of the purification of human cGAS. Recombinant human cGAS from each purification step is resolved on a 10% denaturing SDS-PAGE gel and visualized with Coomassie G-250 stain. Ni-NTA, Nickel-nitrilotriacetic acid affinity purification; IEX, ion exchange chromatography; SEC, size-exclusion chromatography. (C) Crystals of the human cGAS–DNA complex after 3 days growth at 18 °C (2 μL drop size). (D) X-ray diffraction data collected from a crystal of the hcGAS–DNA complex. (E) $2F_o - F_c$ electron density map of the 2.3 Å human cGAS–DNA complex (contoured at 1.0 σ). Human cGAS protein and the double-stranded DNA are shown as sticks in magenta and red-yellow, respectively. (F) Overview of the human cGAS–DNA complex. Human cGAS forms a 2:2 complex with DNA where two cGAS molecules embrace two DNA molecules.

optimized for mouse cGAS crystallography by Patel and colleagues (Gao, Ascano, Wu, et al., 2013). It is likely that further refinement and optimization of DNA ligand selection may improve human cGAS–DNA crystallization and the resolution of X-ray diffraction data. The following section lists a detailed protocol for sample preparation of human cGAS and DNA for crystallization.

3.1.1 Equipment

1. Temperature controlled culture shaker
2. Centrifuges (for 1.5 mL microfuge tubes, 50 mL tubes, and 1 L bottles)
3. Sonicator, French Press or alternative bacterial lysis technology
4. Ni-NTA resin (Qiagen)
5. Gravity column
6. Dialysis tubing (8–10kDa molecular weight cutoff)
7. FPLC System or low-pressure chromatography system with UV detector
8. 5 mL HiTrap Heparin HP ion-exchange column (GE Healthcare)
9. Spin concentrator columns (10 or 30 kDa molecular weight cutoff)
10. 16/600 Superdex S75 column (GE Healthcare)

3.1.2 Reagents

1. MDG media: 0.5% glucose, 25 mM Na_2HPO_4, 25 mM KH_2PO_4, 50 mM NH_4Cl, 5 mM Na_2SO_4, 2 mM $MgSO_4$, 0.25% aspartic acid, 100 μg mL^{-1} ampicillin, 34 μg mL^{-1} chloramphenicol, and trace metals (25 μM $FeCl_3$, 10 μM $CaCl_2$, 5 μM $MnCl_2$, 5 μM $ZnSO_4$, 1 μM $CoCl_2$, 1 μM $CuCl_2$, 1 μM $NiCl_2$, 1 μM Na_2MoO_4, 1 μM Na_2SeO_3, 1 μM H_3BO_3) (Studier, 2005).
2. M9ZB media: 0.5% (v/v) glycerol, 1% Cas-amino acids, 47.8 mM Na_2HPO_4, 22 mM KH_2PO_4, 18.7 mM NH_4Cl, 85.6 mM NaCl, 2 mM $MgSO_4$, 100 μg mL^{-1} ampicillin, 34 μg mL^{-1} chloramphenicol, and trace metals as above (Studier, 2005).
3. Ni lysis buffer: 20 mM HEPES-KOH pH 7.5, 400 mM NaCl, 30 mM imidazole, 10% (v/v) glycerol, 1 mM DTT, 0.5 μg mL^{-1} Pepstatin A protease inhibitor.
4. Ni wash buffer: 20 mM HEPES-KOH pH 7.5, 1 M NaCl, 10% (v/v) glycerol, 30 mM imidazole, 1 mM DTT.
5. Ni elution buffer: 20 mM HEPES-KOH pH 7.5, 400 mM NaCl, 10% (v/v) glycerol, 300 mM imidazole, 1 mM DTT.

6. Dialysis buffer: 20 mM HEPES-KOH pH 7.5, 300 mM NaCl, 1 mM DTT.
7. IEX Buffer A: 20 mM HEPES-KOH pH 7.5, 1 mM DTT.
8. IEX Buffer B: 20 mM HEPES-KOH pH 7.5, 1 M NaCl, 1 mM DTT.
9. SEC Buffer: 20 mM HEPES-KOH pH 7.5, 250 mM KCl, 1 mM TCEP.

Note: Where indicated, add Pepstatin A and DTT immediately before use.

3.1.3 Protocol

1. *E. coli transformation and recombinant protein expression.* Human cGAS (residues 157–522, containing K187N/L195R substitutions) is cloned into a custom pET vector for expression of a 6 × His-SUMO2-human $cGAS_{157-522}$ fusion protein in *E. coli* (Zhou et al., 2018). For expression, we use a previously optimized minimal media system (Studier, 2005). *Day 1*: To start, transform the pET plasmid into *E. coli* BL21-RIL DE3 bacteria (Agilent)—an *E. coli* strain that harbors the pRARE2 tRNA plasmid to express rare tRNAs and increase the level of human cGAS expression. Select transformants by overnight growth at 37 °C on plates prepared with MDG media and antibiotic selection. These MDG plates and all subsequent media steps include 100 $\mu g\,mL^{-1}$ ampicillin and 34 $\mu g\,mL^{-1}$ chloramphenicol. *Day 2*: Prepare a starter culture with 50 mL of MDG media in a 250 mL Erlenmeyer flask and inoculate with three fresh transformant colonies. Cover loosely, and grow the culture overnight at 37 °C with 230 RPM shaking. The resulting OD_{600} of this overnight MDG culture is typically 5.0–6.0. *Day 3*: Prepare 2 × large-scale cultures with 1 L of M9ZB media in a 2 L flask, and inoculate using the overnight MDG culture to a starting OD_{600} of ~0.05. Grow the large M9ZB cultures at 37 °C with 230 RPM shaking until the cultures reach an OD_{600} of 2.5–3.0. RIL bacteria in M9ZB media double every ~60 min and this growth step typically takes 4–5 h. Once cultures have reached the correct OD_{600}, transfer flasks to an ice water bath and cool for 20 min. This step prior to induction is critical to avoid translation initiation and protein aggregation prior to the *E. coli* reaching the optimal temperature for human cGAS protein expression. Next, supplement cultures with IPTG to a final concentration of 0.5 mM and incubate cultures at 16 °C with 230 RPM shaking for an additional 12–16 h.

   Note on media and expression strategy selection: Successful expression of human cGAS has been reported with a variety of plasmid vectors and solubility tags including an N-terminal maltose-binding protein tag. Additionally, multiple medias support robust recombinant protein expression, though rich medias like 2YT or medias supplemented with additional zinc (a structural metal bound in the cGAS zinc-ribbon insertion required for DNA recognition) (Kranzusch et al., 2013) result in higher protein yields. An additional consideration with media systems is loss of the pET plasmid insert due to leaky protein expression and toxicity associated with 2′3′ cGAMP synthesis. While use of MDG and M9ZB is not required for human cGAS expression, this media system offers a reliable method for robust expression of human cGAS and related bacterial cGAS/DncV-like nucleotidyltransferases (Whiteley et al., 2019; Zhou et al., 2018).

2. *E. coli lysis and Ni-NTA chromatography purification.* Harvest *E. coli* by centrifugation in 1 L bottles for 20 min at 3300 × *g*, 4 °C. Perform all subsequent steps with lysate and buffers chilled on ice or working at 4 °C. Re-suspend pellets with 120 mL Ni lysis buffer and lyse by sonication. Collect lysates and centrifuge for 30 min at 47,850 × *g*, 4 °C. Collect the supernatant and pass through packed glass wool or a 0.2–0.5 μm filter. Save a 50 μL aliquot of sample as a total lysate sample for later analysis. Separately, equilibrate ~8 mL of packed Ni-NTA resin by pre-washing with lysis buffer, and then add directly to the clarified *E. coli* supernatant. Mix lysate and resin, and pour into a pre-washed gravity column. Collect the lysate and add back to the top of the column to pass lysate over Ni-NTA resin 2 × total. Save final flow-through for later analysis. Next, wash the resin with 20 mL of Ni lysis buffer followed by 80 mL of Ni wash buffer and then 40 mL of Ni lysis buffer. The high-salt Ni wash buffer is important to remove excess nucleic acid that co-purifies with human cGAS. Elute recombinant human cGAS from the resin with 30 mL of Ni elution buffer and collect as one 30 mL fraction. Analyze elution and all previously collected samples by A_{280} and SDS-PAGE analysis. Typical results of the final elution sample are ~2 mg mL^{-1} protein and ~90% purity.

   Tip: Sonication is a critical parameter for optimization of lysis. Keep cultures cool and avoid excessive sonication to prevent protein denaturation. Following sonication, the suspension should remain cool and turn slightly darker after lysis. An initial recommended starting point for optimization is 3–5 min total time on, 10 s on followed by 20 s off, 6.5 W, on ice.

3. *Dialysis and SUMO-tag removal.* For removal of the SUMO2 solubility tag, add ~250 μg of human SENP2 protease (fragment D364–L589 with M497A mutation) (Reverter & Lima, 2006) directly to the elution sample. Transfer sample into pre-washed dialysis tubing or a dialysis cassette with a molecular-weight cutoff of ~8–10kDa. Dialyze sample against ~1.5–2 L of dialysis buffer at 4 °C with gently spinning overnight for ~12–15 h.
4. *Heparin ion-exchange chromatography.* Centrifuge the post dialysis sample for 10 min at 3220 × *g*, 4 °C to remove any precipitate and save a 50 μL sample for later analysis. Using an FPLC system, load the supernatant onto 2 × 5 mL Heparin HP ion-exchange columns that are connected in tandem. Use a flow rate of ~2 mL min^{-1} for each step. Wash the column with 20 mL of 30% IEX Buffer B and then elute with a gradient of 30–100% IEX Buffer B over ~50 mL total volume while collecting 1 mL fractions. Human cGAS elutes as a single peak near the end of the gradient. Collect and pool all peak fractions, save a 50 μL sample for later analysis and proceed to the next purification step.
5. *Size-exclusion chromatography.* Concentrate the IEX elution sample to ~1.3 mL by centrifugation in a 30 kDa molecular weight cutoff spin concentrator. Centrifuge sample in 5–10 min intervals at 3220 × *g*, 4 °C. Transfer the concentrated sample to a 1.5 mL microfuge tube and remove particulate precipitation by centrifugation for 5 min at 16,000 × *g*, 4 °C. Transfer clarified sample to a new microfuge tube and store on ice until ready for size-exclusion chromatography. Purify the sample using a 16/600 Superdex S75 or S200 column pre-equilibrated in SEC buffer. Human cGAS elutes as two adjacent peaks potentially representing a monomer and dimer conformation. Both size-exclusion peaks are active and capable of synthesizing 2′3′ cGAMP *in vitro* with similar effects. Repeat spin concentration steps to concentrate the final human cGAS protein sample to ~10 mg mL^{-1}. Use protein immediately for crystallography, or flash-freeze aliquots in liquid nitrogen and store at −80 °C. Fig. 2B shows the SDS-PAGE gel analysis of each purification step.

   Tip: Human cGAS stocks at ~10 mg mL^{-1} are stable at −80 °C for at least 1 year, and can typically be used for 2 × freeze-thaw cycles. Lower concentration stocks for biochemistry can be supplemented with 10% (v/v) glycerol and stored long-term at − 80 °C or alternatively supplemented with 50% (v/v) glycerol and stored at −20 °C for up to 4 months.

6. *Preparation of synthetic DNA for crystallography*. A 17 bp (DNA17) with 1 nt overhangs derived from the work of Patel and colleagues (Gao, Ascano, Wu, et al., 2013) is used for formation of the human cGAS–DNA complex for crystallography. Double-stranded DNA17 is prepared by annealing synthetic DNA oligos with the sequences 5′ AAA TTG CCG AAG ACG AA 3′ (DNA17 sense) and 5′ TTT CGT CTT CGG CAA TT 3′ (DNA17 antisense). Likewise, a 45 bp (DNA45) with blunt ends derived from immunostimulatory *Listeria* bacterial DNA (Stetson & Medzhitov, 2006) is used for biochemical reactions and prepared by annealing synthetic DNA oligos with the sequences 5′ TAC AGA TCT ACT AGT GAT CTA TGA CTG ATC TGT ACA TGA TCT ACA 3′ (DNA45 sense) and 5′ TGT AGA TCA TGT ACA GAT CAG TCA TAG ATC ACT AGT AGA TCT GTA 3′ (DNA45 antisense). Anneal oligos by preparing a 50 μL reaction containing 10 mM of each single-stranded DNA oligo. Use a PCR block to slow-anneal oligos by heating to 95 °C for 5 min and then cooling at 0.1 °C s^{-1} to 4 °C. Store dsDNA aliquots at −20 °C for further use.

   Tip: Using salt-based annealing buffers is generally not necessary, we recommend preparing the annealing reactions with nuclease-free dH_2O. Standard de-salted oligos are sufficient for biochemistry and crystallography, though polyacrylamide gel electrophoresis-purification is an option for further optimization.

3.2 Assays to assess human cGAS–DNA complex formation

Human cGAS–DNA complex formation can be monitored biochemically, providing a valuable method to optimize co-complex formation and understand the mechanism of enzyme activation. Wildtype human cGAS is prone to aggregation in the presence of DNA *in vitro*, limiting attempts to form a stable minimal cGAS–DNA complex for structural studies. The human cGAS K187N/L195R variant identified in our bacterial genetic assay allows formation of a minimal human cGAS–DNA complex for crystallization. Complex formation can be verified by adapting a gradient dialysis technique developed for *in vitro* nucleosome assembly (Lee & Narlikar, 2001) and monitoring cGAS–DNA interactions with an agarose electrophoretic mobility shift assay (EMSA) (Fig. 3A).

Note: Gradient dialysis is not necessary for human cGAS K187N/L195R—DNA17 crystallization, but the procedure yields more reliable results for analyzing the impact of cGAS mutations on DNA interactions *in vitro*.

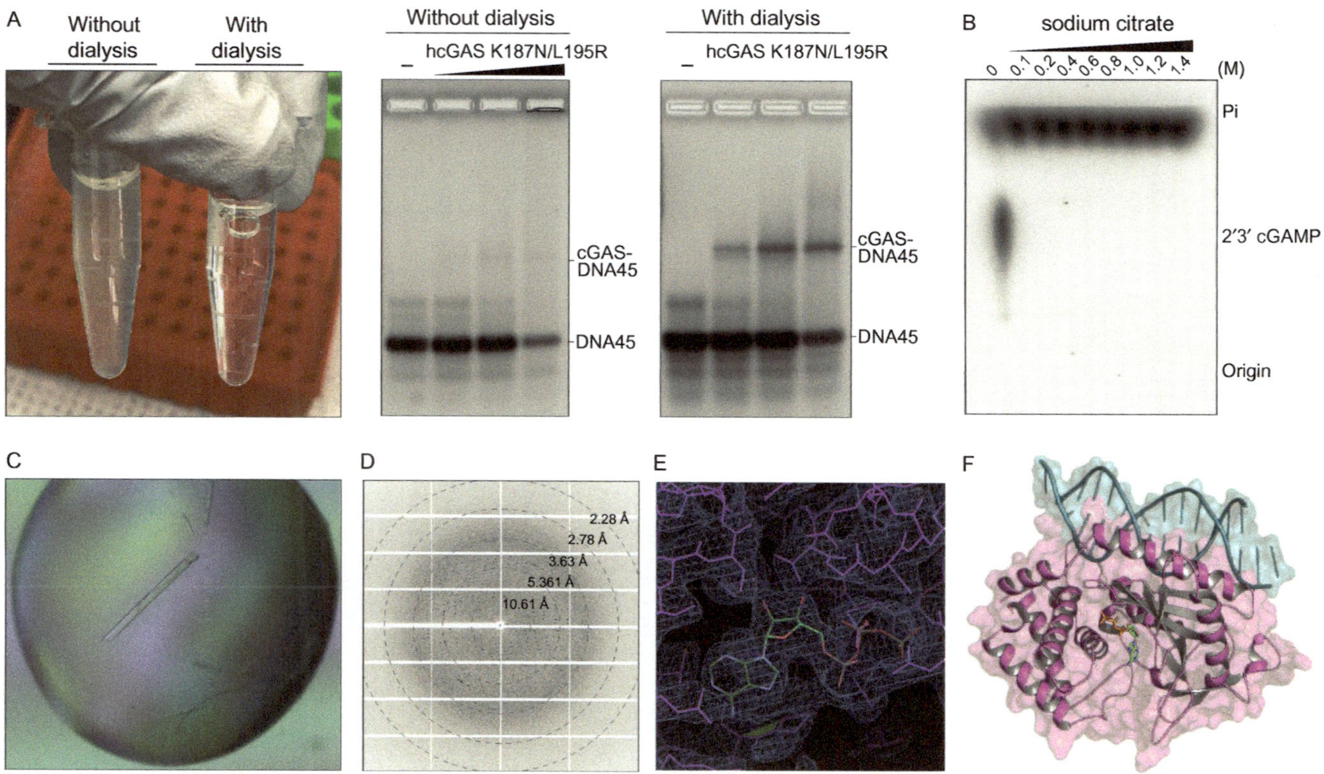

Fig. 3 See legend on next page.

3.2.1 Equipment

1. Dialysis tubing (8–10 kDa molecular-weight cutoff)
2. Peristaltic pump

3.2.2 Reagents

1. High-salt buffer: 20 mM HEPES-KOH pH 7.5, 400 mM KCl, 1 mM DTT.
2. Low-salt buffer: 20 mM HEPES-KOH pH 7.5, 50 mM KCl, 1 mM DTT.
3. 10× TB buffer (for 1000 mL): 108 g Tris, 55 g boric acid, 700 mL Milli-Q dH_2O, qs to 1000 mL, heat into solution by autoclaving for 15–30 min.

3.2.3 Protocol

1. *Gradient dialysis.* Mix human cGAS and double-stranded DNA17 or DNA45 in high-salt buffer at 4 °C. Typically we create a 1 mL solution with a 1:1 ratio of cGAS:DNA at a final concentration of 2 μM. At this high-salt condition (400 mM KCl), human cGAS is unable to bind double-stranded DNA and the solution should remain clear with no precipitation. Next, transfer the protein and DNA mixture into dialysis tubing and place in a ~500 mL beaker of high-salt buffer. Working at 4 °C, setup a second ~500 mL beaker of low-salt buffer, and use one line of a peristaltic pump to slowly transfer liquid from the low-salt solution to

Fig. 3 Assembly of human cGAS–DNA complex by gradient dialysis, and structural determination of the human cGAS–DNA–ATP complex. (A) In the presence of DNA, human cGAS quickly aggregates (left panel, left tube), but human cGAS–DNA complexes remain soluble when assembled with gradient dialysis (left panel, right tube). Right panel: *In vitro* electrophoretic mobility shift assay analysis of human cGAS K187N/L195R–DNA complex formation. Co-complex assembly with gradient dialysis dramatically improves stable complex formation. (B) The impact of sodium citrate on human cGAS production of 2′3′ cGAMP *in vitro.* A concentration gradient of sodium citrate was added to human cGAS reactions in the presence of 45 bp double-stranded DNA, and 2′3′ cGAMP formation was monitored by incorporation of [α-^{32}P] ATP. Reactions were visualized by treating with alkaline phosphatase and separating by thin-layer chromatography. M, Molar concentration. (C) Crystals of the human cGAS–DNA–ATP complex after incubation in the soaking solution (1.5 μL drop size). (D) X-ray diffraction data collected from a crystal of the hcGAS–DNA–ATP complex. (E) $2F_o - F_c$ electron density map of the human cGAS active-site derived from the 2.8 Å human cGAS–DNA–ATP complex (contoured at 1.0 σ). ATP is shown at the center. (F) Overview of a single 1:1 cGAS–DNA unit in the human cGAS–DNA–ATP complex. Human cGAS and DNA are showed in magenta and blue, respectively. ATP is shown in sticks (green-yellow).

the high-salt solution. Use a second line of a peristaltic pump to transfer liquid from the high-salt solution to a waste beaker so that overtime the salt solution surrounding the cGAS and DNA mixture slowly reduces to ~225 mM KCl. Transfer liquid at a rate of ~200 μl min^{-1} over the course of ~48 h at 4 °C to form stable cGAS–DNA complexes. Following incubation, transfer the cGAS–DNA complexes out of the dialysis tubing and store on ice for further biochemical or structural analysis.

2. *EMSA*. Assess final human cGAS–DNA complex formation using a standard agarose EMSA protocol, for example, see instructions from Buratowski and Chodosh (1996). Supplement cGAS–DNA binding reactions with 5% (v/v) glycerol for loading on the gel, and then separate unbound DNA and cGAS-DNA complexes on a 2% agarose gel run in 0.5 × TB buffer at 4 °C. Visualize complex formation by subsequently staining the gel in a bath of 0.5 × TB buffer containing 10 μg mL^{-1} ethidium bromide and imaging with UV light (Fig. 3A). Human cGAS–DNA complex formation typically occurs as a single band corresponding to a minimal 2:2 complex. In addition to the minimal cGAS–DNA complex formed with short DNA, higher-order oligomerization through "DNA laddering" and a cGAS–DNA liquid phase transition are key features of cGAS activation (Andreeva et al., 2017; Du & Chen, 2018). A series of larger cGAS–DNA bands can be observed for mouse cGAS–DNA complexes and likely correspond to higher-order complex formation.

3.3 Human cGAS–DNA crystallization protocol

The human cGAS variant K187N/L195R enables crystallization of the human cGAS–DNA complex using techniques similar to those previously developed to study mouse and other mammalian homolog cGAS enzymes (Andreeva et al., 2017; Gao, Ascano, Wu, et al., 2013; Li et al., 2013; Zhang et al., 2014). For general considerations relating to protein crystallography and X-ray diffraction data collection, please see detailed protocols recently outlined by McPherson (2017) and Moreno (2017). In this section we briefly describe the sparse matrix screening strategy used to identify initial crystallization conditions, and then focus primarily on a step-by-step guide to optimized procedures for crystallization of the human cGAS–DNA complex (Zhou et al., 2018).

3.3.1 Equipment

1. Protein crystallization robot
2. Hanging-drop crystallography trays (96-well and 15-well)

3.3.2 Reagents

1. Reaction buffer: 20 mM HEPES-KOH pH 7.5, 125 mM KCl, 5 mM ATP, 10.5 mM $MgCl_2$, 1 mM TCEP.
2. EasyXtal sparse matrix screening solutions (Qiagen).
3. Classics II B8 buffer: 0.1 M HEPES pH 7.5, 1.4 M sodium citrate.
4. Optimized buffer: 0.1 M HEPES-NaOH pH 7.0, 1.4 M sodium citrate.

3.3.3 Protocol

1. *Sparse matrix identification of crystallization conditions.* Initial crystallization conditions were determined with 96-well sparse matrix screening using a Mosquito protein crystallization robot (TTP LabTech) and EasyXtal screening solutions (Qiagen). Briefly, purified human cGAS K187N/L195R (residues 157–522) was mixed with DNA17 in a molar ratio of 1:1.25 protein:DNA on ice with a final salt concentration of 116 mM KCl. The final concentrations of human cGAS K187N/L195R and DNA17 are 80 and 100 μM, respectively. After mixture on ice in a microfuge tube, the reaction was centrifuged for 5 min at 16,000 × *g*, 4 °C and precipitated protein was removed by transferring the solution to a new microfuge tube. The human cGAS and DNA mixture was then incubated for 30 min at 18 °C and centrifuged again to remove precipitant prior to sparse matrix screening. Crystallization drops were set at 18 °C in a 96-well hanging-drop format with 200 nL of human cGAS–DNA mixture and 200 nL of screen solution using a Mosquito crystallization robot. After 1–2 days of growth at 18 °C, crystals were observed in Qiagen Classics II B8 buffer: 0.1 M HEPES pH 7.5, 1.4 M sodium citrate.
2. *Optimized Human cGAS–DNA Crystallization Conditions.* Optimized human cGAS–DNA crystals can be grown using 15-well Qiagen EasyXtal hanging drop trays and a final reservoir solution of 0.1 M HEPES-NaOH pH 7.0, 1.4 M sodium citrate. First, prepare human cGAS–DNA mixture as outlined above, and then set 2 μL hanging-drops over a 350 μL reservoir solution. Set drops by hand using 1 μL of human cGAS–DNA solution and then 1 μL of reservoir solution. Small rod-shaped crystals should be visible after ~24 h, and crystals reach optimal size after ~3 days at 18 °C (Fig. 2C).

   Tip: It is possible to set three drops on each lid for 45 total drops per plate. Additionally, we recommend using reservoir solution conditions ranging from 0.1 M HEPES-NaOH pH 7.0–7.5, and 1.2–1.6 M sodium citrate. The human cGAS–DNA rod-shaped crystals often grow in clusters. Using a whisker from Kozak or another feline source, gently tease the base of the rods to release individual crystals for harvesting. Alternatively, individual crystals can be grown using standard micro-seeding techniques. Briefly, harvest and crush crystals into reservoir solution and then use 0.2 μL of a 1:1000–1:10,000 dilution to seed drops for crystal growth.

3.4 Human cGAS–DNA structure determination

Following crystal growth, the last experimental steps are to cryoprotect human cGAS–DNA crystals, freeze in liquid nitrogen, and collect X-ray diffraction data. We detail all steps specific for human cGAS–DNA crystals below, and recommend consulting recent reviews by Dauter and Yamamoto and colleagues for general considerations on X-ray diffraction data collection (Dauter, 2017; Yamamoto et al., 2017).

3.4.1 *Equipment*

1. Light microscope
2. Crystal harvesting tools (nylon loops, vials, vial canes or pucks, liquid nitrogen dewar)
3. Synchrotron radiation source

3.4.2 *Reagents*

1. Reaction buffer: 20 mM HEPES-KOH pH 7.5, 125 mM KCl, 5 mM ATP, 10.5 mM $MgCl_2$, 1 mM TCEP.
2. Cryo-protection solution: 0.1 M HEPES-NaOH pH 7.0, 1.4 M sodium citrate, 10% (v/v) glycerol.

3.4.3 *Protocol*

1. *Crystal cryoprotection and harvest.* Human cGAS–DNA crystals grown in high sodium citrate conditions can be successfully cryoprotected using glycerol as an additive. Prepare 500 μL of cryoprotectant solution (0.1 M HEPES-NaOH pH 7.0, 1.4 M sodium citrate, 10% (v/v) glycerol). 10% (v/v) glycerol is sufficient for cryoprotection in this optimized crystallization condition, and crystals grown in alternative conditions can be harvested using 8–25% (v/v) glycerol or as predicted by previously determined experimental tables for crystal cryoprotection (McFerrin & Snell, 2002). After 3 days of growth at 18 °C in 15-well plates, loosen the hanging drop lid and spot a 1.5 μL drop of cryoprotectant solution near the drop containing crystals. Harvest crystals with a nylon loop by gently transferring individual crystals into the cryoprotectant solution. Incubate human cGAS–DNA crystals in cryoprotectant

for 30–180 s and then harvest again with a nylon loop and plunge immediately into liquid nitrogen to freeze for data collection. Length of soaking time of crystals in cryoprotectant solution is a parameter that requires optimization, but we generally found that 30 s was sufficient to prevent ice formation during crystal freezing.

2. *Data collection and structure determination.* X-ray diffraction data are next collected at a synchrotron radiation source. Human cGAS–DNA crystals belong to the hexagonal spacegroup $P\ 6_1\ 2\ 2$ and the asymmetric unit contains one molecule of human cGAS and one molecule of DNA17 (note: the full dimeric 2:2 human cGAS–DNA assembly is generated through crystal symmetry). Crystals typically diffract between 2.5 and 2.8 Å with our best crystals yielding an optimized dataset to ~2.3 Å. Process X-ray diffraction data using standard software, in our case data were processed with XDS and AIMLESS (Kabsch, 2010) prior to structure determination with molecular replacement using Phaser in PHENIX (Adams et al., 2010). An initial phase estimate was obtained using molecular replacement and the *apo* human cGAS structure (PDB 4KM5) (Kranzusch et al., 2013), but we now recommend using the human cGAS–DNA structure (PDB 6CT9) (Zhou et al., 2018). Complete iterative model building and refinement with Coot (Emsley & Cowtan, 2004) and PHENIX (Fig. 2D–F).

4. Crystallization of a human cGAS–DNA–ligand complex

Structural analysis using *apo* human cGAS and mouse cGAS–DNA structures has facilitated development of initial small molecule leads (Hall et al., 2017; Vincent et al., 2017), but these structures are not suitable for studying compounds that specifically target the active human cGAS conformation. Crystallization of human cGAS in an activated state bound to double-stranded DNA now enables the use of X-ray crystallography to determine high-resolution structures of the human cGAS–DNA complex (Zhou et al., 2018). However, the optimized conditions used to crystallize the human cGAS–DNA complex contain 1.4 M sodium citrate and in spite of extensive co-crystallization attempts with ATP, GTP, or non-hydrolyzable NTP ligands included in the drop, no nucleotide density was observed.

Citrate anions can chelate divalent cations like Mg^{2+}, and we reasoned that metal chelation was the most likely explanation for inhibition of stable nucleotide ligand interactions during crystallization. We tested this hypothesis using an experimental thin-layer chromatography assay designed to analyze 2′3′ cGAMP formation (Kranzusch et al., 2014), and found that sodium citrate potently blocks 2′3′ cGAMP synthesis at 10% (0.1 M) of the concentration that was found in the crystallization drop (Fig. 3B). These results strongly suggested it would not be possible to co-crystallize a human cGAS–DNA–ligand ternary complex with the current conditions, and we therefore sought to develop a crystal soaking approach to remove sodium citrate and obtain ligand-bound crystals. The following section outlines the procedure we used to obtain human cGAS–DNA–ATP ligand-bound crystals and future applications for the study of human-specific small molecule inhibitors.

4.1 Crystal soaking and human cGAS–DNA–ligand crystallization

The critical factor for successfully soaking ligands into human cGAS–DNA crystals is to identify a solution that allows replacement of citrate with a precipitant that permits Mg^{2+} and ligand coordination in the enzyme active site. After an exhaustive search of alternative ionic precipitants, we turned to protocols based on PEG-5000 MME that were previously developed to introduce heavy atoms for anomalous dispersion phase determination (Bergfors, 2009). Optimizing these techniques for the human cGAS–DNA crystals allowed us to identify succinate and PEG-5000 MME as a combination that successfully maintains crystal stability while allowing Mg^{2+} and nucleotide ligand coordination.

4.2 Protocol for ligand soaking and crystal harvest

4.2.1 Reagents

1. Soaking solution: 0.1 M sodium succinate, 35% PEG-5000 MME, 3 mM ATP, 6.3 mM $MgCl_2$, 10% (v/v) glycerol.

4.2.2 Protocol

1. Prepare 500 μL of crystal soaking solution and supplement with nucleotide or appropriate small molecule ligand.
2. After 3 days growth of human cGAS–DNA crystals in citrate solution in 15-well plates (see Section 3.3.3), spot a 1.5 μL drop of soaking solution near to the crystal drop. Loop crystals and transfer to soaking drop for ~10 min (Fig. 3C). After soaking, loop the crystals and directly plunge

into liquid nitrogen to freeze for puck storage and X-ray diffraction data collection as outlined in Section 3.4.3.

3. Data collection and structure determination procedures for the human cGAS–DNA–ligand complex are as outlined for the human cGAS–DNA complex (Fig. 3D–F) (Zhou et al., 2018).

   Note: All currently identified small molecule inhibitors of cGAS target the active-site nucleotide donor pocket where ATP resides in the first step of 2′3′ cGAMP synthesis. Our structure of the human cGAS–DNA–ATP complex demonstrates the current procedures are suitable for crystallization of ligands in this part of the active-site. Further crystallization trials and optimizations should allow capture of ligands in the acceptor pocket including GTP and nonhydrolyzable nucleotide analogs.

5. Summary

Following the discovery of cGAS and the mammalian second messenger 2′3′ cGAMP (Chen et al., 2016; Sun et al., 2013), the work of many research groups has combined to explain the basic mechanism of cGAS-dependent recognition of cytosolic DNA. Most recently, major regulatory themes have emerged including a specific role for higher-order cGAS–DNA complex formation with structured DNA laddering and dynamic organization through liquid phase separation (Andreeva et al., 2017; Du & Chen, 2018). Further biochemistry and structural biology are required to connect these two processes and explain how initial detection of cytosolic DNA is transitioned to become higher-order oligomerization and sustained 2′3′ cGAMP synthesis. Additionally, the structural basis for how the cGAS N-terminus and protein co-factors regulate enzyme function remains unclear (Almine et al., 2017; Jonsson et al., 2017; Lian et al., 2018; Liu et al., 2019; Shannon et al., 2018; Tao et al., 2017; Yoh et al., 2015). The assays outlined here, combined with many developments in the field, provide a foundation to explain regulation of human cGAS and support the field-wide effort toward greater biological understanding and therapeutic manipulation of cGAS-STING signaling.

Acknowledgments

The authors are grateful to John Mekalanos, Olga Danilchanka, and all members of the Kranzusch lab for helpful comments and discussion. The work was funded by the Claudia Adams Barr Program for Innovative Cancer Research (P.J.K.), Richard and Susan Smith Family Foundation (P.J.K.), Charles H. Hood Foundation (P.J.K.), a Cancer Research Institute CLIP Grant (P.J.K.), A.T.W. is supported as a fellow of The Jane Coffin Childs Memorial Fund for Medical Research.

References

Ablasser, A., Goldeck, M., Cavlar, T., Deimling, T., Witte, G., Rohl, I., et al. (2013). cGAS produces a 2′-5′-linked cyclic dinucleotide second messenger that activates STING. *Nature*, *498*, 380–384.

Adams, P. D., Afonine, P. V., Bunkoczi, G., Chen, V. B., Davis, I. W., Echols, N., et al. (2010). PHENIX: A comprehensive python-based system for macromolecular structure solution. *Acta Crystallographica Section D, Biological Crystallography*, *66*, 213–221.

Almine, J. F., O'Hare, C. A., Dunphy, G., Haga, I. R., Naik, R. J., Atrih, A., et al. (2017). IFI16 and cGAS cooperate in the activation of STING during DNA sensing in human keratinocytes. *Nature Communications*, *8*, 14392.

Andreeva, L., Hiller, B., Kostrewa, D., Lassig, C., de Oliveira Mann, C. C., Jan Drexler, D., et al. (2017). cGAS senses long and HMGB/TFAM-bound U-turn DNA by forming protein-DNA ladders. *Nature*, *549*, 394–398.

Bergfors, T. M. (2009). *Protein crystallization*. Internat'l University Line.

Biolatti, M., Dell'Oste, V., Pautasso, S., Gugliesi, F., von Einem, J., Krapp, C., et al. (2018). Human cytomegalovirus tegument protein pp65 (pUL83) dampens type I interferon production by inactivating the DNA sensor cGAS without affecting STING. *Journal of Virology*, *92*. e01774-17.

Buratowski, S., & Chodosh, L. A. (1996). Mobility shift DNA-binding assay using gel electrophoresis. *Current Protocols in Molecular Biology*, *36*(1), 12.2.1–12.2.11. chapter 12, Unit 12 12, PMID: 18265085.

Burdette, D. L., Monroe, K. M., Sotelo-Troha, K., Iwig, J. S., Eckert, B., Hyodo, M., et al. (2011). STING is a direct innate immune sensor of cyclic di-GMP. *Nature*, *478*, 515–518.

Chen, Q., Sun, L., & Chen, Z. J. (2016). Regulation and function of the cGAS-STING pathway of cytosolic DNA sensing. *Nature Immunology*, *17*, 1142–1149.

Civril, F., Deimling, T., de Oliveira Mann, C. C., Ablasser, A., Moldt, M., Witte, G., et al. (2013). Structural mechanism of cytosolic DNA sensing by cGAS. *Nature*, *498*, 332–337.

Dauter, Z. (2017). Collection of X-ray diffraction data from macromolecular crystals. *Methods in Molecular Biology*, *1607*, 165–184.

Davies, B. W., Bogard, R. W., Young, T. S., & Mekalanos, J. J. (2012). Coordinated regulation of accessory genetic elements produces cyclic di-nucleotides for *V. cholerae* virulence. *Cell*, *149*, 358–370.

Diner, E. J., Burdette, D. L., Wilson, S. C., Monroe, K. M., Kellenberger, C. A., Hyodo, M., et al. (2013). The innate immune DNA sensor cGAS produces a noncanonical cyclic dinucleotide that activates human STING. *Cell Reports*, *3*, 1355–1361.

Du, M., & Chen, Z. J. (2018). DNA-induced liquid phase condensation of cGAS activates innate immune signaling. *Science*, *361*, 704–709.

Eaglesham, J. B., Pan, Y., Kupper, T. S., & Kranzusch, P. J. (2019). Viral and metazoan poxins are cGAMP-specific nucleases that restrict cGAS-STING signalling. *Nature*, *566*, 259–263.

Emsley, P., & Cowtan, K. (2004). Coot: Model-building tools for molecular graphics. *Acta Crystallographica Section D, Biological Crystallography*, *60*, 2126–2132.

Ferrieres, L., Hemery, G., Nham, T., Guerout, A. M., Mazel, D., Beloin, C., et al. (2010). Silent mischief: Bacteriophage Mu insertions contaminate products of *Escherichia coli* random mutagenesis performed using suicidal transposon delivery plasmids mobilized by broad-host-range RP4 conjugative machinery. *Journal of Bacteriology*, *192*, 6418–6427.

Gao, P., Ascano, M., Wu, Y., Barchet, W., Gaffney, B. L., Zillinger, T., et al. (2013). Cyclic [G(2′,5′)pA(3′,5′)p] is the metazoan second messenger produced by DNA-activated cyclic GMP-AMP synthase. *Cell*, *153*, 1094–1107.

Gao, P., Ascano, M., Zillinger, T., Wang, W., Dai, P., Serganov, A. A., et al. (2013). Structure-function analysis of STING activation by c[G(2′,5′)pA(3′,5′)p] and targeting by antiviral DMXAA. *Cell*, *154*, 748–762.

George, R. D., McVicker, G., Diederich, R., Ng, S. B., MacKenzie, A. P., Swanson, W. J., et al. (2011). Trans genomic capture and sequencing of primate exomes reveals new targets of positive selection. *Genome Research*, *21*, 1686–1694.
Guzman, L. M., Belin, D., Carson, M. J., & Beckwith, J. (1995). Tight regulation, modulation, and high-level expression by vectors containing the arabinose PBAD promoter. *Journal of Bacteriology*, *177*, 4121–4130.
Hall, J., Brault, A., Vincent, F., Weng, S., Wang, H., Dumlao, D., et al. (2017). Discovery of PF-06928215 as a high affinity inhibitor of cGAS enabled by a novel fluorescence polarization assay. *PLoS One*, *12*, e0184843.
Hancks, D. C., Hartley, M. K., Hagan, C., Clark, N. L., & Elde, N. C. (2015). Overlapping patterns of rapid evolution in the nucleic acid sensors cGAS and OAS1 suggest a common mechanism of pathogen antagonism and escape. *PLoS Genetics*, *11*, e1005203.
Hollis, T. (2007). Crystallization of protein-DNA complexes. *Methods in Molecular Biology*, *363*, 225–237.
Huang, Z. F., Zou, H. M., Liao, B. W., Zhang, H. Y., Yang, Y., Fu, Y. Z., et al. (2018). Human cytomegalovirus protein UL31 inhibits DNA sensing of cGAS to mediate immune evasion. *Cell Host & Microbe*, *24*, 69–80.e64.
Ishikawa, H., & Barber, G. N. (2008). STING is an endoplasmic reticulum adaptor that facilitates innate immune signalling. *Nature*, *455*, 674–678.
Jin, L., Waterman, P. M., Jonscher, K. R., Short, C. M., Reisdorph, N. A., & Cambier, J. C. (2008). MPYS, a novel membrane tetraspanner, is associated with major histocompatibility complex class II and mediates transduction of apoptotic signals. *Molecular and Cellular Biology*, *28*, 5014–5026.
Jonsson, K. L., Laustsen, A., Krapp, C., Skipper, K. A., Thavachelvam, K., Hotter, D., et al. (2017). IFI16 is required for DNA sensing in human macrophages by promoting production and function of cGAMP. *Nature Communications*, *8*, 14391.
Kabsch, W. (2010). XDS. *Acta Crystallographica Section D, Biological Crystallography*, *66*, 125–132.
Kato, K., Ishii, R., Goto, E., Ishitani, R., Tokunaga, F., & Nureki, O. (2013). Structural and functional analyses of DNA-sensing and immune activation by human cGAS. *PLoS One*, *8*, e76983.
Khoo, L. T., & Chen, L. Y. (2018). Role of the cGAS-STING pathway in cancer development and oncotherapeutic approaches. *EMBO Reports*, *19*, e46935.
Kranzusch, P. J., Lee, A. S., Berger, J. M., & Doudna, J. A. (2013). Structure of human cGAS reveals a conserved family of second-messenger enzymes in innate immunity. *Cell Reports*, *3*, 1362–1368.
Kranzusch, P. J., Lee, A. S. Y., Wilson, S. C., Solovykh, M. S., Vance, R. E., Berger, J. M., et al. (2014). Structure-guided reprogramming of human cGAS dinucleotide linkage specificity. *Cell*, *158*, 1011–1021.
Lee, K. M., & Narlikar, G. (2001). Assembly of nucleosomal templates by salt dialysis. *Current Protocols in Molecular Biology*, ***54***(1), 21.6.1–21.6.16, chapter 21, Unit 21.26.
Li, X., Shu, C., Yi, G., Chaton, C. T., Shelton, C. L., Diao, J., et al. (2013). Cyclic GMP-AMP synthase is activated by double-stranded DNA-induced oligomerization. *Immunity*, *39*, 1019–1031.
Lian, H., Wei, J., Zang, R., Ye, W., Yang, Q., Zhang, X. N., et al. (2018). ZCCHC3 is a co-sensor of cGAS for dsDNA recognition in innate immune response. *Nature Communications*, *9*, 3349.
Liu, Z. S., Cai, H., Xue, W., Wang, M., Xia, T., Li, W. J., et al. (2019). G3BP1 promotes DNA binding and activation of cGAS. *Nature Immunology*, *20*, 18–28.
McFerrin, M. B., & Snell, E. H. (2002). The development and application of a method to quantify the quality of cryoprotectant solutions using standard area-detector X-ray images. *Journal of Applied Crystallography*, *35*, 538–545.

McPherson, A. (2017). Protein crystallization. *Methods in Molecular Biology*, *1607*, 17–50.
Moreno, A. (2017). Advanced methods of protein crystallization. *Methods in Molecular Biology*, *1607*, 51–76.
Reverter, D., & Lima, C. D. (2006). Structural basis for SENP2 protease interactions with SUMO precursors and conjugated substrates. *Nature Structural & Molecular Biology*, *13*, 1060–1068.
Severin, G. B., Ramliden, M. S., Hawver, L. A., Wang, K., Pell, M. E., Kieninger, A. K., et al. (2018). Direct activation of a phospholipase by cyclic GMP-AMP in El Tor *Vibrio cholerae*. *Proceedings of the National Academy of Sciences of the United States of America*, *115*, E6048–E6055.
Shannon, J. L., Murphy, M. S., Kantheti, U., Burnett, J. M., Hahn, M. G., Dorrity, T. J., et al. (2018). Polyglutamine binding protein 1 (PQBP1) inhibits innate immune responses to cytosolic DNA. *Molecular Immunology*, *99*, 182–190.
Stetson, D. B., & Medzhitov, R. (2006). Recognition of cytosolic DNA activates an IRF3-dependent innate immune response. *Immunity*, *24*, 93–103.
Studier, F. W. (2005). Protein production by auto-induction in high density shaking cultures. *Protein Expression and Purification*, *41*, 207–234.
Sun, W., Li, Y., Chen, L., Chen, H., You, F., Zhou, X., et al. (2009). ERIS, an endoplasmic reticulum IFN stimulator, activates innate immune signaling through dimerization. *Proceedings of the National Academy of Sciences of the United States of America*, *106*, 8653–8658.
Sun, L., Wu, J., Du, F., Chen, X., & Chen, Z. J. (2013). Cyclic GMP-AMP synthase is a cytosolic DNA sensor that activates the type I interferon pathway. *Science*, *339*, 786–791.
Tao, J., Zhang, X. W., Jin, J., Du, X. X., Lian, T., Yang, J., et al. (2017). Nonspecific DNA binding of cGAS N terminus promotes cGAS activation. *Journal of Immunology*, *198*, 3627–3636.
Vincent, J., Adura, C., Gao, P., Luz, A., Lama, L., Asano, Y., et al. (2017). Small molecule inhibition of cGAS reduces interferon expression in primary macrophages from autoimmune mice. *Nature Communications*, *8*, 750.
Whiteley, A. T., Eaglesham, J. B., de Oliveira Mann, C. C., Morehouse, B. R., Lowey, B., Nieminen, E. A., et al. (2019). Bacterial cGAS-like enzymes synthesize diverse nucleotide signals. *Nature*, *567*, 194–199.
Wu, J. J., Li, W., Shao, Y., Avey, D., Fu, B., Gillen, J., et al. (2015). Inhibition of cGAS DNA sensing by a herpesvirus virion protein. *Cell Host & Microbe*, *18*, 333–344.
Yamamoto, M., Hirata, K., Yamashita, K., Hasegawa, K., Ueno, G., Ago, H., et al. (2017). Protein microcrystallography using synchrotron radiation. *IUCrJ*, *4*, 529–539.
Yoh, S. M., Schneider, M., Seifried, J., Soonthornvacharin, S., Akleh, R. E., Olivieri, K. C., et al. (2015). PQBP1 is a proximal sensor of the cGAS-dependent innate response to HIV-1. *Cell*, *161*, 1293–1305.
Zhang, G., Chan, B., Samarina, N., Abere, B., Weidner-Glunde, M., Buch, A., et al. (2016). Cytoplasmic isoforms of kaposi sarcoma herpesvirus LANA recruit and antagonize the innate immune DNA sensor cGAS. *Proceedings of the National Academy of Sciences of the United States of America*, *113*, E1034–E1043.
Zhang, X., Shi, H., Wu, J., Zhang, X., Sun, L., Chen, C., et al. (2013). Cyclic GMP-AMP containing mixed phosphodiester linkages is an endogenous high-affinity ligand for STING. *Molecular Cell*, *51*, 226–235.
Zhang, X., Wu, J., Du, F., Xu, H., Sun, L., Chen, Z., et al. (2014). The cytosolic DNA sensor cGAS forms an oligomeric complex with DNA and undergoes switch-like conformational changes in the activation loop. *Cell Reports*, *6*, 421–430.
Zhang, J., Zhao, J., Xu, S., Li, J., He, S., Zeng, Y., et al. (2018). Species-specific deamidation of cGAS by herpes simplex virus UL37 protein facilitates viral replication. *Cell Host & Microbe*, *24*, 234–248.e235.

Zhong, B., Yang, Y., Li, S., Wang, Y. Y., Li, Y., Diao, F., et al. (2008). The adaptor protein MITA links virus-sensing receptors to IRF3 transcription factor activation. *Immunity*, *29*, 538–550.

Zhou, W., Whiteley, A. T., de Oliveira Mann, C. C., Morehouse, B. R., Nowak, R. P., Fischer, E. S., et al. (2018). Structure of the human cGAS-DNA complex reveals enhanced control of immune surveillance. *Cell*, *174*. 300-311.e311.

CHAPTER THREE

Chemical synthesis, purification, and characterization of 3′-5′-linked canonical cyclic dinucleotides (CDNs)

Changhao Wang[a,*], Min Hao[a], Qianqian Qi[a], Yashao Chen[a], Jörg S. Hartig[b,c,*]

[a]Key Laboratory of Applied Surface and Colloid Chemistry, Ministry of Education, School of Chemistry and Chemical Engineering, Shaanxi Normal University, Xi'an, China
[b]Department of Chemistry, University of Konstanz, Konstanz, Germany
[c]Konstanz Research School Chemical Biology (KoRS-CB), University of Konstanz, Konstanz, Germany
*Corresponding authors: e-mail address: changhaowang@snnu.edu.cn; joerg.hartig@uni-konstanz.de

Contents

Abstract

So far, four cyclic dinucleotides (CDNs) have been discovered as important second messengers in nature, where three canonical CDNs of c-di-GMP, c-di-AMP and c-AMP-GMP were found in bacteria containing two 3′-5′ phosphodiester linkages and one non-canonical CDN 2′3′-c-GMP-AMP was identified in mammals containing mixed 2′-5′ and 3′-5′ phosphodiester linkages. The CDNs are produced by specific cyclases and degraded by phosphodiesterases (PDEs). All of the known CDNs could bind to the

Methods in Enzymology, Volume 625
ISSN 0076-6879
https://doi.org/10.1016/bs.mie.2019.04.022

stimulator of interferon genes (STING) to induce type I interferon (IFN) responses and the three bacterial CDNs are sensed by specific riboswitches to regulate gene expression. The emerging physiological functions of bacterial CDNs lead the motivation to investigate other possible canonical CDNs. In recent years, many endeavors have been devoted to develop fast, convenient and cheap strategies for chemically synthesizing CDNs and their analogues. The phosphoramidite approach using commercial starting materials has attracted much attention. Herein, we describe an adapted one-pot strategy that enables fast synthesis of crude 3′-5′-linked canonical CDNs followed by purification of the obtained CDNs using reversed phase high-performance of liquid chromatography (HPLC). Furthermore, we report the full characterization of CDNs by mass spectrometry (MS) and nuclear magnetic resonance (NMR) techniques.

1. Introduction

Cyclic dinucleotides (CDNs) are natural cyclic RNA molecules bearing a 12-membered ring scaffold and two ribonucleotides. To date, four CDNs have been discovered in nature (Fig. 1), including three canonical CDNs of bis-(3′5′)-cyclic dimeric guanosine monophosphate (c-di-GMP), bis-(3′5′)-cyclic dimeric adenosine monophosphate (c-di-AMP) and bis-(3′5′)-cyclic adenosine monophosphate-guanosine monophosphate (c-AMP-GMP), and

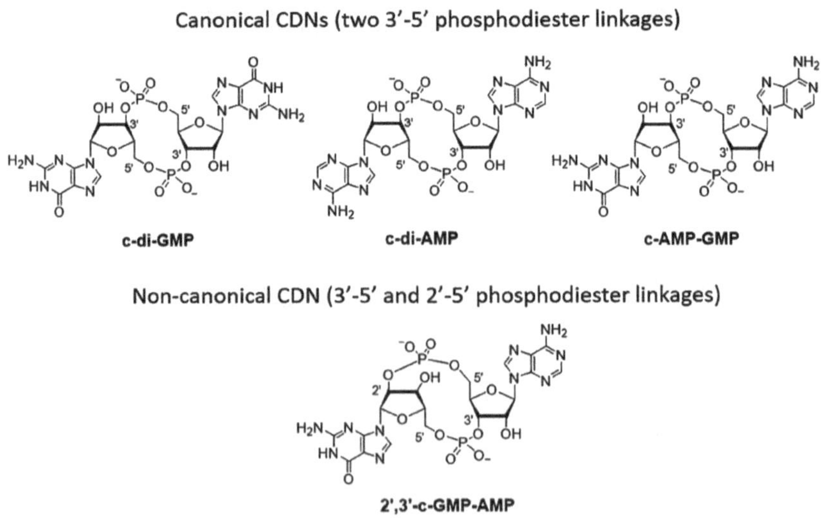

Fig. 1 The chemical structures of naturally occurring canonical CDNs with two 3′-5′ phosphodiester linkages and non-canonical CDN with mixed 3′-5′ and 2′-5′ phosphodiester linkages.

one non-canonical CDN of (2′5′),(3′5′)-cyclic guanosine monophosphate-adenosine monophosphate (2′,3′-c-GMP-AMP). All of the four CDNs have been identified as second messenger signaling molecules in regulation of many cellular processes.

In 1987, c-di-GMP was initially identified as an allosteric activator of a bacterial cellulose synthase in *Gluconacetobacter xylinus* (Ross et al., 1987). C-di-GMP is naturally produced from two guanosine triphosphate (GTP) molecules by diguanylate cyclases (DGCs) containing GGDEF domain, while it could be degraded into 5′-phosphoguanylyl-(3′5′)-guanosine (5′-pGpG) by phosphodiesterases (PDEs) containing EAL or HD-GYP domain and 5′-pGpG is further decomposed into two guanosine monophosphate (GMP) molecules by PDE containing HD-GYP domain. In the past three decades, c-di-GMP has been demonstrated as a regulator in many physiological processes such as cell-cell communication, motility, biofilm formation, biosynthesis of exopolysaccharide, sessility and virulence (Hengge, 2009). In 2008, c-di-AMP was discovered (Witte, Hartung, Buttner, & Hopfner, 2008) and 2 years later its bacterial production was found in *Listeria monocytogenes* (Woodward, Iavarone, & Portnoy, 2010). It has been identified as a crucial intracellular signaling molecule for the regulation of cell wall metabolism, DNA damage, osmotic stress response and bacterial sporulation (Corrigan & Gruendling, 2013). DNA integrity scanning protein A (DisA) that possessing diadenylate cyclase (DAC) activity can convert two adenosine triphosphate (ATP) molecules to c-di-AMP, and the degradation of c-di-AMP into 5′-phosphoadenylyl-(3′5′)-adenosine (5′-pApA) can be conducted by PDEs containing DHH-DHHA or HD domain. In 2012, the third bacterial CDN c-AMP-GMP was discovered in *Vibrio cholerae* (Davies, Bogard, Young, & Mekalanos, 2012). C-AMP-GMP is synthesized from ATP and GTP by a dinucleotide cyclase (DncV) from *Vibrio cholerae*, while it could be degraded into an intermediate of 5′-phosphoadenylyl-(3′5′)-guanosine (5′-pApG) by specific cyclic GMP-AMP PDEs (V-cGAP1, V-cGAP2 or V-cGAP3) and further split into 5′-adenylyl-(3′5′)-guanosine (5′-ApG) by V-cGAP1 (Gao et al., 2015). C-AMP-GMP has been demonstrated to be a crucial signaling molecule for regulating bacterial virulence, promoting intestinal colonization, downregulating chemotaxis and controlling exoelectrogenic process (Davies et al., 2012; Nelson et al., 2015). In 2013, an isomer of c-AMP-GMP (2′,3′-c-GMP-AMP) with mixed 2′-5′ and 3′-5′ phosphodiester linkages was identified as a novel second messenger in metazoan (Gao et al., 2013). 2′,3′-c-GMP-AMP is cyclized from ATP and GTP by a cyclic GMP-AMP synthase (cGAS) upon binding to double-stranded DNA in cytosol, and is degraded to an intermediate of 5′-pApG and fast decomposed into AMP and GMP by Ecto-nucleotide pyrophosphatase phosphodiesterase 1 (ENPP1) (Kato et al., 2018). 2′,3′-c-GMP-AMP has been demonstrated to bind and activate the stimulator of interferon genes (STING) and thereafter induce the production of type I interferons (IFNs) (Ablasser et al., 2013; Danilchanka & Mekalanos, 2013). Beside 2′,3′-c-GMP-AMP, all other three bacterial CDNs could also activate STING to elicit the innate immune response in mammalian cells, resulting in the expression of type I IFNs (Burdette et al., 2011; Gao et al., 2013; Parvatiyar et al., 2012).

The CDN second messengers could be sensed by effectors made of protein or RNA, resulting in the production of specific phenotypic responses for regulating the metabolic pathways. For protein effectors, the four known CDNs could efficiently interact with their dinucleotide cyclases and PDEs. Moreover, c-di-GMP has been identified to bind to PilZ domain proteins, FleQ and PelD transcription factors, I site effectors (Hengge, 2009). C-di-AMP has been found to interact with transcription factor of DarR and domain proteins of KtrA, CpaA, KdpD and PstA (Corrigan & Gruendling, 2013). For RNA effectors, the three bacterial CDNs have been identified to tightly bind to RNA molecules called riboswitches. Riboswitches are noncoding RNA motifs located in the 5′-untranslated region (UTR) of messenger RNA (mRNA) molecules consisting of an aptamer domain and an expression platform. After selectively binding to small molecule ligands, the aptamer domain changes its conformation to further rearrange the expression platform, leading to the regulation of gene expression. C-di-GMP has been found to target two classes of riboswitches of the genes related to environment, membranes, and motility (GEMM) with high affinities (Lee, Baker, Weinberg, Sudarsan, & Breaker, 2010; Sudarsan et al., 2008). In the in-line probing assays, c-di-GMP binds GEMM class I and class II riboswitches with the apparent K_d at 10 pM and 2 nM, respectively. C-di-AMP is recognized by the ydaO riboswitch with an apparent K_d of about 700 pM (Nelson et al., 2013). Very recently, c-AMP-GMP has been discovered to be sensed by a subset of GEMM class Ib riboswitches in *Geobacter* species, where GEMM class Ib is the mutant of c-di-GMP GEMM class I riboswitch with mutation of G20 (Nelson et al., 2015). Bioinformatical analysis of GEMM class I riboswitches revealed that several sequence variations are conserving with the mutation of either G20 or C92, which could potentially sense other unknown CDNs (Nelson et al., 2015).

Besides the basic starting units GTP and ATP of known CDNs, the triphosphates of cytidine triphosphate (CTP) and uracil triphosphate (UTP) are existing in nature in high abundancy, leading us to speculate that potentially novel CDNs with cytidine monophosphate (CMP) or uracil monophosphate (UMP) moieties might also play similar roles in bacterial signaling. Besides the known canonical CDNs, seven more 3′-5′-linked canonical CDNs bearing natural RNA bases are depicted in Fig. 2. Since only the known CDNs have their corresponding dinucleotide cyclases, the novel unknown CDNs could be achieved by chemical synthesis. The chemical approaches for the known CDNs are widely investigated and synthesis strategies for 3′-5′-linked CDNs have been recently reviewed (Clivio, Coantic-Castex, & Guillaume, 2013). Either for CDN synthesis or its

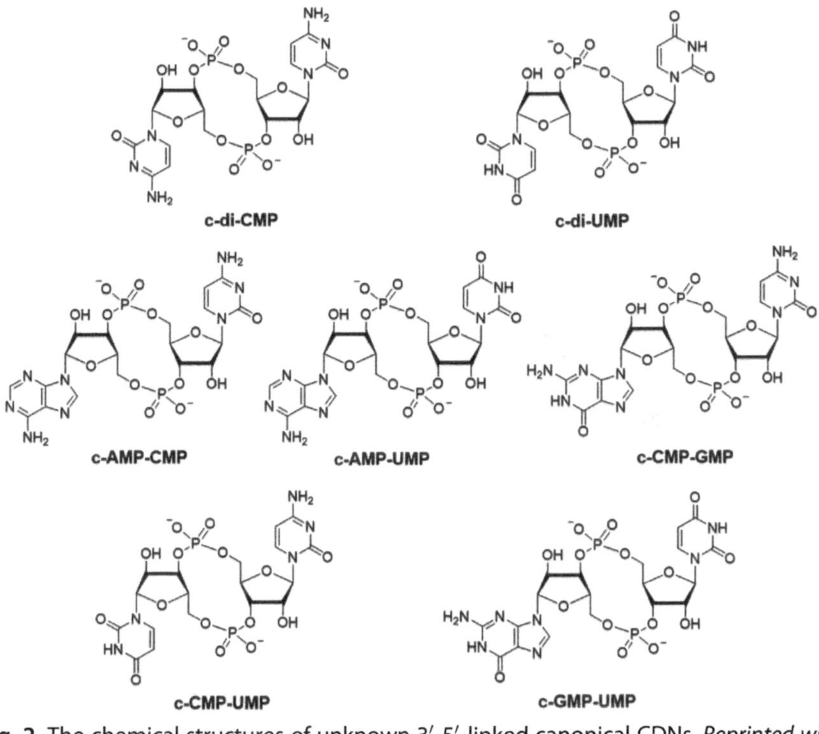

Fig. 2 The chemical structures of unknown 3′-5′-linked canonical CDNs. *Reprinted with permission from Wang, C., Sinn, M., Stifel, J., Heiler, A. C., Sommershof, A. & Hartig, J. S. (2017). Synthesis of all possible canonical (3′-5′-linked) cyclic dinucleotides and evaluation of riboswitch interactions and immune-stimulatory effects.* Journal of the American Chemical Society, 139, *16154–16160. Copyright (2017) American Chemical Society.*

derivatives, the linear precursor of CDN is commonly using phosphoramidite approach and the cyclization step is using phosphotriester approach (Ching, Tan, Chua, & Lam, 2010; Hayakawa, Nagata, Hirata, Hyodo, & Kawai, 2003; Hyodo, Sato, & Hayakawa, 2006; Shanahan, Gaffney, Jones, & Strobel, 2011). Some reports described that c-di-GMP could be produced by H-phosphonate approach (Yan & Aguilar, 2007; Zhang, Gaffney, & Jones, 2004) and c-di-AMP, c-di-CMP and c-di-UMP could be synthesized by phosphodiester approach (Coutsogeorgopoulos & Khorana, 1964; Rammler, Lapidot, & Khorana, 1963; Sawai, Higa, & Kuroda, 1992). An alternative base insertion approach was developed to synthesize CDNs by introducing nucleobases at the last stage to a cyclic sugar-phosphate backbone, yet this method is suitable to synthesize symmetric CDNs other than asymmetric CDNs (Amiot, Heintz, & Giese, 2006). Lam and coworkers employed nucleosides as the starting materials to synthesize c-di-GMP, c-di-UMP and c-GMP-UMP using nearly 20 steps and several times of purifications (Ching et al., 2010). Solid-phase synthesis was applied to synthesize CDNs because they can be easily prepared by automated synthesis using the starting materials with the support linking to either the nucleobase or the 3′-phosphate group (Kiburu, Shurer, Yan, & Sintim, 2008; Shanahan et al., 2011). However, in most cases, the non-commercial starting materials lead to complicated and time-consuming synthesis routes. The most recent phosphoramidite approach is frequently used because of commercial starting materials. In particular, the Jones group has developed a one-pot strategy for synthesizing CDNs and their derivatives, making the synthesis of CDNs become fast and convenient (Gaffney & Jones, 2014; Gaffney, Stebbins, & Jones, 2013; Gaffney, Veliath, Zhao, & Jones, 2010). Very recently, we modified the one-pot strategy to synthesize all possible 3′-5′-linked canonical CDNs and achieved the CDN synthesis within 1 day (Wang et al., 2017). In this chapter, we describe the adapted one-pot strategy for synthesis of the 3′-5′-linked CDNs, purification protocols, and the necessary characterization techniques.

2. Synthesis of crude CDNs

2.1 Chemicals

(a) RNA phosphoramidites: DMT-2′O-TBDMS-rA(Bz) phosphoramidite, DMT-2′O-TBDMS-rC(Ac) phosphoramidite, DMT-2′O-TBDMS-rG(*i*B) phosphoramidite, and DMT-2′O-TBDMS-rU phosphoramidite (*Sigma-Aldrich or Linktech*)

(b) Pyridine trifluoroacetate
(c) *tert*-Butylamine (*t*BuNH$_2$)
(d) Dichloroacetic acid (DCA)
(e) *tert*-Butyl hydroperoxide (*t*BuOOH, ~5.5 M in decane over molecular sieve 4 Å)
(f) 2-Chloro-5,5-dimethyl-1,3,2-dioxaphosphorinane 2-oxide (DMOCP)
(g) Acetonitrile, extra dry over molecular sieve (CH_3CN, *Acros*)
(h) Pyridine, extra dry over molecular sieve (*Acros*)
(i) Sodium bisulfite ($NaHSO_3$)
(j) Sodium bicarbonate (NH_4HCO_3)
(k) Methylamine, 33 wt% in anhydrous ethanol
(l) Triethylamine (TEA)
(m) Triethylamine trihydrofluoride (TEA•3HF)
(n) Dichloromethane (CH_2Cl_2)
(o) Ethyl acetate
(p) Diethyl ether
(q) Acetone

2.2 Equipment

(a) Rotary evaporator equipped with vacuum system
(b) Magnetic stirrer integrated with heating module
(c) 50 μL Hamilton syringe

2.3 Synthesis procedure for crude c-AMP-CMP

As shown in Scheme 1, the synthesis of c-AMP-CMP is taken as an example to describe the typical procedure as follows:

(a) Dissolve DMT-2′O-TBDMS-rA(Bz) phosphoramidite (**1**, 0.46 mmol, 1.0 eq.) and pyridinium trifluoroacetate (107 mg, 0.55 mmol, 1.2 eq.) in dry CH_3CN (4 mL) with a 50 mL round flask under argon atmosphere. Then, add H_2O (17 μL, 0.94 mmol, 2 eq.) to hydrolysis of **1** and stir for 1 min to give **2**.
(b) Add *t*BuNH$_2$ (2.5 mL) to remove the cyanoethyl group (CE) and stir for 1 min to give **3**, concentrate the solution to foam under reduced pressure using dry CH_3CN for three times (5 mL × 3).
(c) Dissolve the resulted foam in CH_2Cl_2 (5.5 mL) under argon atmosphere. Add DCA solution (6% in CH_2Cl_2, 5.5 mL, 4.0 mmol) to remove the 4,4′-dimethoxytriphenylmethyl group (DMT) and H_2O

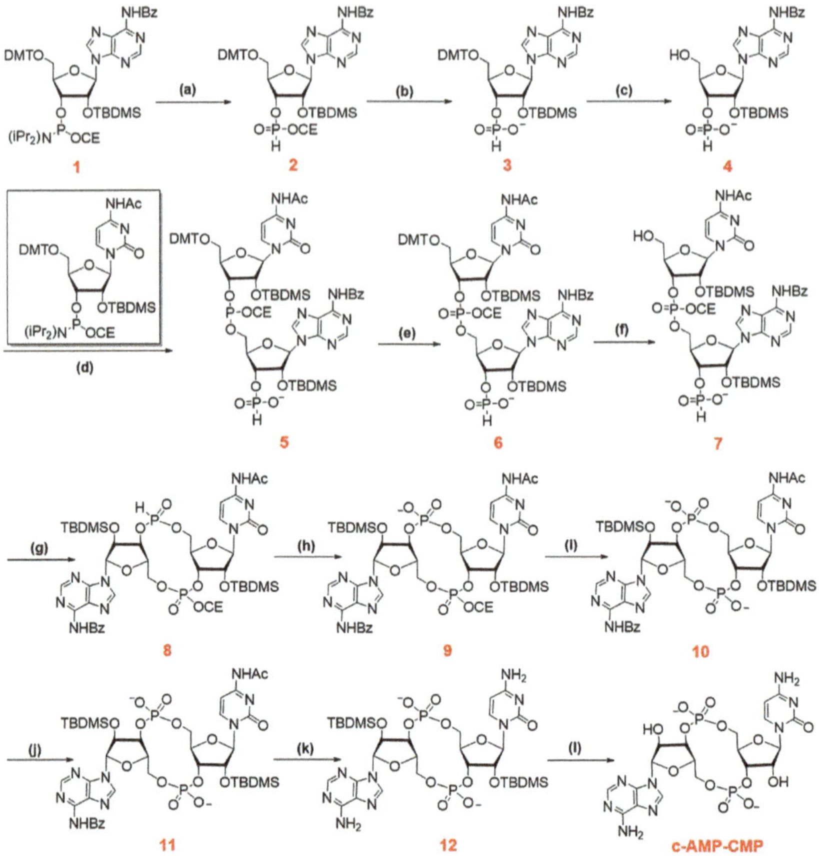

Scheme 1 Synthesis routes for c-AMP-CMP. *Adapted with permission from Wang, C., Sinn, M., Stifel, J., Heiler, A. C., Sommershof, A. & Hartig, J. S. (2017). Synthesis of all possible canonical (3′-5′-linked) cyclic dinucleotides and evaluation of riboswitch interactions and immune-stimulatory effects.* Journal of the American Chemical Society, 139, *16154–16160. Copyright (2017) American Chemical Society.*

(83 μL, 4.6 mmol, 10 eq.) to suppress the retritylation, stir for 10 min to give **4**. Then, add pyridine (0.66 mL, 8.2 mmol) to quench the excess DCA, concentrate the solution to foam under reduced pressure using dry CH_3CN for three times (5 mL × 3) and leave the last solution (A) at about 1 mL.

Tip: If there are some solids precipitated from the solution (A) after evaporation, put the flask in a water bath (ca. 40 °C) to re-dissolve the solids for the next step.

(d) Dissolve DMT-2′O-TBDMS-rC(Ac) phosphoramidite (0.6 mmol, 1.3 eq.) in dry CH_3CN (5 mL), concentrate the solution to foam under reduced pressure doubly using dry CH_3CN (5 mL × 2) and leave the last solution (B) at about 2 mL. Then, mix the solution A from step c and the solution B under argon atmosphere to proceed the coupling reaction, stir for 2 min to give **5**.

(e) Add *t*BuOOH solution (5.5 *M* in decane, 0.25 mL, 1.38 mmol) to oxide phosphite triester to phosphodiester bond and stir for 30 min to give **6**. Then, put the flask into an ice bath, add $NaHSO_3$ aqueous solution (0.5 M, 230 μL, 1.15 mmol, 2.5 eq.) to consume the excess *t*BuOOH and stir for 10 min. At last, remove the ice bath, stir for 5 min at room temperature, and concentrate the solution to oil under reduced pressure.

(f) Dissolve the above oil in CH_2Cl_2 (7.4 mL) under argon atmosphere. Add DCA solution (6% in CH_2Cl_2, 7.4 mL, 5.4 mmol) and H_2O (83 μL, 4.6 mmol, 10 eq.) to remove the 4,4′-dimethoxytriphenylmethyl (DMT) group, stir for 10 min to give **7**.

(g) Add pyridine (1.5 mL, 18.6 mmol) to quench the reaction, and concentrate the solution to oil. Next, add pyridine (15 mL), concentrate the solution under reduced pressure, and leave last solution at about 8 mL. Then, add DMOCP (303 mg, 1.6 mmol, 3.5 eq.) to promote the cyclization and stir for 30 min to give **8**.

(h) Add H_2O (294 μL, 16.3 mmol) to quench the excess of DMOCP and I_2 (152 mg, 0.6 mmol) to transform H-phosphonate diester to phosphodiester, stir for 5 min to give **9**. Next, pour the above mixture into $NaHSO_3$ aqueous solution (10 m*M*, 88 mL, 0.92 mmol, 2.0 eq.) to consume the excess I_2 and stir for 30 min. Then, slowly add $NaHCO_3$ (1.8 g) to neutralize the excess acids and stir for 5 min. At last, triply use the mixture of ethyl acetate and diethyl ether (1:1, 100 mL × 3) to extract the aqueous solution, combine the organic layer and concentrate to oil under reduced pressure.

Tip: The compound **9** could be purified on silica gel column and the compounds obtained after step (i) are hard to be purified using silica gel column.

(i) Dissolve the above oil in CH_3CN (2.5 mL), add *t*$BuNH_2$ (2.5 mL) to remove the 2-cyanoethyl (CE) group and stir for 10 min to give **10**. Concentrate the solution to foam under reduced pressure using dry CH_3CN for three times (5 mL × 3). Next, dissolve the foam with a small volume of CH_2Cl_2. (*Tip:* Add CH_3OH to promote the solubility if the foam cannot be fully dissolved in CH_2Cl_2.) Then, slowly add diethyl ether to precipitate the off-white solid, collect and dry the precipitate.

(j) Dissolve the above crude precipitate (0.1 mmol) in methylamine solution (33 wt% in anhydrous ethanol, 10.0 mL, 80 mmol) under argon to remove the benzoyl group (Bz) and acetyl group (Ac) and stir for 90 min at room temperature to give **11**. Then, concentrate the solution to oil and dry the oil under reduced pressure for three times with a mixture of pyridine (0.4 mL) and TEA (0.2 mL).

(k) Dissolve the above residual oil in pyridine (0.4 mL) under argon and place the reaction flask in an oil bath at 50 °C. Then, spontaneously add TEA (1.4 mL, 10 mmol) and $Et_3N{\cdot}3HF$ (0.9 mL, 5.5 mmol) to remove the *tert*-butyldimethylsilyl groups (TBDMS), and stir for 3 h to give **12**.

(l) Remove the oil bath and cool to room temperature. Then, slowly add acetone to the above solution to form the off-white precipitate and stir for 10 min. At last, collect the precipitate and dry it. The crude c-AMP-CMP was obtained in triethylammonium salt form.

3. Purification of crude CDNs

3.1 Preparation of HPLC buffers

(1) Preparation of triethylammonium acetate buffer (TEAA, 50 m*M*, pH 7.0): Dissolve 2.76 mL glacial acetic acid into a 1.0 L beaker containing ~970 mL of milli-Q water. Next, slowly add 6.95 mL TEA into the above solution and mix the solution thoroughly. Then, adjust the pH value to ~7.0 with diluted acetic acid solution and add the total volume to 1.0 L with milli-Q water. Finally, filter the as-prepared TEAA buffer through a 0.2 μm HPLC-certified Nylon filter for HPLC.

(2) Preparation of ammonium bicarbonate buffer (ABC, 50 m*M*, pH 7.0): Dissolve 3.95 g ammonium bicarbonate into a 1.0 L beaker containing ~970 mL of milli-Q water. Next, adjust the pH value to ~7.0 with diluted formic acid solution and adjust the total volume to 1.0 L with milli-Q water. Finally, filter the as-prepared ABC buffer through a 0.2 μm HPLC-certified Nylon filter for HPLC.

3.2 Purification protocol for crude c-AMP-CMP

For purification of CDN, a set of Shimadzu liquid chromatography (LC) system is employed that equipped with LC 20AT pumps, SPD-M20A

PDA detectors, and FRC-10A fraction collector. Acetonitrile is used as mobile phase A, while either TEAA buffer (50 m*M*, pH 7.0) or ABC buffer (50 m*M*, pH 7.0) is used as mobile phase B. A 250/10 Nucleodur C18 Pyramid, 5 μm column (*Macherey-Nagel*) is used with the gradients of acetonitrile (2–25%) in either TEAA buffer or ABC buffer. The high performance liquid chromatography (HPLC) protocol for purification of crude c-AMP-CMP is shown in Table 1.

Tips:

(1) Compared with TEAA buffer at 50 m*M*, the peak of CDN on HPLC trace becomes sharper when using TEAA buffer at 100 m*M*. The separation efficiency of crude CDN is nearly the same either using 50 m*M* or 100 m*M* TEAA buffer, but high concentration of TEAA needs longer time to remove the TEAA salt from the c-AMP-CMP fraction through lyophilization.

Table 1 HPLC protocol for purification of crude c-AMP-CMP.

Sample preparation	30 mg of crude c-AMP-CMP in 180 μL H_2O	
HPLC instrument	Shimadzu systems equipped with LC 20AT pumps, SPD-M20A PDA detectors, and FRC-10A fraction collector	
Column	VP 250/10 NUCLEODUR Pyramid, 5 μM	
Flow rate	4 mL/min	
Wavelength	260 nm	
Temperature	35 °C	
Mobile phase	A: TEAA buffer (50 m*M*, pH 7.0) B: CH_3CN	
	Time (min)	**CH_3CN (B, %)**
Time program	0	2
	5	2
	25	20
	27.5	100
	30	100
	35	2
	40	2

(2) Compared with TEAA buffer and ABC buffer at the same concentration, using the later one as mobile phase B will bring earlier elution time of CDN and little higher separation efficiency. However, ABC buffer is more difficult to remove than TEAA buffer in lyophilization under the same condition.

(3) For different CDNs, the HPLC gradients of CH_3CN is varied in the range from 14 to 20% at the time point of 25 min.

3.3 HPLC purification of crude c-AMP-CMP

The crude c-AMP-CMP sample solution is filtered in advance through a polytetrafluoroethylene (PTFE) filter (25 mm × 0.45 μm) before injection. First, a small volume of injection (1–5 μL) is injected to roughly check the purity of the crude c-AMP-CMP, the separation efficiency and the sample concentration. When we inject 1 μL of crude c-AMP-CMP sample (Fig. 3A), the main fraction 1 is at 17.6 min with the amount ratio beyond 90%, which is the c-AMP-CMP evidenced by mass spectrometry (MS) and nuclear magnetic resonance (NMR) in Section 4. We observed the fine separation efficiency between fraction 1 at 17.6 min and fraction 2 at 18.1 min. Once 20 μL of the crude c-AMP-CMP sample is injected (Fig. 3B), we found that the remaining time of fraction 1 is maintained although the intensity of the fraction is out of the maximum range. However, the retention time of the faction 2 shifts to 19.5 min with injection volume of 20 μL. The fractions at 17.5 and 19.5 min are respectively collected for further MS and NMR characterizations.

4. Characterization of CDNs

4.1 Materials and equipment

Liquid chromatography mass spectrometry (LC-MS) is conducted on a Shimadzu LCMS-2020 single quadrupole mass spectrometer. High resolution mass spectra (HRMS) are recorded on a Bruker Daltronics microTOF II mass spectrometer. All of the CDN fractions for NMR are dissolved in deuterium water. One dimensional (1-D) and two dimensional (2-D) NMR spectra are recorded at 300 K on Bruker series instruments, e.g. Bruker Avance III 400 MHz spectrometers with a BBFO plus probe for N to F/H or F, Bruker Avance III 600 MHz spectrometer with a TCI-H/C/N triple resonance cryoprobe and Bruker Avance III 600 MHz spectrometer with QXI-H/C/N/P quadruple resonance probe. The data are processed and analyzed by MestReNova software.

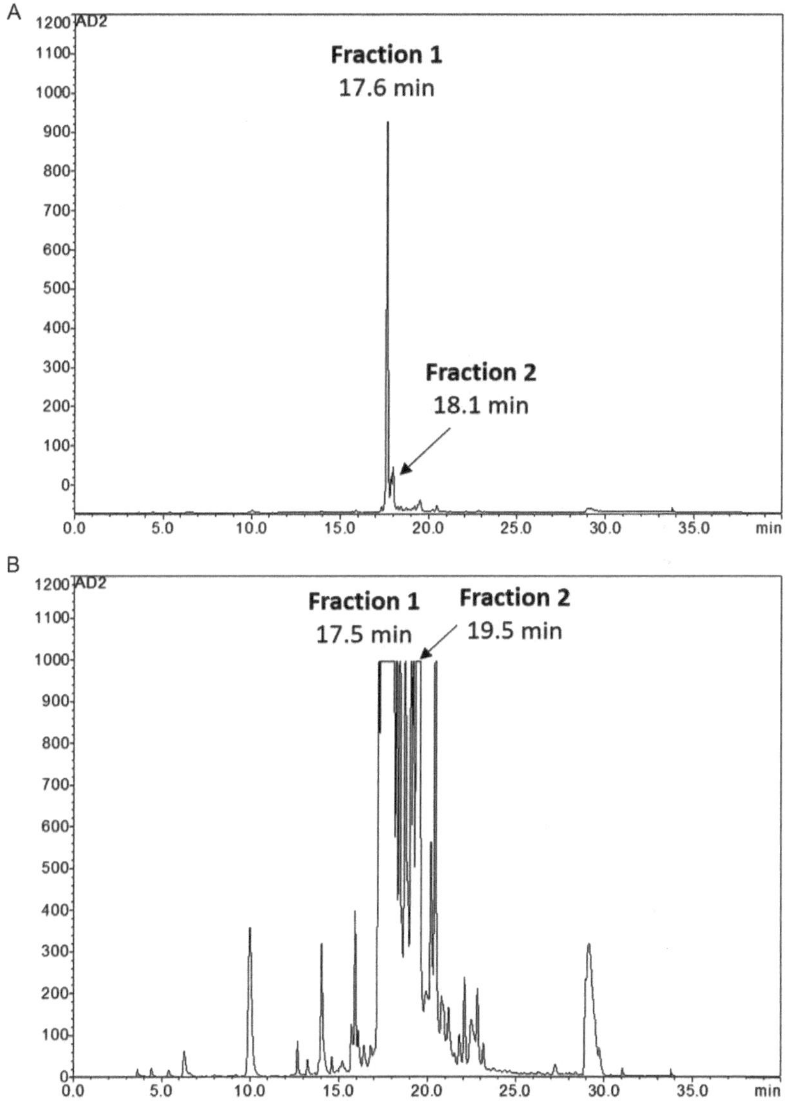

Fig. 3 HPLC traces for purification of crude c-AMP-CMP using protocol from Table 1. (A) 1 μL and (B) 20 μL injection volume of crude c-AMP-CMP sample.

4.2 MS analysis of c-AMP-CMP fractions

The fractions from purification of crude c-AMP-CMP in Section 3.3 could be directedly analyzed by LC-MS. The LC-MS data are listed in Table 2. Compared to the calculated molecular weight of c-AMP-CMP (634.1), the LC-MS data of fraction 1 corresponds to c-AMP-CMP under either positive mode ($M+1=635.2$) or negative mode ($M-1=633.1$). For the

Table 2 LC-MS analysis for the c-AMP-CMP fractions.

LC-MS data	**Fraction 1 (17.5 min)**	**Fraction 2 (19.5 min)**
Positive mode (M + 1)	635.2	573.2
Negative mode (M − 1)	633.1	571.2
Possible M.W.	634.1	572.1
Possible structure	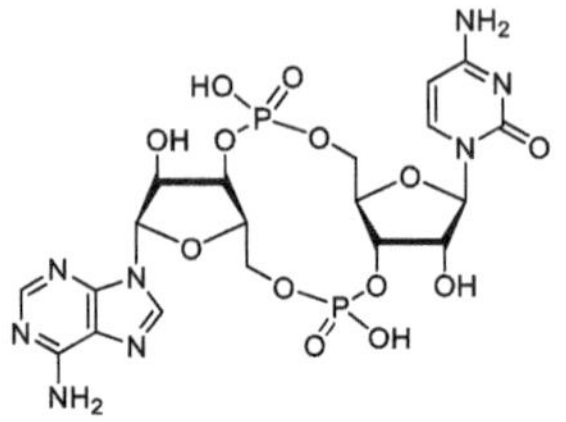	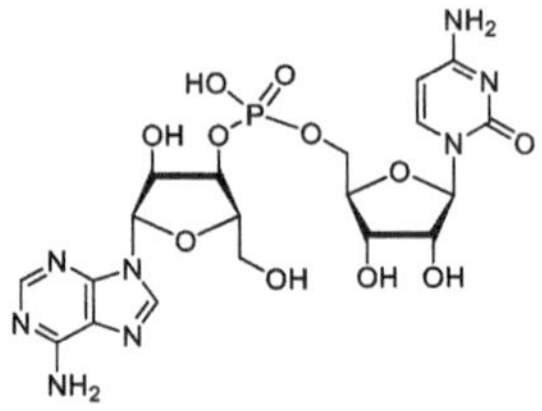

fraction 2, the experimental molecular weight is 572.1 from the LC-MS results, indicating that fraction might be the non-cyclic dinucleotide c-AMP-CMP (Table 2). The aim fraction 1 at 17.5 min is further confirmed by HRMS to ensure the mass analysis of c-AMP-CMP.

4.3 NMR analysis of c-AMP-CMP

After the fractions are analyzed by LC-MS, the aim fraction is lyophilized for NMR characterization. If for an unknown CDN, the fine structure of CDN should be identified by 1-D and 2-D NMR techniques. Protons and carbons in c-AMP-CMP are assigned by ^{1}H NMR, ^{13}C NMR, ^{1}H–^{13}C heteronuclear single quantum correlation (HSQC), ^{1}H–^{1}H correlation spectroscopy (COSY), and ^{1}H–^{13}C heteronuclear multiple-bond correlation (HMBC), see Fig. 4 for details. ^{1}H NMR and ^{13}C NMR spectra are first to confirm whether the number of protons or carbons is correct for c-AMP-CMP (Fig. 4B and C).

Based on the chemical shifts, we can roughly assign the protons and carbons to the structure of c-AMP-CMP (Fig. 4A). ^{1}H–^{1}H COSY spectrum reflects the correlations between *J*-coupled protons that can help to revise the initial assignment of protons (Fig. 4D). In combination of ^{1}H–^{13}C HSQC and ^{1}H–^{13}C HMBC (Fig. 4E and F), the precise assignment of protons and carbons could be confirmed, where ^{1}H–^{13}C HSQC reveals the correlations between the proton and its bounding carbon and ^{1}H–^{13}C HMBC reflects ^{1}H–^{13}C correlations over longer ranges of about 2–4 bonds. There are two different phosphorus in c-AMP-CMP, that could be distinguished by the spectra of ^{1}H NMR, ^{31}P NMR, and ^{1}H–^{31}P HMBC (Fig. 5). In view

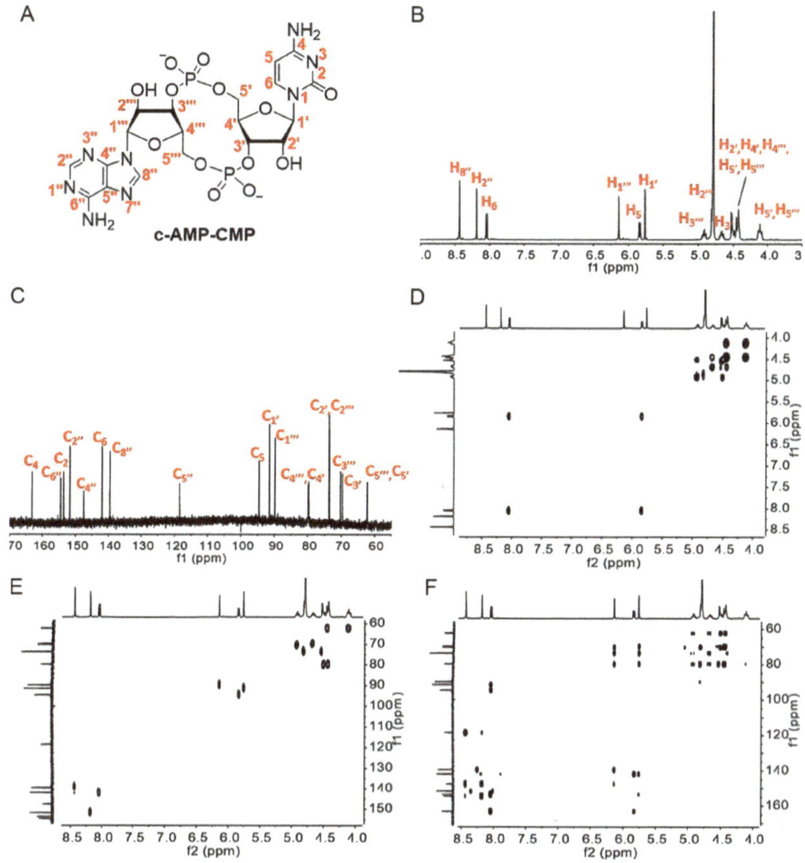

Fig. 4 Assignment of protons and carbons in c-AMP-CMP. (A) Chemical structure of c-AMP-CMP; NMR spectra of c-AMP-CMP in D_2O: (B) ^1H NMR, (C) ^{13}C NMR, (D) ^1H–^1H COSY, (E) ^1H–^{13}C HSQC, and (F) ^1H–^{13}C HMBC. *Adapted with permission from Wang, C., Sinn, M., Stifel, J., Heiler, A. C., Sommershof, A. & Hartig, J. S. (2017). Synthesis of all possible canonical (3′-5′-linked) cyclic dinucleotides and evaluation of riboswitch interactions and immune-stimulatory effects.* Journal of the American Chemical Society, 139, *16154–16160. Copyright (2017) American Chemical Society.*

of the assignment of protons in c-AMP-CMP (Fig. 5B), ^1H–^{31}P HMBC spectrum provides the additional correlations information between phosphorus and the protons nearby, especially for H3′, H3‴, H5′ and H5‴ (Fig. 5D). From ^1H NMR, ^{31}P NMR and ^1H–^{31}P HMBC spectra, the two phosphorus in c-AMP-CMP are accurately deciphered. Based on the 1-D and 2-D NMR spectra, we can precisely assign the protons, carbons and phosphorus to an unknown CDN, which would be helpful to investigate the interaction between CDN and receptors of either proteins or riboswitches.

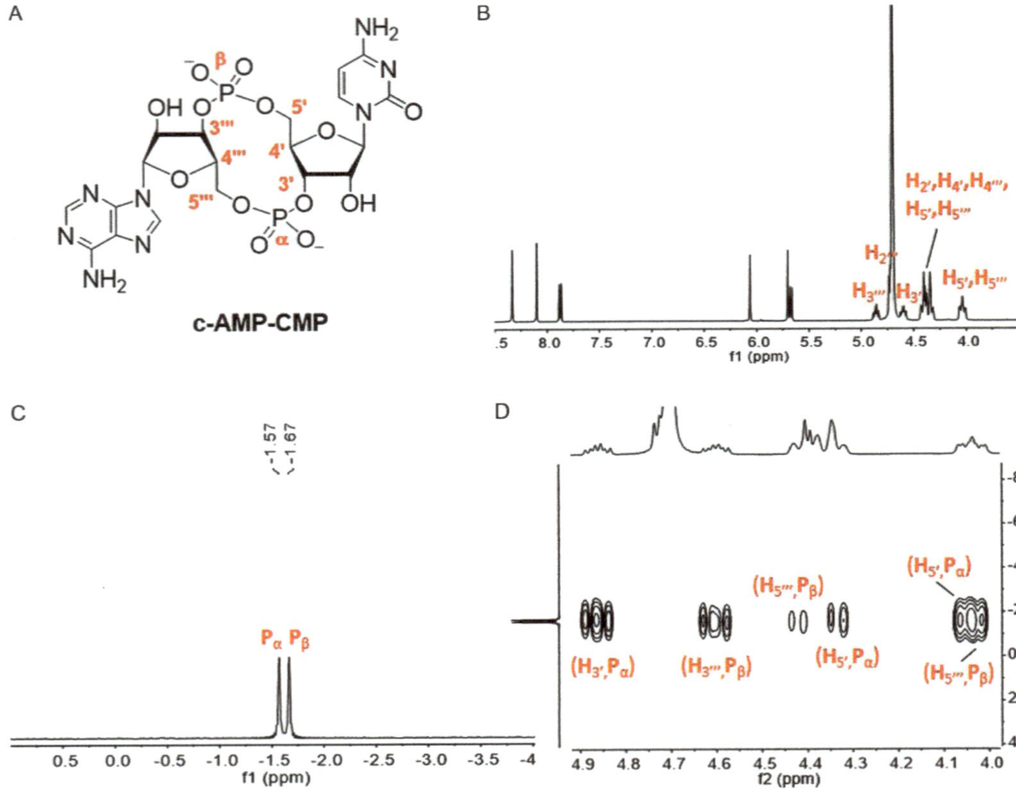

Fig. 5 Assignment of phosphorus in c-AMP-CMP. (A) Chemical structure of c-AMP-CMP; NMR spectra of c-AMP-CMP in D_2O: (B) ^{1}H NMR, (C) ^{31}P NMR, and (D) ^{1}H–^{31}P HMBC. *Adapted with permission from Wang, C., Sinn, M., Stifel, J., Heiler, A. C., Sommershof, A. & Hartig, J. S. (2017). Synthesis of all possible canonical (3′-5′-linked) cyclic dinucleotides and evaluation of riboswitch interactions and immune-stimulatory effects.* Journal of the American Chemical Society, 139, *16154–16160. Copyright (2017) American Chemical Society.*

5. Summary and prospects

In summary, we describe the detailed synthesis procedure for the 3′-5′-linked canonical CDNs using an adapted one-pot strategy and provide helpful tips for some key steps. This typical procedure is suitable for synthesis of CDNs with any combination of nucleobases and enables fast access to the crude CDNs within 1 day. Furthermore, we present the detailed purification protocol for crude CDNs and introduce the techniques for full characterization of the purified CDNs by MS and NMR. This chapter provides a fast and convenient way to obtain canonical CDNs and will promote the investigations for biological functions of possible unknown CDNs.

Until now, four CDNs have been discovered in nature with the building blocks GTP and ATP. All of the known CDNs are serving as second messengers for signal transduction and regulation of many physiological processes. The possibility that all four canonical RNA building blocks could be used to synthesize naturally occurring CDNs has motivated us to explore whether further CDNs might play so-far unknown roles in bacterial signaling. Utilizing the described CDNs as analytical standards we have already tried to detect novel CDNs in 12 different bacterial extracts by LC-MS, although we so far did not find novel CDNs in these bacteria (Wang et al., 2017). However, more sensitive detection methods could be applied and extracts from more diverse species should be investigated in efforts to detect additional CDNs in nature.

Acknowledgments

We are grateful for the financial supports of the Natural Science Foundation of China (21703132), the Fundamental Research Funds for the Central Universities (GK201802001) to C.W. and European Research Council Consolidator Grant (ERC CoG project "RiboDisc") to J.S.H.

References

Ablasser, A., Goldeck, M., Cavlar, T., Deimling, T., Witte, G., Roehl, I., et al. (2013). cGAS produces a 2′-5′-linked cyclic dinucleotide second messenger that activates STING. *Nature*, *498*, 380–384.

Amiot, N., Heintz, K., & Giese, B. (2006). New approach for the synthesis of c-di-GMP and its analogues. *Synthesis*, 4230–4236.

Burdette, D. L., Monroe, K. M., Sotelo-Troha, K., Iwig, J. S., Eckert, B., Hyodo, M., et al. (2011). STING is a direct innate immune sensor of cyclic di-GMP. *Nature*, *478*, 515–518.

Ching, S. M., Tan, W. J., Chua, K. L., & Lam, Y. (2010). Synthesis of cyclic di-nucleotidic acids as potential inhibitors targeting diguanylate cyclase. *Bioorganic & Medicinal Chemistry, 18*, 6657–6665.

Clivio, P., Coantic-Castex, S., & Guillaume, D. (2013). (3′-5′)-Cyclic dinucleotides: Synthetic strategies and biological potential. *Chemical Reviews, 113*, 7354–7401.

Corrigan, R. M., & Gruendling, A. (2013). Cyclic di-AMP: Another second messenger enters the fray. *Nature Reviews. Microbiology, 11*, 513–524.

Coutsogeorgopoulos, C., & Khorana, H. (1964). Studies on polynucleotides. XXXI. The specific synthesis of C3′-C5′-linked ribopolynucleotides (6). A further study of the synthesis of uridine polynucleotides. *Journal of the American Chemical Society, 86*, 2926–2932.

Danilchanka, O., & Mekalanos, J. J. (2013). Cyclic dinucleotides and the innate immune response. *Cell, 154*, 962–970.

Davies, B. W., Bogard, R. W., Young, T. S., & Mekalanos, J. J. (2012). Coordinated regulation of accessory genetic elements produces cyclic di-nucleotides for V. cholerae virulence. *Cell, 149*, 358–370.

Gaffney, B. L., & Jones, R. A. (2014). Synthesis of c-di-GMP analogs with thiourea, urea, carbodiimide, and guanidinium linkages. *Organic Letters, 16*, 158–161.

Gaffney, B. L., Stebbins, N. D., & Jones, R. A. (2013). Synthesis of biotinylated c-di-GMP and c-di-AMP using click conjugation. *Nucleosides, Nucleotides & Nucleic Acids, 32*, 1–16.

Gaffney, B. L., Veliath, E., Zhao, J., & Jones, R. A. (2010). One-flask syntheses of c-di-GMP and the [Rp,Rp] and [Rp,Sp] thiophosphate analogues. *Organic Letters, 12*, 3269–3271.

Gao, P., Ascano, M., Wu, Y., Barchet, W., Gaffney, B. L., Zillinger, T., et al. (2013). Cyclic G(2′,5′)pA(3′,5′)p is the metazoan second messenger produced by DNA-activated cyclic GMP-AMP synthase. *Cell, 153*, 1094–1107.

Gao, P., Ascano, M., Zillinger, T., Wang, W., Dai, P., Serganov, A. A., et al. (2013). Structure-function analysis of STING activation by c G(2′,5′)pA(3′,5′)p and targeting by antiviral DMXAA. *Cell, 154*, 748–762.

Gao, J., Tao, J., Liang, W., Zhao, M., Du, X., Cui, S., et al. (2015). Identification and characterization of phosphodiesterases that specifically degrade 3′3'-cyclic GMP-AMP. *Cell Research, 25*, 539–550.

Hayakawa, Y., Nagata, R., Hirata, A., Hyodo, M., & Kawai, R. (2003). A facile synthesis of cyclic bis(3′-5′)diguanylic acid. *Tetrahedron, 59*, 6465–6471.

Hengge, R. (2009). Principles of c-di-GMP signalling in bacteria. *Nature Reviews. Microbiology, 7*, 263–273.

Hyodo, M., Sato, Y., & Hayakawa, Y. (2006). Synthesis of cyclic bis(3′-5′)diguanylic acid (c-di-GMP) analogs. *Tetrahedron, 62*, 3089–3094.

Kato, K., Nishimasu, H., Oikawa, D., Hirano, S., Hirano, H., Kasuya, G., et al. (2018). Structural insights into cGAMP degradation by Ecto-nucleotide pyrophosphatase phosphodiesterase 1. *Nature Communications, 9*, 4424.

Kiburu, I., Shurer, A., Yan, L., & Sintim, H. O. (2008). A simple solid-phase synthesis of the ubiquitous bacterial signaling molecule, c-di-GMP and analogues. *Molecular BioSystems, 4*, 518–520.

Lee, E. R., Baker, J. L., Weinberg, Z., Sudarsan, N., & Breaker, R. R. (2010). An allosteric self-splicing ribozyme triggered by a bacterial second messenger. *Science, 329*, 845–848.

Nelson, J. W., Sudarsan, N., Furukawa, K., Weinberg, Z., Wang, J. X., & Breaker, R. R. (2013). Riboswitches in eubacteria sense the second messenger c-di-AMP. *Nature Chemical Biology, 9*, 834.

Nelson, J. W., Sudarsan, N., Phillips, G. E., Stav, S., Lunse, C. E., McCown, P. J., et al. (2015). Control of bacterial exoelectrogenesis by c-AMP-GMP. *Proceedings of the National Academy of Sciences of the United States of America, 112*, 5389–5394.

Parvatiyar, K., Zhang, Z., Teles, R. M., Ouyang, S., Jiang, Y., Iyer, S. S., et al. (2012). The helicase DDX41 recognizes the bacterial secondary messengers cyclic di-GMP and cyclic di-AMP to activate a type I interferon immune response. *Nature Immunology, 13*, 1155–1161.

Rammler, D., Lapidot, Y., & Khorana, H. (1963). Studies on polynucleotides. XIX. The specific synthesis of C3″-C5″ inter-ribonucleotidic linkage. A new approach and its use in the synthesis of C3″-C5″-linked uridine oligonucleotides. *Journal of the American Chemical Society, 85*, 1989–1997.

Ross, P., Weinhouse, H., Aloni, Y., Michaeli, D., Weinbergerohana, P., Mayer, R., et al. (1987). Regulation of cellulose synthesis in acetobacter-xylinum by cyclic diguanylic acid. *Nature, 325*, 279–281.

Sawai, H., Higa, K., & Kuroda, K. (1992). Synthesis of cyclic and acyclic oligocytidylates by uranyl ion catalyst in aqueous solution. *Journal of the Chemical Society, Perkin Transactions, 1*, 505–508.

Shanahan, C. A., Gaffney, B. L., Jones, R. A., & Strobel, S. A. (2011). Differential analogue binding by two classes of c-di-GMP riboswitches. *Journal of the American Chemical Society, 133*, 15578–15592.

Sudarsan, N., Lee, E. R., Weinberg, Z., Moy, R. H., Kim, J. N., Link, K. H., et al. (2008). Riboswitches in eubacteria sense the second messenger cyclic di-GMP. *Science, 321*, 411–413.

Wang, C., Sinn, M., Stifel, J., Heiler, A. C., Sommershof, A., & Hartig, J. S. (2017). Synthesis of all possible canonical (3′-5′-linked) cyclic dinucleotides and evaluation of riboswitch interactions and immune-stimulatory effects. *Journal of the American Chemical Society, 139*, 16154–16160.

Witte, G., Hartung, S., Buttner, K., & Hopfner, K. P. (2008). Structural biochemistry of a bacterial checkpoint protein reveals diadenylate cyclase activity regulated by DNA recombination intermediates. *Molecular Cell, 30*, 167–178.

Woodward, J. J., Iavarone, A. T., & Portnoy, D. A. (2010). c-di-AMP secreted by intracellular listeria monocytogenes activates a host type I interferon response. *Science, 328*, 1703–1705.

Yan, H., & Aguilar, A. L. (2007). Synthesis of 3′,5′-cyclic diguanylic acid (cdiGMP) using 1-(4-chlorophenyl)-4-ethoxypiperidin-4-yl as a protecting group for 2′-hydroxy functions of ribonucleosides. *Nucleosides, Nucleotides & Nucleic Acids, 26*, 189–204.

Zhang, Z. Y., Gaffney, B. L., & Jones, R. A. (2004). c-di-GMP displays a monovalent metal ion-dependent polymorphism. *Journal of the American Chemical Society, 126*, 16700–16701.

CHAPTER FOUR

An SPR-based analysis of cGAS substrate K_D and steady-state K_M values

Erik C. Ralph*, Justin Hall
Worldwide Medicinal Chemistry, Pfizer, Groton, CT, United States
*Corresponding author: e-mail address: erik.ralph@pfizer.com

Contents

Abstract

Surface plasmon resonance (SPR) is a standard method for evaluating direct protein-small molecule binding. While studying the catalytic mechanism of cyclic GMP-AMP synthase (cGAS), we developed an SPR-based method to measure steady-state K_M values that complements traditional SPR affinity measurements. The method relies on refractive changes to detect protein interaction with substrates and products, and takes advantage of stimulator of type 1 interferon genes (STING) binding to the cGAS product, 2′,3′-cGAMP. The specific method described here uses co-immobilization of cGAS and double-stranded DNA through a biotin tag; it should be generally applicable to other proteins and protein complexes.

Methods in Enzymology, Volume 625
ISSN 0076-6879
https://doi.org/10.1016/bs.mie.2019.04.023

1. Introduction

Cyclic GMP-AMP synthase (cGAS) catalyzes the reaction of ATP and GTP to form a cyclic dinucleotide with mixed 2′,5′- and 3′,5′-phosphodiester bonds (2′,3′-cGAMP). This enzymatic activity requires binding of cGAS to double stranded DNA (ds-DNA); 2′,3′-cGAMP subsequently binds and activates stimulator of type 1 interferon genes (STING). Thus, cGAS serves as a DNA-sensor protein, producing a secondary messenger for the innate immune response (Junt & Barchet, 2015; Schlee & Hartmann, 2016; Wang, Liu, Zhou, & Wang, 2015). ds-DNA sensing can become dysregulated, and cGAS activation has been observed in autoimmune disorders (An et al., 2017; Gao et al., 2015; Gray, Treuting, Woodward, & Stetson, 2015). Therefore, chemical inhibitors of cGAS may have therapeutic efficacy for some autoimmune diseases.

As noted by a recent publication (Hooy & Sohn, 2018), published methods for monitoring enzymatic activity of cGAS, such as HPLC or mass spectroscopy, are not ideal for mechanistic studies. To address this concern, those authors developed an assay to detect the pyrophosphate released during ATP and GTP linkage. However, 2′,3′-cGAMP is not the only enzymatic product of cGAS; cGAS also releases pyrophosphate when it produces linear homo- and hetero-dinucleotides (AMP-2′-GTP, AMP-3′-ATP and GMP-2′-GTP), which complicates this method of reaction monitoring (Gao et al., 2013; Hall, Ralph, et al., 2017; Kranzusch et al., 2014). We therefore developed an alternative approach that takes advantage of the surface plasmon resonance (SPR) instrumentation we had already implemented in cGAS-ligand binding studies.

SPR is an information-rich technique commonly used in drug discovery for the direct measurement of affinity constants (k_{ON}, k_{OFF}, and K_d) (e.g., Su & Wang, 2018). To perform enzymatic studies, we generated cGAS and STING constructs suitable for immobilization on the SPR surface (Hall, Ralph, et al., 2017). We then took advantage of our SPR instrument's serpentine flow to expose analytes in series to our different proteins. cGAS was immobilized first, with STING immobilized downstream from cGAS (see Section 3), where it could bind 2′,3′-cGAMP produced from ATP and GTP substrates exposed to activated cGAS. The STING response proved to be sensitive (<10 nM, Zhang et al., 2013) and highly selective for 2′,3′-cGAMP (Hall, Ralph, et al., 2017), and could be analyzed as a function of the substrate concentrations injected using standard enzymology equations (see Section 4.6).

Most SPR approaches require protein immobilization; the resulting deliquefaction can change protein behavior compared to solution state. Protein stability, oligomerization, ligand binding, and catalytic ability can be affected. Careful consideration for the site(s) of immobilization are the best start to limit immobilization-specific artifacts. Although specific considerations were required for immobilization of DNA-activated cGAS (see Section 2), the considerations for immobilization of an enzyme on an SPR chip are the same if attempting an enzymatic assay or the traditional SPR ligand binding assays. Comparison of affinity and catalytic results between immobilized and solution states is the best standard to determine if immobilization has affected protein behavior. The K_M values determined with immobilized cGAS agreed well with solution values obtained through orthogonal approaches (see Section 5 and Hall, Ralph, et al., 2017).

The Michaelis constant (K_M) is the equilibrium of free enzyme to all other enzyme species (i.e., the sum of enzyme-substrate and enzyme-product complexes). Therefore, it is theoretically possible to directly measure the response change on the active surface to determine K_M; indeed, such an approach was utilized to study deoxyadenosine kinase using SPR (Pol & Wang, 2006). In those studies, the authors were able to see a positive response change as a function of substrate (or product) injected into the system. Rapid quench studies with a tritium labeled substrate showed product formation was rate limiting, and the authors therefore interpreted the response increase from the immobilized enzyme as corresponding to the enzyme-product complex. Although their interpretation may have been an oversimplification of the results (see Section 3), the authors were nonetheless able to elucidate the kinetic mechanism for this enzyme using this approach. Unfortunately, for cGAS bound to ds-DNA, there was a near-zero SPR response to substrate despite 2′,3′-cGAMP formation (see Section 3 and Hall, Ralph, et al., 2017). Additionally, this approach would not have been selective for 2′,3′-cGAMP over other potential products, as the response change would be the sum of all enzyme-substrate and enzyme-product complexes, including linear homo-dinucleotides.

To the best of our knowledge, these two references (Hall, Ralph, et al., 2017; Pol & Wang, 2006) are the only published attempts to monitor catalytic activity of an immobilized enzyme by SPR. In our opinion, if one is already doing SPR with an active enzyme for ligand binding, it takes effectively no more effort to flow substrate over the enzyme to test its catalytic competence. Thus, SPR-based enzymatic studies may turn out to be a generally useful and robust approach complimenting ligand binding studies.

2. Consideration for the immobilization of proteins, multimers, and the cGAS·ds-DNA complex

Immobilizing a protein multimer can cause strain between components of the complex when multiple components are tagged. Forming the complex in solution under optimum conditions (e.g., stoichiometry, temperature, cofactors) prior to immobilization can help achieve a high fraction of competent multimer. Long linkers on the construct, and long matrices in the solid phase, can help reduce strain and mimic solution conditions. In general, the considerations for immobilization of an enzyme on an SPR chip (e.g., site of immobilization, compatibility of protein with immobilization conditions) are not different if one is planning to attempt an enzymatic assay in addition to ligand binding studies. Given the wealth of reviews detailing these considerations on this subject (e.g., Giannetti, 2011), we make no further attempt to review the literature here. For our studies, we first immobilized neutravidin using standard amine-coupling reagents, which enabled subsequent capture of biotinylated cGAS, DNA, and STING (see Section 4).

cGAS is activated by binding to ds-DNA. ds-DNA length is important for achieving full activation of cGAS in solution (Gao et al., 2013; Kranzusch, Lee, Berger, & Doudna, 2013). At the time of these studies, there were conflicting reports of the effect of ds-DNA length on cGAS activation. The science was confounded by insensitive activity assays (e.g., product sensing by TLC), and what has subsequently been shown to be a difference between human and mouse activation profiles (Zhou et al., 2018). It is now known that in addition to cGAS forming a ds-DNA-dependent dimer (Li et al., 2013), it can form an ordered ladder-like network in the presence of sufficiently long ds-DNA (Andreeva et al., 2017). When we immobilized cGAS bound ds-DNA, both cGAS and ds-DNA were biotin tagged, which may have created strain in the complex. Indeed, identical repeats of immobilization resulted in variable activity amplitudes. These results were seemingly stochastic, and suggested a random element in our attempts to successfully immobilize the final active complex instead of the inactive component parts. We now speculate the stochastic effect may be due to the formation of the competent complex, possibly the ladder superstructure. If so, these experiments may be helped by a stoichiometry that favors cGAS over ds-DNA (>2:1), and ds-DNA that is longer than the 42-mer we used. In support of this hypothesis, the Kranzusch lab has recently shown that the human form of cGAS has a DNA-length dependence to its DNA activation while the mouse cGAS does not, consistent with the need of DNA-ladders for full activation of human cGAS. They were able to trace the human and mouse difference to two residues (K187N and L195R), which allow human cGAS to become fully activated without long ds-DNA (Zhou et al., 2018). In addition to cGAS:DNA stoichiometry or DNA length, the K187N and L195N mutations may help to normalize the variable active fraction we observed in our original experiments.

3. Considerations for SPR-based enzymology

We used a Biacore T200 SPR instrument for these experiments (Hall, Ralph, et al., 2017). It has a serpentine microfluidic path where injected samples pass in series over each surface (Fig. 1A). In a series system, substrate that is converted to product by an immobilized enzyme will be passed to the next surface as a mixture of product amid residual unreacted substrate. If subsequent surfaces are capable of binding only the product (through a product sensor protein), or only the substrate (through a substrate sensor protein), then the mixture of product and unreacted substrate can be parsed to determine the amount of substrate converted as a function of concentration, and exposure time to the catalytic surface. To simplify these experiments, a constant flow rate (e.g., 5 μL min^{-1}) and injection time (e.g., 60 s) should be used to create a constant time of exposure to the catalytic surface. Sensors of product are preferred over sensors of substrates; the former will give an increase of signal over a small background while the latter attempts to measure depletion of signal when the experimental design intentionally minimizes the percentage of depletion that takes place.

2′,3′-cGAMP is produced through a hetero-dinucleotide intermediate, but cGAS also forms homo-dinucleotides that do not contribute to 2′,3′-cGAMP production. cGAS catalyzes the conversion of ATP and GTP to form the linear hetero-dinucleotide intermediate, AMP-2′-GTP, which must then undergo substantial adjustments in the active site before 2′,3′-cGAMP can be formed. It is possible for AMP-2′-GTP to rearrange in the active site, and it is possible for AMP-2′-GTP to instead dissociate from the active site and then rebind to cGAS in the correct position for 2′,3′-cGAMP formation. In a closed reaction vessel, such as a tube, dissociation of AMP-2′-GTP will not result in the irrevocable loss of AMP-2′-GTP from the system; however, in the SPR system, dissociation can result in loss of reaction intermediates to the fluidics. Thus, apart from using a uniform flow rate for ease of analysis, a minimal flow rate is important to minimize reaction intermediate loss.

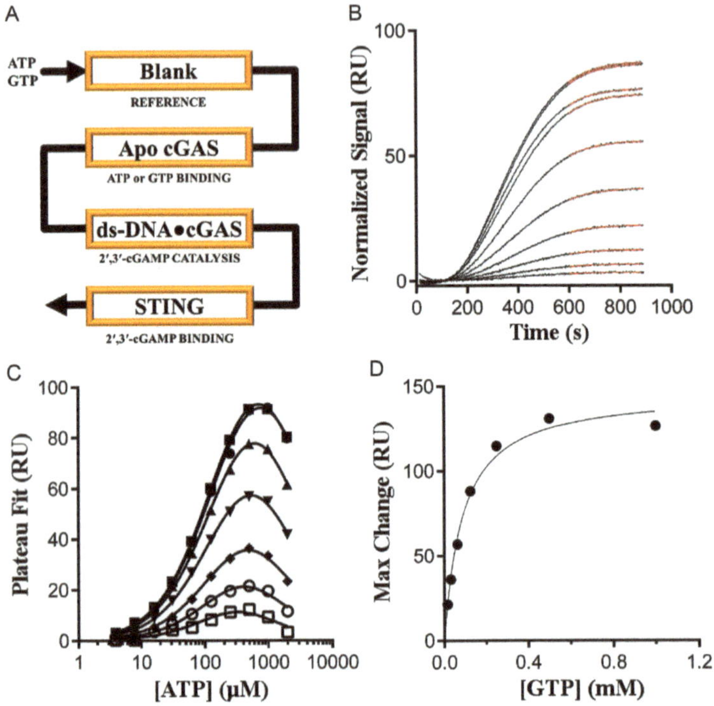

Fig. 1 An SPR-based enzymatic assay. (A) Schematic for SPR sensor layout. Individual channels are indicated by yellow boxes and the flow-path is indicated by the black arrow. (B) Representative data for normalized $STING_{155-341}$ response over 900 s injections of varied substrate concentrations. The red lines show the fit of the data to a single exponential association from 600 to 900 s. (C) Extrapolated plateau response values from panel B. Solid lines show the fit of the data to Eq. (2) using a GTP-concentration independent $K_{M,ATP}$ value with substrate inhibition observed at high nucleotide concentrations. (D) Calculated maximum response values from panel C were plotted as a function of the GTP concentration. The black line shows the fit of the data using the Michaelis–Menten equation. *Figure and legend reprinted from Hall, Ralph, et al. (2017).*

In our experiments, we immobilized apo cGAS first, followed by cGAS that was activated by ds-DNA on the next channel, and finally STING on the last channel. Thus, when we injected mixtures of ATP and GTP, we could measure binding to apo cGAS, generate 2′,3′-cGAMP on the active cGAS surface, and then monitor the amount of 2′,3′-cGAMP formed from the binding response on the STING surface.

The Michaelis constant (K_M) is the equilibrium of free enzyme to all other enzyme species. As demonstrated by Pol and Wang (2006), it is possible to directly measure the response on the active surface alone to determine K_M. However, Pol and Wang (2006) showed the active enzyme channel response was still increasing after more than 10 min, in contrast to their rapid quench data (with free enzyme) and product analysis by MS (from the immobilized enzyme) suggested they were at steady state within 3 s. This underscores the potential complexities with monitoring the enzyme directly. Given the potential of their enzymatic system to bind product, it would have been interesting to see the impact of changing flow rate on their results. If they were monitoring product formation on the enzyme as intended, the rate of response change should have been independent of flow rate. If they were monitoring product formation, with release and recapture of product by immobilized protein, then increasing the flow rate may have diminished rebinding, and thereby diminished time required to reach a constant response level, but also the magnitude of the final response.

For cGAS bound to ds-DNA, there was a near-zero SPR response to substrate despite 2′,3′-cGAMP formation. Lack of response is most simply explained by both positive (substrate mass addition) and negative (catalytic conformation changes) responses on the SPR surface (Crauste et al., 2014). The near-zero catalysis effect seen may be particular to cGAS, and other proteins may not have this same behavior. However, given the lack of a direct response from the activated cGAS channel, we chose to monitor catalysis using STING as a sensor protein for the formation of 2′,3′-cGAMP. STING is a high-affinity (<10 nM) binder of the cGAS product, but does not bind the cGAS substrates under the conditions studied, which is imperative for analysis as all concentrations of substrate tested were mixtures of both substrates and product. Response of the sensor protein to analyte is dependent upon the rates of association and dissociation (k_{on}, k_{off}), and the rates of catalysis (k_{cat}); it is therefore important to wait long enough after the injection for equilibrium to be reached so that steady-state assumptions (and simplifications) can be applied. In SPR, steady state is reached after the response signals have reached a maximum that does not change further during the remainder of the injection (Fig. 1B). The magnitude of the response will be a function of the flow rate after the reaction has reached equilibrium; thus, apart from flow rate affecting loss of intermediates to the fluidics, if a sensor protein is used, the amplitude of the binding signal will be a function of the rate of flow since higher flow rates effectively cause a dilution of the product. For both these reasons a slow flow rate is advised. Lastly, the response on the catalytic surface, or the sensor protein, is intrinsic to these proteins with their analytes; it is therefore necessary to normalize the response by establishing a dose dependence and saturation with the analyte studied.

4. Protocols

4.1 Protein reagents

- *H. sapiens* cGAS with an N-terminal truncation of the first 160 residues, and *H. sapiens* STING residues 155 through 341, were cloned into pET28 containing an N-terminal SUMO-HIS_6 tag and a BirA recognition sequence. Both proteins were expressed in *E. coli* BL21(DE3), purified using nickel-affinity chromatography, and treated with BirA ligase to generate biotin-tagged protein (Avidity). Additional purification was obtained using a HiLoad Superdex75 column. Complete details for the protein purification are available (Hall, Ralph, et al., 2017).

4.2 Chemical reagents

- Neutravidin (Pierce)
- N-terminal biotinylated cGAS
- 3′-Biotinylated Interferon Stimulatory ds-DNA (ISD) (5′-TAC AGA TCT ACT AGT GAT CTA TGA CTG ATC TGT ACA TGA TCT ACA-3′) (integrated ds-DNA Technologies) (Stetson & Medzhitov, 2006)
- Amine-coupling reagents (N-hydroxysuccinimide (NHS) and N-(3-Dimethylaminopropyl)-N′-ethylcarbodiimide hydrochloride (EDC))
- Protein ligands (e.g., nucleotides, dinucleotides, small molecule inhibitors)
- Common buffer reagents (e.g., HEPES, KCl, $MgCl_2$, 1 mM TCEP, DMSO)

4.3 Equipment

- BIACORE T200 (GE Healthcare)
- CM5 sensor chips (GE Healthcare)
- GraphPad Prism v7.04

4.4 Chip preparation and protein capture

- System was primed using the manufactures HEPES buffered saline at 25 °C.
- CM5 chips were warmed to room temperature on the bench before being added to the system.
- SPR surfaces were cleaned and conditions (separate injections of 50 mM NaOH, 1 M NaCl, 0.1% SDS and 10 mM HCl, each for 60 s at 10 μL min^{-1}) prior to normalization using the manufacturer's 70% glycerol solution (GE Healthcare).

- Neutravidin was prepared at 50 μg mL^{-1} in pH 5.5 acetate buffer. EDC and NHS were prepared according to the manufacturer's protocol (GE Healthcare); all surfaces were activated using a 1:1 EDC-NHS mixture for 300 s at 10 μL min^{-1}, followed by immediate neutravidin injection for 600 s, followed by ethanolamine (GE Healthcare) for 300 s. Final neutravidin density was typically between 25,000 and 30,000 resonance units (RUs).
- System was primed with SPR running buffer with no DMSO, and brought to 4 °C.
- cGAS was prepared at 100 μg mL^{-1} in SPR running buffer with no DMSO and injected twice onto channel 2 for 600 s at 5 μL min^{-1}. Final density ca. 7000 RU.
- A 1:1 cGAS•ds-DNA mixture was prepared at 100 μg mL^{-1} cGAS in SPR running buffer with no DMSO, and injected twice onto channel 3 for 600 s at 5 μL min^{-1}. Final density ca. 6000 RU.
- STING was prepared at 100 μg mL^{-1} in SPR running buffer with no DMSO and injected twice onto channel 4 for 600 s at 5 μL min^{-1}. Final density ca. 10,000 RU.

4.5 cGAS-ligand binding studies

- This section is intended to guide the reader through our methods for SPR-based binding assays.

4.5.1 Preparation of samples

- Nucleotides and dinucleotides were prepared from powders to 100 mM stocks in SPR running buffer. Stock pH was adjusted to 7.5 using 5 M NaOH.
- Samples (320 μL) were made at their top concentration in SPR running buffer, then diluted 1:1 (155 μL each) in running buffer to make a 12-concentrations dilution series, including a blank containing no substrate. Samples were made and diluted in a 96-well polypropylene Greiner plate, and then transferred by multichannel pipette to a 384-well polypropylene Greiner plate. Plates were spun at 5000 RCF for 10 min and then placed at 4 °C until experimentation.

4.5.2 Details of injections and data collection

- The system was equilibrated at 4 °C, 60 μL min^{-1} for 4 h prior to experiments.
- Experiments were performed at 60 μL min^{-1} with data collected every 0.1 s.

- A 5-point solvent correction (0.5%, 0.75%, 1.0%, 1.25%, and 1.5% DMSO) was run at the beginning, end, and every 48 samples.
- 2′,3′-cGAMP controls were run to establish the 2′,3′-cGAMP-dependence of the STING response.
- The typical sample injection time for ligand binding studies was 60 s, with dissociation measured for 60 s.

4.5.3 Data analysis

- Sample-dependent responses on channels 2 and 3 were subtracted from channel 1 to account for interactions with neutravidin, and data were further corrected by subtracting a zero-concentration blank from all compounds to account for mismatches in the sample buffer and running buffer. Dissociation equilibrium constants (K_d) were determined using a binary association model ($A + B \leftrightarrow AB$) within the manufacture's software (T200 BIA Eval software). Although slight differences in software may exist, this standard analysis approach is discussed in more detail by Bergsdorf and Wright (2018) using the Biacore 8K Evaluation software.

4.6 cGAS activity assays

- This section is intended to guide the reader through our methods for SPR-based activity assays.

4.6.1 Preparation of samples

- A matrix of ATP and GTP concentrations, including conditions with only ATP or only GTP was made. ATP concentrations were between 2.0 mM and 2.0 μM (11 concentration points and a zero-concentration blank over two-fold dilutions); GTP concentrations were between 1.0 mM and 1.0 μM (11 concentration points and a zero-concentration blank over two-fold dilutions).
- A single 2:1 M ratio of ATP:GTP was made. Top and bottom ATP concentrations were 5 mM and 90 μM (11 concentration points and a zero-concentration blank over 1.5-fold dilutions).
- To ensure the $STING_{155-341}$ response was specific to 2′,3′-cGAMP and not influenced by any of the linear products present in the enzymatic samples (e.g., AMP-2′-GTP), reactions were prepared in SPR running buffer with 2 μM cGAS, 3 μM ISD, and a 2:1 M ratio of GTP to ATP at 0.25, 0.5, 1.0, and 4 mM ATP. Reactions were quenched with 85 mM EDTA after 3 h at 23 °C and analyzed by NMR to determine the concentration of 2′,3′-cGAMP and linear AMP-2′-GTP dinucleotide

compared to concentration standards of 2′,3′-cGAMP. These samples were diluted four-fold into SPR running buffer to minimize noise associated with running buffer and sample buffer mismatch and analyzed at 60 μL min^{-1}, 4 °C, with a 60 s injection and dissociation.

4.6.2 Details of injections and data collection

- System was equilibrated at 4 °C, 5 μL min^{-1} for 4 h prior to experiments.
- Experiments were performed at 4 °C, 5 μL min^{-1} with data collected every 0.1 s.
- A 5-point solvent correction (0.5%, 0.75%, 1.0%, 1.25%, and 1.5% DMSO) was run at the beginning, end, and every 48 samples.
- 2′,3′-cGAMP controls were run to establish the 2′,3′-cGAMP-dependence of the STING response.
- The typical sample injection time for activity assays was 900 s, with dissociation measured for 60 s.

4.6.3 Data analysis

- Sample-dependent responses on channel 4 were subtracted from channel 1 to account for interactions with neutravidin, and data were further corrected by subtracting a zero concentration blank from all compounds to account for mismatches in the sample buffer and running buffer, as described above.
- Data were exported into a format compatible with Microsoft Excel.
- Depending on the data collection rate used, an optional step at this point is to reduce the size of the data set. In Excel, data can be reduced to every *N*th row. A dataset of one to two data points per second is sufficient for further analysis.
- The exported datasets often showed an initial negative deflection, especially at higher nucleotide concentrations, which is suspected to be an artifact of buffer mismatch. In Excel, data were normalized to have zero absorbance at the signal minimum (typically occurred near 100 s), by subtracting the minimum from the corresponding run's dataset.
- Normalized data were copied and pasted into GraphPad Prims as a new *XY* table with run time on the *X*-axis. Because all run data were collected for the same length of time, and we are not attempting to use the rate information in the channel 4 response, several datasets can be plotted and analyzed as a function of any one run's time data.
- To obtain the equilibrium RU level, normalized data were fit to a single exponential association $[Y = Y_0 + (\text{Plateau} - Y_0) \times (1 - \exp(-K \times x))]$

from 600 to 900 s. This equation is preloaded in the software under **Exponential/One-phase association**. The 600 s starting point was chosen empirically by looking at the goodness of the fit; data did not fit well to a single exponential function if too early a starting point was used. Representative data are shown in Fig. 1B.

- The extrapolated plateau values were replotted in a new *XY* table with the ATP concentration used in a given run in the *X* column and the plateau values (and fitting errors) entered in the *Y* column. Note that datasets generated in the presence of different GTP concentrations should be entered into separate columns with the GTP concentrations entered as the column names.
- Compiled data sets were simultaneously fit to a substrate inhibition equation: $Y = V_{max} \times X/(K_m + X \times (1 + X/K_i))$. This equation is preloaded in the software under **Enzyme kinetics – inhibition/Substrate inhibition**. To check for cooperativity in GTP and ATP binding, an extra sum-of-squares *F*-test was used. On the **Compare** tab of the **Parameters:Nonlinear Regression** window, select to see if "the best-fit values of selected unshared parameters differ between datasets." Select the "sum-of-squares F test" and select the K_M parameter. Because the resulting *P*-value ($P = 0.18$) supported a shared K_M value, the data set was then constrained to a shared value. On the **Constrain** tab of the **Parameters:Nonlinear Regression** window, find the K_M parameter and select "Shared value for all datasets" from the drop-down menu. The graph should look like Fig. 1C.
- The software provides V_{max} and K_i values for each data set with a shared K_M value for all datasets analyzed. The V_{max} values and fitting errors were entered into a new *XY* table with GTP concentrations in the *X* column. The data were fit as a function of the GTP concentration using the Michaelis–Menten equation (**Enzyme kinetics – Substrate vs Velocity/Michaelis Menten**): $Y = V_{max} \times X/(K_m + X)$ (Fig. 1D). This analysis provides the GTP K_M.

5. Summary

SPR is an increasingly available technology and has become a popular method for characterizing ligand binding interactions in the pharmaceutical industry. Many pharmaceutical targets are catalytic enzymes, for which substrate and ligand binding studies must be performed. We therefore present here an extension of SPR technology for use in characterizing catalytically competent enzymes to compliment traditional ligand binding studies. A straightforward analysis of these methods yields the steady-state K_M parameter; if the K_M is already known through alternative methods, this approach can serve as an excellent quality control for catalytic competence after the enzyme has been immobilized on the SPR surface. In the case of the enzyme cGAS, this technique served not only as a quality control for the immobilized enzyme, but also as the first published continuous assay for the enzymatic production of 2,3′-cGAMP.

The method took advantage of the known 2,3′-cGAMP-binding protein, STING. Although logic suggests one may be able to monitor an enzyme-labeled channel directly to determine K_M, a sensor protein was required for cGAS. A broader application of this approach is required to determine if this is a peculiarity of the cGAS system or a more general occurrence for enzyme catalysis monitored by SPR. Regardless, utilization of a sensor protein (e.g., STING) can simplify the data analysis and is recommended. A sensor protein may seem like a daunting limitation if a naturally occurring receptor protein is not known for the analyte of interest. However, an increasing number of enzymes of commercial or pharmaceutical interest have antibodies available to detect their catalytic products. Additionally, one may be able to use another enzyme as a reporter protein given sufficient differences in the catalytic requirements. For example, in a pathway where a hydroxylase can be followed by a dehydrogenase, the dehydrogenase may serve as a reporter protein for the hydroxylated product in the absence of its electron acceptor.

Using a sensor protein does introduce limitations. First, it requires instrumentation designed with a serpentine flow path (e.g., Biacore T100 or T200). A serpentine path allows the detection of materials produced on previous flow cells; without this there is no way to couple enzyme and sensor proteins in one injection. Second, when working with a sensor protein, the physics of a flow system must be considered when tuning the response change on the sensor protein channels. The concentration of the enzymatic product as it enters the flow cell for the sensor protein is a function of the rate of production, the total volume of the cell, and the rate of flow; the last of these parameters can be controlled by the experimenter. Too fast of a flow can dilute the reaction product below the detection limit of the sensor protein. A slower flow rate will require a longer time to come to equilibrium, thus increasing the experimental duration. Finally, in the case of somewhat complicated reaction mechanisms, such as with cGAS, a fast flow may lead to stripping of the reaction intermediates from the enzyme, and thus prevent catalysis to the final product.

Table 1 cGAS steady-state apparent K_M values.[a,b]

$K_{M,GTP}$ (μM)	$K_{IS,GTP}$ (μM)[c]	$K_{M,ATP}$ (μM)	$K_{IS,ATP}$ (μM)[c]	Enzyme	Method
90 (12)	ND[d]	140 (4)	2700 (100)	$cGAS_{161}$	SPR
90 (20)	ND	190 (20)	2500 (500)	$cGAS_{161}$	SPR
190 (50)	ND	120 (30)	ND	Full length	NMR
160 (50)[e]	1500 (500)[f]	ND	ND	Full length	HPLC
40 (10)	1500 (600)	50 (10)	700 (200)	Full length	MS
22 (2)	1700 (200)	24 (1)	3600 (500)	Full length	MS

[a]Table reprinted from Hall, Ralph, et al. (2017).
[b]Parenthetical values are errors of the fit.
[c]As presented in the results, K_{IS} values were dependent upon the concentration of the non-varied nucleotide. Presented values were determined at *ca* 1 mM ATP or 0.3 mM GTP.
[d]Data not determined (ND).
[e]Value is the GTP IC_{50} value for inhibition of AMP-3′-ATP formation.
[f]Value was determined from Eq. (2) using a set $K_{M,GTP}$ value of 160 μM.

This method presented here can be used to generate higher confidence in protein quality of an immobilized system, as well as additional steady-state kinetic parameters from only a little additional effort compared to traditional SPR-binding assays. The determined K_M values for cGAS using this method were in good agreement to values determined through orthogonal, non-continuous assays (Table 1) (Hall et al., 2017) and values recently published for an alternative continuous assay for the production of pyrophosphate (Hooy & Sohn, 2018). Therefore, given the commonality of SPR use for ligand binding studies, and the minimal additional work required to use the methods presented here, it is our expectation that this method could be broadly applied for the study of other enzymes.

References

An, J., Durcan, L., Karr, R. M., Briggs, T. A., Rice, G. I., Teal, T. H., et al. (2017). Expression of cyclic GMP-AMP synthase in patients with systemic lupus erythematosus. *Arthritis & Rhematology*, 69(4), 800–807. https://doi.org/10.1002/art.40002.

Andreeva, L., Hiller, B., Kostrewa, D., Lassig, C., de Oliveira Mann, C. C., Jan Drexler, D., et al. (2017). cGAS senses long and HMGB/TFAM-bound U-turn DNA by forming protein-DNA ladders. *Nature*, 549(7672), 394–398. https://doi.org/10.1038/nature23890.

Bergsdorf, C., & Wright, S. K. (2018). A guide to run affinity screens using differential scanning fluorimetry and surface plasmon resonance assays. *Methods in Enzymology*, 610, 135–165. https://doi.org/10.1016/bs.mie.2018.09.015.

Crauste, C., Willand, N., Villemagne, B., Flipo, M., Willery, E., Carette, X., et al. (2014). Unconventional surface plasmon resonance signals reveal quantitative inhibition of transcriptional repressor EthR by synthetic ligands. *Analytical Biochemistry*, 452, 54–66. https://doi.org/10.1016/j.ab.2014.02.011.

Gao, P., Ascano, M., Wu, Y., Barchet, W., Gaffney, B. L., Zillinger, T., et al. (2013). Cyclic [G(2′,5′)pA(3′,5′)p] is the metazoan second messenger produced by DNA-activated cyclic GMP-AMP synthase. *Cell*, 153(5), 1094–1107. https://doi.org/10.1016/j.cell.2013.04.046.

Gao, D., Li, T., Li, X. D., Chen, X., Li, Q. Z., Wight-Carter, M., et al. (2015). Activation of cyclic GMP-AMP synthase by self-DNA causes autoimmune diseases. *Proceedings of the National Academy of Sciences of the United States of America*, 112(42), E5699–E5705. https://doi.org/10.1073/pnas.1516465112.

Giannetti, A. M. (2011). From experimental design to validated hits: A comprehensive walk-through of fragment lead identification using surface plasmon resonance. *Methods in Enzymology*, 493, 169–218.

Gray, E. E., Treuting, P. M., Woodward, J. J., & Stetson, D. B. (2015). Cutting edge: cGAS is required for lethal autoimmune disease in the Trex1-deficient mouse model of Aicardi–Goutieres syndrome. *Journal of Immunology*, 195(5), 1939–1943. https://doi.org/10.4049/jimmunol.1500969.

Hall, J., Brault, A., Vincent, F., Weng, S., Wang, H., Dumlao, D., et al. (2017). Discovery of PF-06928215 as a high affinity inhibitor of cGAS enabled by a novel fluorescence polarization assay. *PLoS One*, 12(9), e0184843. https://doi.org/10.1371/journal.pone.0184843.

Hall, J., Ralph, E. C., Shanker, S., Wang, H., Byrnes, L. J., Horst, R., et al. (2017). The catalytic mechanism of cyclic GMP-AMP synthase (cGAS) and implications for innate immunity and inhibition. *Protein Science*, 26(12), 2367–2380. https://doi.org/10.1002/pro.3304.

Hooy, R. M., & Sohn, J. (2018). The allosteric activation of cGAS underpins its dynamic signaling landscape. *eLife*, 7, e39984. https://doi.org/10.7554/eLife.39984.

Junt, T., & Barchet, W. (2015). Translating nucleic acid-sensing pathways into therapies. *Nature Reviews. Immunology*, 15(9), 529–544. https://doi.org/10.1038/nri3875.

Kranzusch, P. J., Lee, A. S., Berger, J. M., & Doudna, J. A. (2013). Structure of human cGAS reveals a conserved family of second-messenger enzymes in innate immunity. *Cell Reports*, 3(5), 1362–1368. https://doi.org/10.1016/j.celrep.2013.05.008.

Kranzusch, P. J., Lee, A. S., Wilson, S. C., Solovykh, M. S., Vance, R. E., Berger, J. M., et al. (2014). Structure-guided reprogramming of human cGAS dinucleotide linkage specificity. *Cell*, 158(5), 1011–1021. https://doi.org/10.1016/j.cell.2014.07.028.

Li, X., Shu, C., Yi, G., Chaton, C. T., Shelton, C. L., Diao, J., et al. (2013). Cyclic GMP-AMP synthase is activated by double-stranded DNA-induced oligomerization. *Immunity*, 39(6), 1019–1031. https://doi.org/10.1016/j.immuni.2013.10.019.

Pol, E., & Wang, L. (2006). Kinetic mechanism of deoxyadenosine kinase from mycoplasma determined by surface plasmon resonance technology. *Biochemistry*, 45(2), 513–522. https://doi.org/10.1021/bi0515523.

Schlee, M., & Hartmann, G. (2016). Discriminating self from non-self in nucleic acid sensing. *Nature Reviews Immunology*, 16(9), 566–580. https://doi.org/10.1038/nri.2016.78.

Stetson, D. B., & Medzhitov, R. (2006). Recognition of cytosolic DNA activates an IRF3-dependent innate immune response. *Immunity*, 24(1), 93–103. https://doi.org/10.1016/j.immuni.2005.12.003.

Su, Y. W., & Wang, W. (2018). Surface plasmon resonance sensing: From purified biomolecules to intact cells. *Analytical and Bioanalytical Chemistry*, 410(17), 3943–3951. https://doi.org/10.1007/s00216-018-1008-8.

Wang, Q., Liu, X., Zhou, Q., & Wang, C. (2015). Cytosolic sensing of aberrant DNA: Arming STING on the endoplasmic reticulum. *Expert Opinion on Therapeutic Targets*, 19(10), 1397–1409. https://doi.org/10.1517/14728222.2015.1067303.

Zhang, X., Shi, H., Wu, J., Zhang, X., Sun, L., Chen, C., et al. (2013). Cyclic GMP-AMP containing mixed phosphodiester linkages is an endogenous high-affinity ligand for STING. *Molecular Cell*, 51(2), 226–235. https://doi.org/10.1016/j.molcel.2013.05.022.

Zhou, W., Whiteley, A. T., de Oliveira Mann, C. C., Morehouse, B. R., Nowak, R. P., Fischer, E. S., et al. (2018). Structure of the human cGAS-DNA complex reveals enhanced control of immune surveillance. *Cell*, 174(2), 300–311 e311. https://doi.org/10.1016/j.cell.2018.06.026.

CHAPTER FIVE

A pyrophosphatase-coupled assay to monitor the NTase activity of cGAS

Richard Hooy*, Jungsan Sohn*
Department of Biophysics and Biophysical Chemistry, Johns Hopkins University School of Medicine, Baltimore, MD, United States
*Corresponding authors: e-mail address: rhooy1@jhu.edu; jsohn@jhmi.edu

Contents

Abstract

Cyclic GMP-AMP synthase, cGAS, converts ATP and GTP into a cyclic dinucleotide second messenger, cyclic GMP-AMP or cGAMP, through its enzymatic, nucleotidyl transferase (NTase) activity. Although many methods are available to directly measure cGAMP production, these assays often have high cost of implementation and/or experimental limitations. This chapter details how to implement an alternative approach that is relatively inexpensive, accurate and medium-throughput. The assay measures cGAS NTase activity by quantifying pyrophosphate production, a byproduct of the cGAS reaction. A coupling enzyme, pyrophosphatase, catalyzes the hydrolysis of pyrophosphate into inorganic phosphate, which enables facile detection of cGAS activity through conventional phosphomolybdate-malachite green absorbance methodology. This method is amenable for conventional steady-state kinetic measurements as well as high-throughput compound screening.

Methods in Enzymology, Volume 625
ISSN 0076-6879
https://doi.org/10.1016/bs.mie.2019.06.005

1. Introduction

Cyclic GMP-AMP synthase (cGAS), a pattern recognition receptor of the innate immune system, induces the expression of type-I interferon (IFN-I) genes upon detecting cytosolic double-stranded (ds)DNA and synthesizing the unique second messenger, cyclic GMP-AMP (cGAMP), from ATP and GTP (Sun, Wu, Du, Chen, & Chen, 2013). cGAMP binds and activates stimulator of interferon genes (STING) (Ablasser et al., 2013; Gao et al., 2013; Zhang et al., 2013), which recruits and activates Tank-binding kinase 1 (TBK1) leading to activation of transcription factors, IRF3 and NF-κB (Ishikawa, Ma, & Barber, 2009). The IFN-I response has variable outcomes and depends heavily on the strength of cGAS signaling (Gulen et al., 2017; Li & Chen, 2018; Vanpouille-Box, Demaria, Formenti, & Galluzzi, 2018). Hence, understanding how cGAS activity is regulated at the molecular level, has the potential to inform observations made at the cell, tissue and organism level.

Numerous methods have been developed to track cGAS activity (Ablasser et al., 2013; Andreeva et al., 2017; Bose, Su, Marcus, Raulet, & Hammond, 2016; Hall et al., 2017; Liu et al., 2017; Vincent et al., 2017; Wang, Sooreshjani, Mikek, Opoku-Temeng, & Sintim, 2018). Most methods rely on the ability to directly measure cGAMP, the biological signaling molecule. However, an equally reliable proxy for nucleotidyl transferase (NTase) activity is the production of pyrophosphate, the by-product of cGAMP synthesis. Indeed, pyrophosphate detection has been employed to study numerous systems including NTases (Cestari & Stuart, 2013; Chang, Pan, Lin, & Wang, 1984; Lloyd, Thomann, Ibba, & Soll, 1995; McMillan, Hepowit, & Maupin-Furlow, 2016; Seamon & Stivers, 2015; Yi, Velasquez, Holler, & Woodard, 2011). Pyrophosphate is difficult to be measured directly; however, its degradation product, inorganic phosphate can be easily quantified colorimetrically (Baykov, Evtushenko, & Avaeva, 1988). For instance, pyrophosphate is relatively stable under physiological conditions (neutral pH), but its degradation into inorganic phosphate is facilitated under low pH, by divalent cations and/or by enzymes that specifically catalyze the hydrolysis of pyrophosphates, e.g., pyrophosphatases (Stockbridge & Wolfenden, 2011). Enzyme-mediated hydrolysis of pyrophosphate is far superior to non-enzymatic alternatives as evidenced by the 10^6-fold faster hydrolysis rate produced by pyrophosphatases (Stockbridge & Wolfenden, 2011). The high efficiency of pyrophosphatases is particularly critical for ensuring accurate rate measurement of target enzymes.

The specific measurement of cGAS NTase activity employs the coupled hydrolysis of pyrophosphate by *E. coli* pyrophosphatase (Fig. 1A). Reactions are quenched with divalent cation chelators, such as EDTA, and the

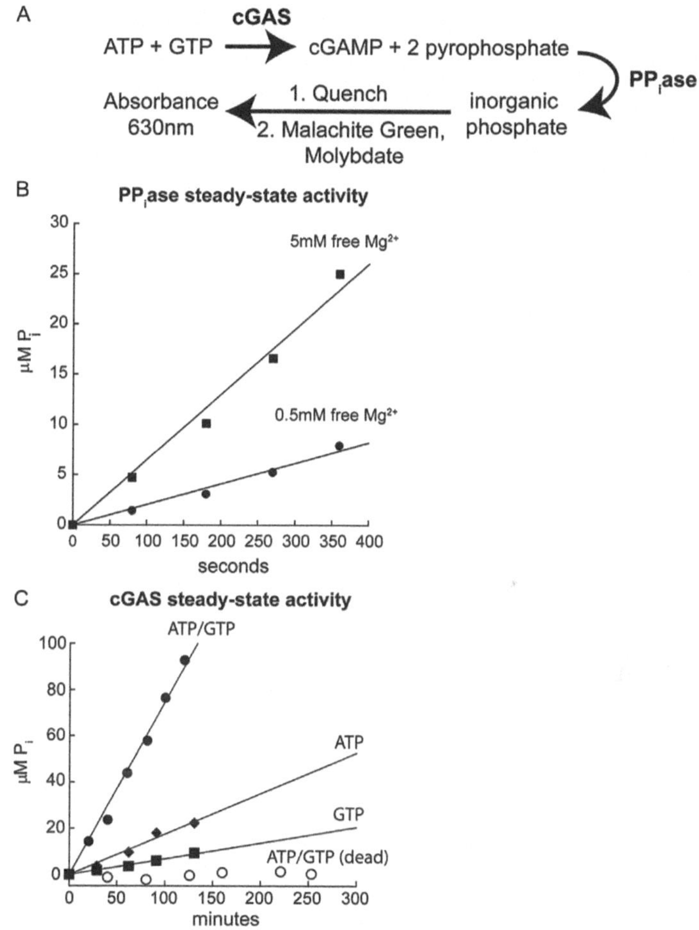

Fig. 1 (A) Scheme of the pyrophosphatase-coupled activity assay. (B) Raw sample data of *E. coli* pyrophosphatase steady-state turnover of inorganic pyrophosphate in the presence of different concentrations of free divalent metal. Each reaction contains 80 µM pyrophosphate pre-mixed with equimolar Mg^{2+}. (C) Raw sample data of cGAS (125 nM) steady state activity in the presence of different substrates and dsDNA. "dead"—catalytically inert cGAS variant with the mutations E225A/D227A.

phosphate concentration is determined colorimetrically by measuring the absorbance of malachite green- phosphomolybdate complexes (Baykov et al., 1988). Absolute values of phosphate are determined by comparing enzyme reactions to an internal, inorganic phosphate standard curve. This strategy has multiple advantages over direct detection of cGAMP. Phosphate detection is cost effective and can be implemented with a plate reader equipped with a UV–vis detector. The assay is also essentially label-free, which relieves restrictions on substrate concentration and substrate identity, and is scalable for high-throughput applications (Cestari & Stuart, 2013; Seamon & Stivers, 2015). Importantly, the assay lends itself to conventional steady-state enzymatic measurements, enabling accurate and rigorous measurement of cGAS enzymatic properties. Lastly, given the relatively slow turnover of cGAS (Vincent et al., 2017; Zhou et al., 2018), phosphate detection provides the benefit of immediate signal amplification per turnover event due to two phosphates being produced per catalytic event.

This chapter details how to successfully implement the PP_iase-coupled NTase activity assay for monitoring the catalytic activity of cGAS in vitro.

2. Materials

2.1 Chemicals, materials and enzymes

Nucleotides: adenosine triphosphate (ATP), guanosine triphosphate (GTP)
Divalent cations: magnesium chloride (anhydrous)
Chemicals: sodium molybdate, malachite green, concentrated sulfuric acid, Tween-20, potassium acetate, Tris, acetic acid, potassium hydroxide, Tris(2-carboxyethyl)phosphine (TCEP), glycerol, ethylenediaminetetraacetic acid (EDTA)
Enzymes: *E. coli* pyrophosphatase and cGAS
Materials: 0.22 μm filter, syringe, 384-well flat-bottom, transparent assay plate (Corning)

2.2 Buffers and assay reagents

2.2.1 Solution stocks

1. 1 M Tris acetate pH 7.4, adjust the pH with acetic acid/KOH, filter and de-gas, store at 4 °C
2. 1 M potassium acetate pH 7.4, adjust the pH with acetic acid/KOH, filter and de-gas, store at 4 °C
3. 0.5 M Tcep, pH with acetic acid/KOH, aliquot and store at −20 °C

4. Malachite Green Solution (MGS): 0.4 g of malachite green, 300 mL ddH_2O, 60 mL concentrated sulfuric acid
5. 7.5% sodium molybdate
6. 11% Tween-20
7. 0.5 M EDTA pH 8.0
8. *Nucleotides*: hydrate ATP and GTP separately in deionized water (ddH_20) supplemented with 20 mM HEPES buffer pH 7.4. Target concentrations are 100 mM for ATP and 50 mM for GTP. Adjust the pH to ~7–8 using KOH. It is best to do this quickly to avoid hydrolysis of either NTP. Measure the concentration of each nucleotide spectrophotometrically. λ_{max} (ATP) = 259 nm, $\varepsilon = 15{,}400\,M^{-1}\,cm^{-1}$; λ_{max} (GTP) = 252 nm, $\varepsilon = 13{,}700\,M^{-1}\,cm^{-1}$. Optional: Divalent metal cations can be added to the nucleotide stocks but should be done so prior to pH adjustment. Metals are typically equimolar or in 10% molar excess.
9. 800 μM Phosphate standard—hydrate sodium phosphate monobasic in ddH_20. Calibrate the phosphate concentration using a commercial stock of aqueous phosphate and the protocol described below for quantifying phosphate.
10. 1 M $MgCl_2$–0.22 μm filtered.

2.2.2 Buffers and assay solutions

NTase Reaction Buffer—25 mM Tris Acetate pH 7.4, 125 mM potassium acetate pH 7.4, 0.5 mM Tcep pH 7.4, 5 mM magnesium chloride, 5% glycerol.

2× *Quench Buffer*—25 mM Tris Acetate pH 7.4, 125 mM potassium acetate pH 7.4, 0.5 mM Tcep pH 7.4, 25 mM EDTA pH 8.0, 5% glycerol.

Color Development Solution (CDS)—200 μL 7.5% (w/v) sodium molybdate (ddH_20), 16 μL Tween-20, 800 μL MGS

2.2.3 Recombinant enzymes

cGAS—(See other chapters within this Volume and Hooy & Sohn, 2018).

E. coli pyrophosphatase—see (Seamon & Stivers, 2015) for full details, or can be purchased from commercial sources (e.g., New England BioLabs).

2.2.4 Equipment

Plate reader equipped with UV–vis spectrophotometer (e.g., TECAN M1000)

Nutating shaker

3. Implementing the PP_iase-coupled NTase activity assay

3.1 Measuring steady-state NTase activity

This protocol describes how to measure cGAS substrate turnover as a function of time under steady-state conditions.

3.1.1 Protocol

1. Prepare a 10 × stock of nucleotides. Combine reaction buffer, ATP, GTP and 2M equivalents of $MgCl_2$. Mix thoroughly. *Note*: Additional equivalents of metal are only required if the nucleotide stocks were not hydrated with metal.
2. Combine reaction buffer, PP_iase and $cGAS^{FL}$ in "cGAS reaction" tubes. Repeat this for "control" reactions (i.e., minus cGAS). *Note*: As described in Section 3.3.1, the concentration of PP_iase will depend on reaction conditions. One must ensure that PP_iase activity is not rate limiting.
3. Initiate each reaction by adding 10 × nucleotide(s). Mix thoroughly. *Note*: It is important to time-match each "cGAS reaction" with its "control" reaction. To achieve this, use a multi-channel to simultaneously initiate reactions, or stagger the initiation of each reaction.
4. Add 20 μL of 2 × Quench Buffer to the appropriate number of wells on a 384-well assay plate. For each time point, remove 20 μL of each reaction and add it to the corresponding well containing 2 × Quench buffer. Ensure that all "cGAS reaction" and "control" reaction pairs run for the same duration by using a multi-channel or by staggering quench times, exactly as in step 3.
5. Repeat step 4 for a minimum of three additional time points. *Note*: The exact time course is not as crucial; however, we recommend collecting evenly spaced time points to facilitate data fitting. Also, the duration of the experiment will be dictated by the apparent rate of the target enzyme; ensure that the collected time points provide adequate signal-to-noise yet remain within the linear range of the assay.
6. In the assay plate, prepare 20 μL volumes of 90 μM, 60 μM, 30 μM, and 15 μM inorganic phosphate diluted in reaction buffer. Add 20 μL of 2 × Quench Buffer to each.
7. Prepare the Color Development Solution (CDS) when all time points have been collected. Add 10 μL of CDS to each well. Set the plate on

a nutating shaker at room temperature for 30–60 min. *Note*: To minimize time discrepancies between reaction pairs, use a multi-channel or automated machine/robot to distribute the CDS.

8. Scan the absorbance of each well between 610 nm and 650 nm in 2 nm steps. Calculate the maximum absorbance value for each well.

3.2 Data processing and analysis

Subtract the absorbance value of control reactions from time-matched cGAS reactions. Convert the subtracted value to phosphate concentration using the slope of the phosphate standard curve. Calculate the rate of phosphate production by fitting the phosphate concentration over time to a line. Divide the slope by two to account for each pyrophosphate produced. Normalize the rate by dividing by the concentration of $cGAS^{FL}$ used in the assay to obtain apparent rates of PP_i production min^{-1} $[cGAS]^{-1}$.

3.3 Additional experimental notes

3.3.1 Measuring the specific activity of PP_iase

To accurately measure the NTase activity of a target enzyme, it is crucial to ensure that the coupling step, PP_i hydrolysis, is not rate-limiting (i.e., changing PP_iase concentrations must not affect observed rates). Although PP_iase is an efficient enzyme (Stockbridge & Wolfenden, 2011), reaction conditions may reduce its efficiency. Measure the specific activity of PP_iase against purified PP_i under the exactly same conditions as those with cGAS as described in Section 3.1. Use this measurement as a guide to select appropriate PP_iase concentrations.

3.3.2 Steady-state activity

Under steady-state conditions, the rate of product formation over time should be linear. Moreover, control-subtracted cGAS reactions typically exhibit virtually no burst-like behavior, i.e., the y-intercept is approximately zero, even if the enzyme is pre-bound to dsDNA.

3.3.3 Measuring the effectiveness of allosteric ligands

cGAS NTase activity is allosterically regulated by dsDNA binding. To accurately and quantitatively determine how effective an allosteric ligand modulates cGAS activity, it is crucial that the cGAS concentration be at least an order of magnitude lower than the measured apparent binding affinity for the ligand, e.g., $[cGAS] \ll K_{act}$; K_{act} is the concentration of allosteric ligand required to achieve half-maximal stimulation. This will ensure that the measured K_{act} reflects an accurate binding affinity instead of a stoichiometric interaction. If the enzyme concentration required to achieve this condition becomes impractical, modify the buffer composition or report the binding affinities as upper limits.

4. Summary and prospects

The NTase activity assay described here is inexpensive to implement, facile to perform and is highly reproducible. It's utility stems from the straightforward and established approach of monitoring inorganic phosphate concentration by absorbance instead of cyclic dinucleotide accumulation by some alternative observable. The assay benefits from an inherent signal amplification that occurs due to each PP_i giving rise to two inorganic phosphates, and its label-free nature circumvents constraints imposed by substrate analogs. The assay lends itself to medium-throughput steady-state kinetic measurements and enables accurate determination of enzymatic properties such as maximal turnover rates, k_{cat}, and the functional binding affinity for substrate, K_m. Importantly, nucleotides other than ATP and GTP can also be assayed with ease. Lastly, allosteric effectors can be easily and quantitatively compared by measuring NTase activity as a function of allosteric ligand concentration, or by re-measuring substrate utilization parameters in the presence of allosteric ligands.

Although the assay does not directly measure the more biologically relevant signaling molecule, cGAMP, the measured NTase activity profile mirrors the cGAS signaling response in cells and recapitulates rate measurements made in vitro by orthogonal methods (Hooy & Sohn, 2018). This justifies using the assay to screen libraries of small molecule compounds for allosteric and active site modulators of cGAS. Indeed, the assay can be implemented in 96-well and 384-well formats and demonstrates Z' values >0.5 (Cestari & Stuart, 2013; Seamon & Stivers, 2015; Zhang, Chung, & Oldenburg, 1999), although this value can range depending on reaction conditions.

The main disadvantage of the assay is its discontinuous nature. This makes kinetic measurements rather tedious. One alternative strategy is to combine this PP_iase-coupled assay with the real-time, spectrophotometric assay that couples the conversion of MesG ($\lambda_{max}=330$ nm) and inorganic phosphate to the free base moiety ($\lambda_{max}=360$ nm) and 1-phosphate-ribose via the enzyme, purine nucleotide phosphorylase (PNP) (Lloyd et al., 1995; Upson, Haugland, Malekzadeh, & Haugland, 1996). Appropriate considerations should be made to ensure the additional coupling enzymes and reagents do not interfere with cGAS or PP_iase, and that the coupling step is not rate-limiting. However, such a modification could enable facile pre-steady-state kinetic measurements and would improve the detection limit.

References

Ablasser, A., Goldeck, M., Cavlar, T., Deimling, T., Witte, G., Rohl, I., et al. (2013). cGAS produces a 2′-5′-linked cyclic dinucleotide second messenger that activates STING. *Nature*, *498*, 380–384.

Andreeva, L., Hiller, B., Kostrewa, D., Lassig, C., de Oliveira Mann, C. C., Jan Drexler, D., et al. (2017). cGAS senses long and HMGB/TFAM-bound U-turn DNA by forming protein-DNA ladders. *Nature*, *549*, 394–398.

Baykov, A. A., Evtushenko, O. A., & Avaeva, S. M. (1988). A malachite green procedure for orthophosphate determination and its use in alkaline phosphatase-based enzyme immunoassay. *Analytical Biochemistry*, *171*, 266–270.

Bose, D., Su, Y., Marcus, A., Raulet, D. H., & Hammond, M. C. (2016). An RNA-based fluorescent biosensor for high-throughput analysis of the cGAS-cGAMP-STING pathway. *Cell Chemical Biology*, *23*, 1539–1549.

Cestari, I., & Stuart, K. (2013). A spectrophotometric assay for quantitative measurement of aminoacyl-tRNA synthetase activity. *Journal of Biomolecular Screening*, *18*, 490–497.

Chang, G. G., Pan, F., Lin, Y. H., & Wang, H. Y. (1984). Continuous spectrophotometric assay for aminoacyl-tRNA synthetases. *Analytical Biochemistry*, *142*, 369–372.

Gao, P., Ascano, M., Wu, Y., Barchet, W., Gaffney, B. L., Zillinger, T., et al. (2013). Cyclic [G(2′,5′)pA(3′,5′)p] is the metazoan second messenger produced by DNA-activated cyclic GMP-AMP synthase. *Cell*, *153*, 1094–1107.

Gulen, M. F., Koch, U., Haag, S. M., Schuler, F., Apetoh, L., Villunger, A., et al. (2017). Signalling strength determines proapoptotic functions of STING. *Nature Communications*, *8*, 427.

Hall, J., Brault, A., Vincent, F., Weng, S., Wang, H., Dumlao, D., et al. (2017). Discovery of PF-06928215 as a high affinity inhibitor of cGAS enabled by a novel fluorescence polarization assay. *PLoS One*, *12*, e0184843.

Hooy, R. M., & Sohn, J. (2018). The allosteric activation of cGAS underpins its dynamic signaling landscape. *eLife*, *7*.

Ishikawa, H., Ma, Z., & Barber, G. N. (2009). STING regulates intracellular DNA-mediated, type I interferon-dependent innate immunity. *Nature*, *461*, 788–792.

Li, T., & Chen, Z. J. (2018). The cGAS-cGAMP-STING pathway connects DNA damage to inflammation, senescence, and cancer. *The Journal of Experimental Medicine*, *215*(5), 1287–1299.

Liu, B., Tang, L., Zhang, X., Ma, J., Sehgal, M., Cheng, J., et al. (2017). A cell-based high throughput screening assay for the discovery of cGAS-STING pathway agonists. *Antiviral Research*, *147*, 37–46.

Lloyd, A. J., Thomann, H. U., Ibba, M., & Soll, D. (1995). A broadly applicable continuous spectrophotometric assay for measuring aminoacyl-tRNA synthetase activity. *Nucleic Acids Research*, *23*, 2886–2892.

McMillan, L. J., Hepowit, N. L., & Maupin-Furlow, J. A. (2016). Archaeal inorganic pyrophosphatase displays robust activity under high-salt conditions and in organic solvents. *Applied and Environmental Microbiology*, *82*, 538–548.

Seamon, K. J., & Stivers, J. T. (2015). A high-throughput enzyme-coupled assay for SAMHD1 dNTPase. *Journal of Biomolecular Screening*, *20*, 801–809.

Stockbridge, R. B., & Wolfenden, R. (2011). Enhancement of the rate of pyrophosphate hydrolysis by nonenzymatic catalysts and by inorganic pyrophosphatase. *The Journal of Biological Chemistry*, *286*, 18538–18546.

Sun, L., Wu, J., Du, F., Chen, X., & Chen, Z. J. (2013). Cyclic GMP-AMP synthase is a cytosolic DNA sensor that activates the type I interferon pathway. *Science*, *339*, 786–791.

Upson, R. H., Haugland, R. P., Malekzadeh, M. N., & Haugland, R. P. (1996). A spectrophotometric method to measure enzymatic activity in reactions that generate inorganic pyrophosphate. *Analytical Biochemistry*, *243*, 41–45.

Vanpouille-Box, C., Demaria, S., Formenti, S. C., & Galluzzi, L. (2018). Cytosolic DNA sensing in organismal tumor control. *Cancer Cell*, *34*, 361–378.

Vincent, J., Adura, C., Gao, P., Luz, A., Lama, L., Asano, Y., et al. (2017). Small molecule inhibition of cGAS reduces interferon expression in primary macrophages from autoimmune mice. *Nature Communications*, *8*, 750.

Wang, M., Sooreshjani, M. A., Mikek, C., Opoku-Temeng, C., & Sintim, H. O. (2018). Suramin potently inhibits cGAMP synthase, cGAS, in THP1 cells to modulate IFN-beta levels. *Future Medicinal Chemistry*, *10*, 1301–1317.

Yi, L., Velasquez, M. S., Holler, T. P., & Woodard, R. W. (2011). A simple assay for 3-deoxy-d-manno-octulosonate cytidylyltransferase and its use as a pathway screen. *Analytical Biochemistry*, *416*, 152–158.

Zhang, J. H., Chung, T. D., & Oldenburg, K. R. (1999). A simple statistical parameter for use in evaluation and validation of high throughput screening assays. *Journal of Biomolecular Screening*, *4*, 67–73.

Zhang, X., Shi, H., Wu, J., Zhang, X., Sun, L., Chen, C., et al. (2013). Cyclic GMP-AMP containing mixed phosphodiester linkages is an endogenous high-affinity ligand for STING. *Molecular Cell*, *51*, 226–235.

Zhou, W., Whiteley, A. T., de Oliveira Mann, C. C., Morehouse, B. R., Nowak, R. P., Fischer, E. S., et al. (2018). Structure of the human cGAS-DNA complex reveals enhanced control of immune surveillance. *Cell*, *174*, 300–311 e11.

CHAPTER SIX

Tracking the polymerization of DNA sensors and inflammasomes using FRET

Zachary Mazanek, Jungsan Sohn*
Department of Biophysics and Biophysical Chemistry, Johns Hopkins University School of Medicine, Baltimore, MD, United States
*Corresponding author: e-mail address: jsohn@jhmi.edu

Contents

Abstract

Upon activation, DNA sensors and inflammasomes often polymerize into filamentous oligomers, and transduce the incoming pathogenic signal via inducing the assembly of downstream filaments. Given the complexity of these supramolecular structures, kinetics and thermodynamics that govern their assembly mechanisms remain poorly understood. Here, we present a simple yet robust assay that can track the assembly of these filaments in a quantitative manner. This FRET-based measurement is cost effective and also amenable to high-throughput screening.

1. Introduction

The assembly of supramolecular signaling platforms has emerged as a new paradigm in innate immune signaling pathways (Wu, 2013). Several DNA sensors and virtually all inflammasomes form filaments in order to be signaling competent (Lu et al., 2014; Matyszewski, Morrone, & Sohn, 2018; Morrone et al., 2014, 2015; Shen et al., 2019; Stratmann, Morrone, van Oijen, & Sohn, 2015). For instance, Absent-In-Melanoma-2 (AIM2)-like

Methods in Enzymology, Volume 625
ISSN 0076-6879
https://doi.org/10.1016/bs.mie.2019.06.006

receptors assemble filaments on cytoplasmic double-stranded (ds)DNA (Lu et al., 2014; Matyszewski et al., 2018; Morrone et al., 2014, 2015; Stratmann et al., 2015), which then induce various inflammatory responses including the activation of inflammasome pathways and boosting interferon production (Guo, Callaway, & Ting, 2015). Of note, AIM2, the eponymous member of the family, directly induces the polymerization of a central hub in inflammasome pathways-ASC (Apoptosis associated Speck-forming protein containing Caspase recruiting domain (CARD)). ASC then induces the polymerization of caspase-1, activating the protease via proximity induced auto-proteolysis (Fernandes-Alnemri, Yu, Datta, Wu, & Alnemri, 2009; Guo et al., 2015; Hornung et al., 2009; Roberts et al., 2009). Caspase-1, also known as the interleukin converting enzyme, mediates the maturation of interleukin-1β and 18, and also induces cell death via activating pore-forming gasdermin-D (Guo et al., 2015; Liu et al., 2016).

A unique aspect of inflammasome/ALR-mediated DNA sensing is that the signal is transduced by the sequential assembly of filamentous oligomers, unlike more conventional enzyme-mediated cascades (e.g., phosphorylation relay) (Lu et al., 2014; Matyszewski et al., 2018). To understand how DNA-sensing/inflammasome pathways operate at the molecular level, it is imperative to devise a quantitative assay to track the assembly in a controlled manner. Here, we describe how we have used Förster Resonance Energy Transfer (FRET) to monitor the assembly of several key components involved in DNA-sensing/inflammasome pathways (Matyszewski et al., 2018; Morrone et al., 2014, 2015; Stratmann et al., 2015).

2. Protocols

PYrin Domains (PYDs) are major warheads in inflammasome signaling. For instance, upon binding ligands (e.g., DNA), the PYD of upstream receptors assemble into filaments, which then induces the assembly of the PYD filament of ASC (ASC^{PYD}) (Lu et al., 2014; Matyszewski et al., 2018; Morrone et al., 2014, 2015; Stratmann et al., 2015). ASC is a bipartite protein that contains an N-terminal PYD and a C-terminal CARD (ASC^{CARD}). The ASC^{PYD} filament induces the oligomerization of ASC^{CARD}, subsequently inducing the polymerization of caspase-1(Lu et al., 2014). Although we have used the FRET method in both full-length AIM2 and its isolated PYD, given that signal transduction is exclusively mediated by PYD-PYD interactions, we will focus on how to track the polymerization of PYDs using FRET. We will break down our protocols into three sections: protein prep, labeling, and assay.

2.1 Preparing PYDs for labeling

Maleimide linked fluorophores work well for labeling inflammasomes because they contain very few cysteines. In order to use maleimide-linked fluorophores, a target protein must contain at least one solvent exposed cysteine residue. In cases where a protein lacks a cysteine or has no solvent exposed cysteines one can be easily inserted via mutagenesis PCR. We recommend adding the cysteine at either terminus to avoid disrupting the structural integrity of the protein. Alternatively, if there are several solvent exposed cysteine residues, one can add a Sortase tag to the C-terminus of the protein and use it as a method for linking a fluorophore to the protein (Guimaraes et al., 2013). PYDs can be expressed in *E. coli* and purified without much issue (Lu et al., 2014; Matyszewski et al., 2018; Morrone et al., 2014, 2015; Stratmann et al., 2015) (see the chapter "Preparation of filamentous proteins for electron microscopy visualization and reconstruction" in this volume).

Note: PYDs tend to auto-assemble/aggregate. To prevent this, we fuse a Maltose Binding Protein tag (MBP) at the N-terminus of the protein (Fig. 1A). This not only increases solubility and production of PYDs,

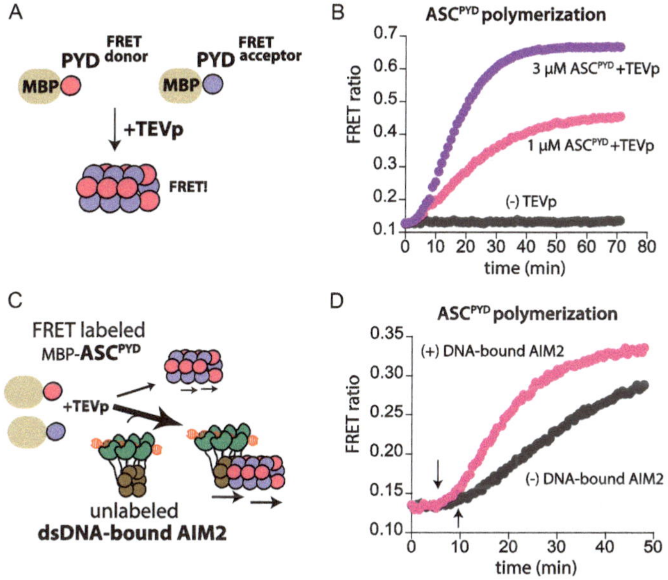

Fig. 1 FRET assay scheme and examples. (A) A cartoon description of FRET-based polymerization assays. (B) An example of concentration dependent polymerization of ASC^{PYD} tracked by FRET. (C) A cartoon description of a FRET-based polymerization assay that monitors "seeding" effect. (D) An example of seeding reactions. Arrows indicate the shift from the lag phase to the exponential phase.

but also keeps them largely monomeric, thus providing a handle for controlling the polymerization activity. A protease site should also be included between the tag and the protein of interest so that it can be cleaved later (i.e., triggering polymerization). Tobacco Etch Virus protease (TEVp) is a good option because it is stable, and it has a specific sequence that leaves a minimal scar after cleavage. Fluorescent proteins or other tags can also be used, however, MBP has the added benefit of being an affinity tag. Fluorescent proteins are generally undesirable for doing FRET based assays since they can interfere with filament formation and therefore FRET. The affinity tag should be left intact prior to labeling as it will be used to isolate the monomer protein after the labeling reaction is complete. *Important*: avoid dithiolthreitol (DTT) or beta-mercaptoethanol (BME) as they will compete with target cysteines. Tris(2-carboxyethyl)phosphine (TCEP) is a good alternative reducing agent.

2.2 Labeling strategy

Maleimide linked fluorophores are widely available from several vendors. We generally use a FRET donor that has Excitation λ_{max} around 550 nm (e.g., $Alexa_{568}$) and an acceptor that has Excitation λ_{max} around 650 nm (e.g., $Alexa_{647}$). *Note*: we find that using a donor Excitation λ_{max} around 488 nm and using an acceptor Excitation λ_{max} 550 nm do not work well, as it mostly results in quenching of the donor than increasing the emission of the acceptor. *Note*: label one batch of target protein with a FRET donor and another batch (preferably equimolar) with a FRET acceptor. One limitation of FRET assays is that they do not directly confirm the formation of filaments. This problem can be alleviated with quick nsEM experiments to confirm the presence of structures such as filaments (see the chapter "Preparation of filamentous proteins for electron microscopy visualization and reconstruction" in this volume).

Labeling protocol

1. Labeling should start immediately after protein preparation is finished. Although labeling can be done at any concentration, we find not exceeding 50 μM usually helps in preventing aggregation. For example, combined fractions from size-exclusion chromatography (SEC) can be taken and used for labeling without concentrating as long as the resulting protein is free of other protein contaminants.
2. Dissolve maleimide dyes in 100% DMSO to a sufficiently high concentration. 10 mM works well for labeling micromolar amounts of protein. Dye tubes should be stored wrapped in foil at −20 °C.

3. Determine the concentration of protein in the starting solution.
4. Add dye at 5 × the protein concentration to the protein containing solution.
5. Close the tubes tightly and seal with parafilm to prevent leaks. Wrap each tube in aluminum foil to prevent excessive photobleaching during the labeling process. Or one can use dark brown tubes.
6. Allow tubes to sit on a rotator or nutator for 2 h at room temperature. Do not use a shaker for small volumes as it does not provide sufficient mixing and will result in lower labeling efficiency.
7. After 2 h, spin tubes down using (5 min 10 × *g*), add DTT to 2 mM, BME to 5 mM if using Nickel NTA resin for the next step, to deactivate remaining maleimide dye. Place back on the rotator or nutator for 30–60 min at room temperature.
8. While dye is being deactivated prepare about 300 μL of packed resin (either amylose or Nickel NTA).
9. Spin down tubes again and run the solution over affinity resin matching the tag on the protein
10. Make a buffer containing 40 mM HEPES at pH 7.4, 150 mM KCl, 5% Glycerol, 5 mM DTT and 1 mM EDTA. *Note:* Replace DTT with BME and omit EDTA if using Nickel NTA resin.
11. Wash the column with buffer until the flow-through is clear. The resin should maintain a bright color which indicates your labeled protein is bound to the resin. If the resin lacks color, either the protein failed to bind or, more likely, labeling efficiency was poor.
12. Elute the protein from the column slowly using only 200–300 μL (e.g., buffer + maltose or imidazole) at a time. Watch carefully and collect fractions in 100 μL volumes. This ensures that the protein will be concentrated by combining only the fractions with the most intense color. *Note*: After labeling, concentrating using centrifugal devices is not recommended as they can cause aggregation in filament forming proteins.
13. Measure the labeling efficiency of the protein by dividing the concentration of the dye, adjusted for the number of solvent accessible cysteines, by the protein concentration. *Note:* 100% labeling efficiency is not required to achieve a sufficient FRET signal in filament forming species. We have found 50% labeling efficiency to be optimal.

Alternatives to labeling protocol: Proteins can be labeled at 4 °C overnight if they are not stable at room temperature or if labeling efficiency is low.

2.3 FRET-based assays

FRET-based assays are traditionally used to measure proximity of two molecules in solution. In the case of filamentous protein complexes, it can be used to measure the kinetics of filament formation. As monomeric units assemble the FRET ratio between donor and acceptor fluorophores increases due to an increasing number of subunits coming into proximity with each other over time (Fig. 1A and B).

To set up a FRET assay, 384 well plates are a good option because small volumes (typically 15 μL) can be used, and reactions with varying conditions can be run and compared simultaneously. A suitable buffer is required. Filament formation and stability is sensitive to pH, ionic strength, and additives such as glycerol and detergent. To replicate a cellular environment we use: 40 mM HEPES at pH 7.4, 160 mM KCl, 5% Glycerol, 0.1% Triton X100, 1 mM EDTA and 5 mM DTT. These conditions can be adjusted as needed in order to optimize behavior of filaments in solution.

1. Add buffer to each well.
2. Add donor and acceptor labeled proteins at 1:1 M ratio.
3. To initiate reaction, add TEVp to each well and mix rapidly and thoroughly (~2 μL). *Note*: Make sure that the amount of TEVp is not rate-limiting by trying a few different conditions (1:10 or 1:5 target: TEVp).
4. Once TEVp is added, immediately start the data collection over 60–120 min (depending on conditions), recording time points every 1–2 min (Fig. 1). *Note:* Upon exciting the FRET donor, we monitor the emission of the FRET donor and the acceptor.

Several types of assays can be done using this setup. Filaments can be grown and monitored simply by cleaving the solubility tag at different concentrations to determine the critical concentration at which filaments will form. Once a suitable concentration is found at which PYD filaments assemble efficiently and consistently, conditions can be varied to study the effect changing conditions have on PYDs (Fig. 1B).

Experiments can also be done by seeding a PYD filament with an existing PYD filament (i.e., signal transduction between upstream and downstream assemblies). First, filaments are formed using unlabeled protein. This is done by incubating the PYDs with TEVp for a period of time that will allow filaments to fully form. *Note*: Pre-establish this condition by monitoring the cleavage by SDS-PAGE. It is important to make sure filaments have fully formed before introducing labeled protein to the reaction because unlabeled protein can dilute the FRET signal and give false results.

To initiate a FRET reaction under seeding conditions simply add the premixed FRET pair to the mixture and monitor the change in FRET values. Different seed filaments can be used to determine their effect on the PYD of interest. For example, a preformed AIM2 filament can be used to seed an ASC^{PYD} filament and compared to an ASC^{PYD} filament seeding ASC^{PYD} (Fig. 1C and D). *Note*: to monitor ligand-mediated oligomerization (e.g., DNA), pre-cleave the FRET-labeled proteins separately below their critical concentrations for auto-assembly (Matyszewski et al., 2018). Add ligand to initiate reaction and monitor the FRET ratios in either time- or [ligand]-dependent manner (Matyszewski et al., 2018; Morrone et al., 2014, 2015).

Once *FRET ratios* (= *divide acceptor emission by donor emission at each time point*) have been calculated at a number of time points curves can be fit to a variant of the Hill equation to estimate the half-time (t_h) of polymerization:

$$\text{Observed FRET ratio at time } (t) = (\text{max observed FRET ratio})/(1 + (t_h/t)^n),$$

where n allows the curve to fit the lag phase (i.e., rate-limiting nucleation Matyszewski et al., 2018) of the FRET curves. The apparent polymerization rate (k_{poly}) can be calculated as $\ln(2)/t_h$.

Note: A double exponential equation can also be used to fit the data, but the apparent polymerization rate is a good approximation that is easier to conceptualize and compare. Nevertheless, pay attention to how the shape of reaction curves changes (Fig. 1D arrows), as the initial lag-phase (nucleation) will be shorted in the presence of "seeds." Some difficulties can arise in doing FRET assays. One is temperature. Slight variations in room temperature can affect the kinetics of polymerization and give inconsistent results. This can be solved by using a plate reader with a temperature control feature to ensure consistent results. It is also noteworthy that our assay is broadly applicable for tracking any proteins that oligomerize.

Acknowledgment

This work was supported by NSF CAREER Award (1845003) to J.S.

Reference

Fernandes-Alnemri, T., Yu, J. W., Datta, P., Wu, J., & Alnemri, E. S. (2009). AIM2 activates the inflammasome and cell death in response to cytoplasmic DNA. *Nature*, *458*, 509–513.

Guimaraes, C. P., Witte, M. D., Theile, C. S., Bozkurt, G., Kundrat, L., Blom, A. E., et al. (2013). Site-specific C-terminal and internal loop labeling of proteins using sortase-mediated reactions. *Nature Protocols*, *8*, 1787–1799.

Guo, H., Callaway, J. B., & Ting, J. P. (2015). Inflammasomes: Mechanism of action, role in disease, and therapeutics. *Nature Medicine*, *21*, 677–687.

Hornung, V., Ablasser, A., Charrel-Dennis, M., Bauernfeind, F., Horvath, G., Caffrey, D. R., et al. (2009). AIM2 recognizes cytosolic dsDNA and forms a caspase-1-activating inflammasome with ASC. *Nature*, *458*, 514–518.

Liu, X., Zhang, Z., Ruan, J., Pan, Y., Magupalli, V. G., Wu, H., et al. (2016). Inflammasome-activated gasdermin D causes pyroptosis by forming membrane pores. *Nature*, *535*, 153–158.

Lu, A., Magupalli, V. G., Ruan, J., Yin, Q., Atianand, M. K., Vos, M. R., et al. (2014). Unified polymerization mechanism for the assembly of ASC-dependent inflammasomes. *Cell*, *156*, 1193–1206.

Matyszewski, M., Morrone, S. R., & Sohn, J. (2018). Digital signaling network drives the assembly of the AIM2-ASC inflammasome. *Proceedings of the National Academy of Sciences of the United States of America*, *115*, E1963–E1972.

Morrone, S. R., Matyszewski, M., Yu, X., Delannoy, M., Egelman, E. H., & Sohn, J. (2015). Assembly-driven activation of the AIM2 foreign-dsDNA sensor provides a polymerization template for downstream ASC. *Nature Communications*, *6*, 7827.

Morrone, S. R., Wang, T., Constantoulakis, L. M., Hooy, R. M., Delannoy, M. J., & Sohn, J. (2014). Cooperative assembly of IFI16 filaments on dsDNA provides insights into host defense strategy. *Proceedings of the National Academy of Sciences of the United States of America*, *111*, E62–E71.

Roberts, T. L., Idris, A., Dunn, J. A., Kelly, G. M., Burnton, C. M., Hodgson, S., et al. (2009). HIN-200 proteins regulate caspase activation in response to foreign cytoplasmic DNA. *Science*, *323*, 1057–1060.

Shen, C., Lu, A., Xie, W. J., Ruan, J., Negro, R., Egelman, E. H., et al. (2019). Molecular mechanism for NLRP6 inflammasome assembly and activation. *Proceedings of the National Academy of Sciences of the United States of America*, *116*, 2052–2057.

Stratmann, S. A., Morrone, S. R., van Oijen, A. M., & Sohn, J. (2015). The innate immune sensor IFI16 recognizes foreign DNA in the nucleus by scanning along the duplex. *eLife*, *4*, e11721.

Wu, H. (2013). Higher-order assemblies in a new paradigm of signal transduction. *Cell*, *153*, 287–292.

CHAPTER SEVEN

Monitoring gasdermin pore formation in vitro

Shiyu Xia[a,b], **Jianbin Ruan**[a,b,*], **Hao Wu**[a,b,*]
[a]Program in Cellular and Molecular Medicine, Boston Children's Hospital, Boston, MA, United States
[b]Department of Biological Chemistry and Molecular Pharmacology, Harvard Medical School, Boston, MA, United States
*Corresponding authors: e-mail address: jianbin.ruan@childrens.harvard.edu; wu@crystal.harvard.edu

Contents

Abstract

The gasdermin (GSDM) family consists of gasdermin A (GSDMA), B (GSDMB), C (GSDMC), D (GSDMD), E or DNFA5 (GSDME), and DFNB59 in human. Expressed in the skin, gastrointestinal tract, and various immune cells, GSDMs mediate homeostasis and inflammation upon activation by caspases and unknown proteases. In particular, GSDMD is activated by inflammasome-activated caspases-1/-4/-5/-11 as well as a caspase-8-mediated pathway during Yersinia infection. These caspases cleave GSDMD to release its functional N-terminal fragment (GSDMD-NT) from its auto-inhibitory C-terminal fragment (GSDMD-CT). GSDMD-NTs bind to acid lipids in mammalian cell membranes and bacterial membranes, oligomerize, and insert into the membranes to form large transmembrane pores. Consequently, cellular contents including inflammatory cytokines are released and cells can undergo pyroptosis, a highly inflammatory form of cell death. In this chapter, we summarize recent research findings and present experimental procedures to obtain pure recombinant GSDMs for biochemical studies. We highlight a liposome-based assay that yields robust fluorescence signals for characterizing GSDM activities in vitro and may be applicable to other pore-forming proteins and ion channels in general.

Methods in Enzymology, Volume 625
ISSN 0076-6879
https://doi.org/10.1016/bs.mie.2019.04.024

Abbreviations

2ME	β-mercaptoethanol
3C	human rhinovirus 3C protease
CL	Cardiolipin
Cryo-EM	cryo-electron microscopy
CT	C-terminal fragment
DPA	dipicolinic acid
GSDM	Gasdermin
FL	full-length
NT	N-terminal fragment
PBS	phosphate buffered saline
PC	Phosphatidylcholine
PE	Phosphatidylethanolamine
PIP	Phosphatidylinositol phosphate
PS	Phosphatidylserine
$TbCl_3$	terbium(III) chloride
TEV	tobacco etch virus protease
TM	transmembrane

1. Introduction to the gasdermin family

1.1 Pyroptosis, inflammasomes, and gasdermins

Eukaryotic cells can undergo various types of programmed cell death when triggered by certain metabolic events, pathogens, and environmental factors. In many types of cells including myeloid and epithelial cells in the immune system, supramolecular complexes known as inflammasomes can sense pathogens and danger to initiate pyroptosis, a form of inflammatory cell death that features cell swelling, lysis, and release of pro-inflammatory cytokines including IL-1β and IL-18 (Bergsbaken, Fink, & Cookson, 2009; Martinon, Burns, & Tschopp, 2002; Schroder & Tschopp, 2010; Zhang, Chen, Gueydan, & Han, 2018). While this inflammatory response has a protective role in recruiting immune cells to remove infected cells, its dysregulation can also cause pathology such as inflammatory bowel disease, gout, and sepsis (Jorgensen & Miao, 2015; Maloy & Powrie, 2011; Martinon, Petrilli, Mayor, Tardivel, & Tschopp, 2006).

Inflammasomes are categorized based on their activating signals and caspases involved. Canonical inflammasomes recognize various infection and danger signals including flagellin, K+ efflux, and cytosolic DNA to activate caspase-1 (Franchi, Eigenbrod, Munoz-Planillo, & Nunez, 2009; Hornung et al., 2009; Lu et al., 2014; Petrilli et al., 2007; Zhao et al.,

2011). On the other hand, the assembly of the non-canonical inflammasome is induced by lipopolysaccharides (LPS) and oxidized lipids and results in the activation of murine caspase-11 and human caspase-4 and -5 (Kayagaki et al., 2013; Shi et al., 2014; Zanoni et al., 2016). The pore-forming protein gasdermin D (GSDMD) was recently identified as a substrate of these inflammasome-activated caspases, which cleave GSDMD into a functional N-terminal fragment (NT) and an autoinhibitory C-terminal fragment (CT) (Kayagaki et al., 2015; Shi et al., 2015). GSDMD-NTs subsequently translocate to the inner leaflet of mammalian cell membranes and bacterial membranes and bind acid lipids phosphatidylserine (PS), such as phosphatidylinositol phosphates (PIPs), phosphatidic acid (PA), and cardiolipin (CL). Upon lipid binding, GSDMD-NTs oligomerize and carry out membrane insertion to form transmembrane (TM) pores (Aglietti et al., 2016; Ding et al., 2016; Liu et al., 2016).

Although GSDMD pores were discovered in the context of pyroptosis, they also play crucial roles in NETosis (Chen et al., 2018; Sollberger et al., 2018) and can also be present in living cells (Evavold et al., 2018), which may have evolved mechanisms to survive GSDMD pores. For example, caspase-3 and -7 can degrade the pore-forming GSDMD-NT into smaller, non-functional fragments (Orning et al., 2018; Taabazuing, Okondo, & Bachovchin, 2017). In addition, GSDMD-perforated membranes may be shed from cells via an ESCRT-dependent mechanism (Ruhl et al., 2018). It has also been discovered recently that caspase-8-mediated cleavage of GSDMD occurs under Yersinia infection (Orning et al., 2018; Sarhan et al., 2018), challenging the notion that GSDMD activation is only carried out by inflammatory caspases.

The mechanistic elucidation of GSDMD kindled investigations into the GSDM protein family. In humans, the family consists of GSDMA, GSDMB, GSDMC, GSDMD, GSDME/DFNA5, and DFNB59; in mice, there are 10 GSDMs including three isoforms of GSDMA (GSDMA1–3) and four of GSDMC (GSDMC1–4), and GSDMB is absent. With the exception of DFNB59, all GSDMs adopt a two-domain architecture, with GSDM-NT as the pore-forming fragment and GSDM-CT autoinhibitory. Functionally, it has been reported that the mouse GSDMA3 (mGSDMA3) can modulate mitochondrial homeostasis (Lin, Lin, Kuo, & Yang, 2015), while GSDME is activated by caspase-3 and serves as a mediator of pyroptosis and a tumor suppressor (Rogers et al., 2017; Wang et al., 2017). Knowledge is lacking on the physiological roles of most GSDMs and the stimuli and proteases that activate them.

1.2 Pore-forming mechanism of gasdermins

While GSDMD-NT drives pore formation in cells, full-length (FL) GSDMD lacks pore-forming activity and cytotoxicity due to the auto-inhibitory function of GSDMD-CT (Aglietti et al., 2016; Ding et al., 2016; Liu et al., 2016). The structural basis for the autoinhibition in GSDMs has been elucidated by the crystal structure of FL mGSDMA3, a mouse homolog of human GSDMA (Ding et al., 2016). In comparison to GSDM-CTs, which are almost exclusively α-helical, mGSDMA3-NT contains an extended β-sheet formed by nine strands (β3–β11) in addition to several α-helices (α1–α4). In the auto-inhibited inactive state, mGSDMA3-CT folds back onto mGSDMA3-NT to form extensive contacts through hydrophobic, hydrogen bonding and charged interactions, leaving the NT-CT linker accessible to cleavage by proteases. Human and mouse GSDMD-CT and human GSDMB-CT are structurally similar to mGSDMA3-CT, suggesting a conserved auto-inhibitory mechanism across the GSDM family (Kuang et al., 2017; Liu et al., 2018).

The structural transition from autoinhibition to the mature GSDM TM pore has been illustrated by the cryo-electron microscopy (cryo-EM) structure of the mGSDMA3 pore, a 26- to 28-subunit mGSDMA3-NT oligomer with a molecular weight of approximately 0.8 MDa (Ruan, Xia, Liu, Lieberman, & Wu, 2018). The pore features an anti-parallel β-barrel, with each subunit contributing two β-hairpins or four β-strands, as the transmembrane (TM) region and a soluble rim formed by a globular domain of each subunit. The inner diameter of the pore is around 180 Å, large enough to allow the passage of inflammatory cytokines such as IL-1β. Structural alignment of pore-form mGSDMA3-NT to its auto-inhibited state shows that the globular domain remains largely unaltered whereas drastic conformational changes occur at the TM region. More specifically, the first β-hairpin is formed by the β3-β4-β5 region in auto-inhibited mGSDMA3-NT, and the second β-hairpin by the β7-α4-β8 region that is in close contact with mGSDMA3-CT. Adjacent to the inner leaflet of the bilayer membrane, the positively charged α1 helix of mGSDMA3-NT directly binds the negatively charged head of acid lipid CL. The α1 helix is masked by mGSDMA3-CT in the auto-inhibited conformation, consistent with the observation that FL GSDMs do not bind lipids and form pores (Aglietti et al., 2016; Ding et al., 2016; Liu et al., 2016; Ruan et al., 2018).

2. Generation of recombinant gasdermins and proteases

2.1 Molecular cloning, expression, and purification of gasdermins from *E. coli*

To avoid the issues that GSDM-NTs are toxic to cells and tend to form aggregates in lipidic environments even in the presence of GSDM-CTs, we recommend expressing and purifying FL GSDMs. We found that FL GSDMs can be yielded in considerable quantities, about 10 mg pure protein from 1 L culture of *E. coli*, as monomers when they are appended a solubility tag such as a maltose-binding protein (MBP) or a small ubiquitin-like modifier (SUMO). The following steps describe the detailed procedure for cloning, expressing, and purifying recombinant human GSDMD (hGSDMD) for in vitro experiments (Ruan et al., 2018).

Gasdermin cloning and expression

1. Following standard restriction enzyme cloning protocols, clone the full-length hGSDMD sequence into the pDB.His.MBP vector with a tobacco etch virus (TEV)-cleavable N-terminal 6xHis-MBP tag. We call this construct 6xHis-MBP-hGSDMD. For mGSDMA3, clone the full-length sequence into a pET28a vector with an N-terminal 6xHis-SUMO tag. As the protease that cleaves and activates mGSDMA3 is unknown, insert a human rhinovirus (HRV) 3C protease cleavage site (LEVLFQGP) after residue E262 using standard site-directed mutagenesis methods such as overlap extension PCR. We call this construct 6xHis-SUMO-mGSDMA3-3C.
2. Using 6xHis-MBP-hGSDMD and 6xHis-SUMO-mGSDMA3-3C as template vectors, introduce mutations using the QuikChange Mutagenesis kit (Agilent Technologies).
3. Verify all cloning results by sequencing. Transform the 6xHis-MBP-hGSDMD and 6xHis-SUMO-mGSDMA3-3C plasmids into BL21 (DE3) *E. coli* competent cells and grow on LB plates containing kanamycin (50 μg/mL) overnight at 37 °C.
4. Pick a single colony into 10 mL LB medium containing kanamycin (50 μg/mL) and culture at 37 °C under shaking at 210 rpm until $O.D._{600}$ reaches 0.6–1.0.

5. Induce protein expression by adding 0.2 mM isopropyl-β-D-thiogalactopyranoside (IPTG) to the culture. Lower the temperature to 18 °C and continue growing the cells for 18 h. The expression level can be checked on SDS-PAGE by comparing uninduced and induced cells.
6. Harvest the bacterial cells by centrifugation at 3500 rpm for 30 min and transfer the pellet into a 50 mL tube. Wash with pellet by resuspending it in PBS, redoing centrifugation, and discarding the supernatant. Freeze the pellet with liquid nitrogen and store at −80 °C for future use.

Gasdermin purification

1. Fully thaw and resuspend a pellet in 10 mL of lysis buffer (50 mM Tris-HCl at pH 8.0, 150 mM NaCl, 5 mM imidazole, and 2 mM β-mercaptoethanol (2ME) per gram of pellet.
2. Ultrasonicate the pellet suspension for 10 min on an ice-water bath to fully lyse the cells.
3. Transfer the lysate to a centrifuge tube and spin at 15000 rpm for 30 min at 4 °C. Collect the supernatant, which contains most of the recombinant GSDMs.
4. Equilibrate Ni-NTA beads (Qiagen) with lysis buffer. For 1 L of bacterial culture, we recommend using 3 mL of Ni-NTA beads. The following steps are based on 1 L of culture and the quantities of reagents can be scaled accordingly.
5. Incubate the equilibrated Ni-NTA beads with the supernatant at 4 °C for 1 h.
6. Load the supernatant-bead mixture onto a glass Econo-Column and collect flow-through fractions.
7. Wash the beads with at least 20 column volumes (CVs) of wash buffer (50 mM Tris-HCl at pH 8.0, 150 mM NaCl, 20 mM imidazole, and 2 mM 2ME). Collect wash fractions.
8. Elute recombinant GSDMs off the column with 5 CVs of elution buffer (50 mM Tris-HCl at pH 8.0, 150 mM NaCl, 300 mM imidazole, and 2 mM 2ME). Collect elution fractions. Analyze the supernatant, flow-through, wash, and elution fractions using SDS-PAGE. The elution fractions should contain recombinant GSDMs at around 100 kD for 6xHis-MBP-hGSDMD and around 70 kD for 6xHis-SUMO-mGSDMA3-3C.
9. Pool elution fractions containing pure recombinant GSDMs. Spin the mixture at 15,000 rpm for 10 min at 4 °C to remove insoluble aggregates. Load the sample into a Superdex 200 (10/300) size exclusion column equilibrated with gel filtration buffer (50 mM Tris-HCl at pH 8.0,

150 mM NaCl). Pool and save monomer peak fractions. This step removes soluble aggregates that hinder complete enzymatic cleavage in the following step.

10. Add TEV to the pool to cleavage the MBP tag off the recombinant hGSDMD, or add ULP1 to cleave the SUMO tag off the recombinant mGSDMA3. The protease to GSDM concentration ratio should be around 1:50. We observe nearly complete cleavage after 24 h at 4 °C. TEV, ULP1, and the 3C protease used for cleaving engineered mGSDMA3, are available commercially or can be generated using bacterial expression systems and straightforward Ni-NTA purification procedures. We recommend using TEV and ULP1 tagged with 6xHis for easier removal of these enzymes later.
11. After the 24-h cleavage, incubate the reaction mixture with 1 mL of Ni-NTA beads equilibrated with gel filtration buffer for 1 h. Then, Load the mixture onto a glass Econo-Column and collect flow-through fractions. MBP, SUMO, ULP1 and TEV will be bound to the beads and stay in the column due to the presence of 6xHis tags.
12. Pool the flow-through fractions containing GSDMs without MBP or SUMO, which should show a molecular weight of around 50 kD on SDS-PAGE. Spin the mixture at 15,000 rpm for 10 min at 4 °C to remove insoluble aggregates. Load the sample into a Superdex 200 (10/300) size exclusion column equilibrated with gel filtration buffer. Pool and concentrate monomer peak fractions to 5 μM. Aliquot the proteins, freeze with liquid nitrogen, and store at −80 °C for future use.

2.2 Overproduction of caspase-11 using insect cells

Caspase-11, an upstream activating protease for GSDMD, can be overexpressed in insect cell systems and purified for in vitro cleavage of hGSDMD. We employed baculovirus-infected Sf9 cells to generate caspase-11 following the established protocols for the Bac-to-Bac Baculovirus Expression System (Invitrogen), briefly described as following.

1. Full-length caspase-11 sequence was cloned into the pFastBac-HTa vector with a TEV-cleavable N-terminal 6xHis tag. Verify results by sequencing.
2. Transform the pFastBac-HTa-caspase-11 plasmid into *E. coli* DH10Bac and the bacteria cultured on an LB plate containing kanamycin (50 μg/mL), gentamycin (7 μg/mL), tetracycline (10 μg/mL), Bluo-gal (100 μg/mL), and IPTG (40 μg/mL) for 48 h at 37 °C.

3. Pick a white colony and grow in a miniprep LB culture containing kanamycin (50 μg/mL), gentamycin (7 μg/mL), and tetracycline (10 μg/mL) for 12 h at 37 °C under shaking at 210 rpm.
4. Extract bacmids from the miniprep culture using the isopropanol precipitation method.
5. Transfect Sf9 insect cells with the bacmid to generate P1, P2, and P3 baculoviruses following the manufacturer's instructions.
6. Culture 1 L of Sf9 cells to a density of 2–3 million cells/mL. Then, infect the cells with 20 mL of P3 viruses and culture the infected cells for 48 h at 27 °C.
7. Collect the cells by spinning the culture at 2500 rpm for 15 min at 4 °C. Wash the pellet with PBS. Freeze with liquid nitrogen and store at −80 °C for future use.
8. Purify 6xHis-caspase-11 following the same protocol as for 6xHis-MBP-hGSDMD and 6xHis-SUMO-mGSDMA3. Eluate the 6xHis-caspase-11 from the Ni-NTA resin, concentrate to 2 μM, and aliquot for storage. Note that caspase-11 undergoes autoproteolysis to generate active fragments, and therefore storage at 4 °C over time can enhance its enzymatic activity (Yang, Chang, & Baltimore, 1998).

3. Liposome leakage assay

3.1 Liposome formulation

GSDM-NTs bind acidic lipids, including PS, PIPs, and CL, via charge-charge interactions. Liposomes that contain these acidic lipids can recruit GSDM-NTs, which form pores and enable the exchange of materials across the liposomal membranes. Based on this principle, we designed a liposome leakage assay that monitors GSDM activities in vitro (Ruan et al., 2018). Briefly, we encapsulate terbium ions (Tb^{3+}) into the liposomes and provide dipicolinic acid (DPA) outside the liposomes. GSDM pore formation on the liposomal membranes induces the leakage of Tb^{3+} into the external buffer containing DPA, leading to a fluorogenic reaction between Tb^{3+} and DPA. The following steps describe the process to generate liposomes for the leakage assay. The lipids (Avanti Polar Lipids) we used were dissolved in chloroform to 25 mg/mL, upon which the calculations below are based.

1. Mix 80 μL phophatidylcholine (PC), 128 μL phosphatidylethanolamine (PE), and 64 μL CL in a glass tube. Note that other lipid combinations can be used provided that acidic lipids such as PIPs, PS, and CL are included.

2. Evaporate the chloroform under a steady stream of nitrogen gas while rotating the tube to create a lipid film on the glass surface.
3. Using a vortex mixer, suspend the lipid film in 1 mL buffer A (20 mM HEPES at pH 7.4, 150 mM NaCl, 50 mM sodium citrate, and 15 mM $TbCl_3$) to generate the liposomes.
4. Push the liposome suspension through a 100-nm Whatman® Nuclepore™ Track-Etched Membrane 30 × using an extruder to obtain liposomes more homogeneous in size.
5. Equilibrate a Superose 6 (10/300 GL) gel filtration column with buffer B (20 mM HEPES at pH 7.4, 150 mM NaCl). Load the liposomes onto the column and collect 5 mL of void fractions to produce a stock of PC-PE-CL liposomes at 1.6 mM lipid concentration.

3.2 In vitro pore formation, assay protocol, and data interpretation

The leakage of Tb^{3+} from GSDM-perforated liposomes can be monitored by an increase in fluorescence intensity when Tb^{3+} binds DPA in the external buffer (Fig. 1A). Below is a protocol for the liposome leakage assay using purified hGSDMD and caspase-11 as an example. The combination of purified mGSDMA3-3C and the 3C protease will yield similar results.

1. Dilute the PC-PE-CL liposomes with buffer C (20 mM HEPES at pH 7.4, 150 mM NaCl, and 50 μM DPA) to a final lipid concentration of 50 μM for the leakage assay.
2. In a 50 μL reaction system, add 30 μL liposomes (final lipid concentration: 30 μM), 5 μL hGSDMD (final concentration: 0.5 μM), 5 μL

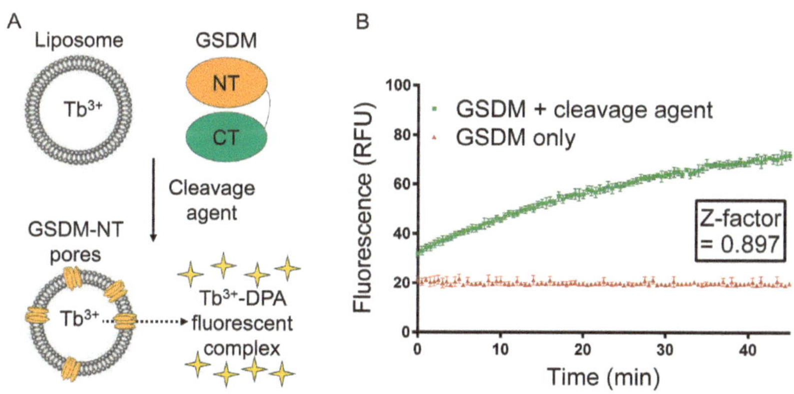

Fig. 1 A liposome leakage assay for monitoring gasdermin activity. (A) A scheme of the liposomal assay. (B) Statistical analysis using the Z-factor. Data are plotted as mean ± standard deviation.

caspase-11 (final concentration: 0.2 μM). To set up a negative control, replace the caspase-11 with buffer C.

3. Mix the reactions well using a pipette, and transfer the reactions to Corning 3820 384-well assay plate.
4. Using a Molecular Devices SpectraMax M5 plate reader, continuously record fluorescence at 545 nm after excitation at 276 nm for 45 min at 30 s intervals.

The fluorescence data can be plotted with respect to time. A robust increase in fluorescence, which indicates active GSDM pore formation, is expected when both GSDM and the cleavage agent are present, in comparison to no or little fluorescence in the absence of the cleavage agent. To quantify the separation between the positive signal and the negative control, we employ a statistical measure known as the Z-factor, which is frequently used in high-throughput screening (Zhang, Chung, & Oldenburg, 1999). The Z-factor is calculated based on the means (μ) and standard deviations (σ) of the experimental (exp) and control groups (ctr) as follows.

$$\text{Z-factor} = 1 - \frac{3\left(\mu_{\text{exp}} + \mu_{\text{ctr}}\right)}{\left|\sigma_{\text{exp}} - \sigma_{\text{ctr}}\right|}$$

A Z-factor of less than 0.5 indicates an insignificant separation between the two groups, whereas a Z-factor between 0.5 and 1 suggests significant separation. Using the assay protocol above, we are able to reproducibly achieve a Z-factor of around 0.9 (Fig. 1B).

4. Conclusions and discussion

GSDMs represent a new family of β-barrel pore-forming proteins that play vital roles in innate immunity and cell death. While the functions and activation mechanisms of GSDMD and GSDME have been more extensively studied, the physiological significance and the pathways that regulate the generation of other GSDM-NTs remain largely elusive. In addition to aiding the identification of new upstream activating enzymes for GSDMs, the liposome leakage assay that monitors GSDM activities in vitro can also characterize the effects of GSDM mutations, some of which are related to human diseases. Similar liposomal assays have been employed to study other membrane proteins including pore-forming proteins and ion channels

(Faudry, Perdu, & Attree, 2013; Posson, Rusinova, Andersen, & Nimigean, 2018; Wang et al., 2014), suggesting the plasticity and wide applicability of this liposome-based method.

Acknowledgments

This work was supported by US NIH grants DP1HD087988, R01A139914, and R01AI124491 to H.W., a Charles A. King Trust Postdoctoral Fellowship to J.R., and an Albert J. Ryan Fellowship to S.X.

References

Aglietti, R. A., Estevez, A., Gupta, A., Ramirez, M. G., Liu, P. S., Kayagaki, N., et al. (2016). GsdmD p30 elicited by caspase-11 during pyroptosis forms pores in membranes. *Proceedings of the National Academy of Sciences of the United States of America*, *113*, 7858–7863.

Bergsbaken, T., Fink, S. L., & Cookson, B. T. (2009). Pyroptosis: Host cell death and inflammation. *Nature Reviews. Microbiology*, *7*, 99–109.

Chen, K. W., Monteleone, M., Boucher, D., Sollberger, G., Ramnath, D., Condon, N. D., et al. (2018). Noncanonical inflammasome signaling elicits gasdermin D-dependent neutrophil extracellular traps. *Science Immunology*, *3*, aar6676.

Ding, J., Wang, K., Liu, W., She, Y., Sun, Q., Shi, J., et al. (2016). Pore-forming activity and structural autoinhibition of the gasdermin family. *Nature*, *535*, 111–116.

Evavold, C. L., Ruan, J., Tan, Y., Xia, S., Wu, H., & Kagan, J. C. (2018). The pore-forming protein Gasdermin D regulates interleukin-1 secretion from living macrophages. *Immunity*, *48*, 35–44 e6.

Faudry, E., Perdu, C., & Attree, I. (2013). Pore formation by T3SS translocators: Liposome leakage assay. *Methods in Molecular Biology*, *966*, 173–185.

Franchi, L., Eigenbrod, T., Munoz-Planillo, R., & Nunez, G. (2009). The inflammasome: A caspase-1-activation platform that regulates immune responses and disease pathogenesis. *Nature Immunology*, *10*, 241–247.

Hornung, V., Ablasser, A., Charrel-Dennis, M., Bauernfeind, F., Horvath, G., Caffrey, D. R., et al. (2009). AIM2 recognizes cytosolic dsDNA and forms a caspase-1-activating inflammasome with ASC. *Nature*, *458*, 514–518.

Jorgensen, I., & Miao, E. A. (2015). Pyroptotic cell death defends against intracellular pathogens. *Immunological Reviews*, *265*, 130–142.

Kayagaki, N., Stowe, I. B., Lee, B. L., O'Rourke, K., Anderson, K., Warming, S., et al. (2015). Caspase-11 cleaves gasdermin D for non-canonical inflammasome signalling. *Nature*, *526*, 666–671.

Kayagaki, N., Wong, M. T., Stowe, I. B., Ramani, S. R., Gonzalez, L. C., Akashi-Takamura, S., et al. (2013). Noncanonical inflammasome activation by intracellular LPS independent of TLR4. *Science*, *341*, 1246–1249.

Kuang, S. Y., Zheng, J., Yang, H., Li, S. H., Duan, S. Y., Shen, Y. F., et al. (2017). Structure insight of GSDMD reveals the basis of GSDMD autoinhibition in cell pyroptosis. *Proceedings of the National Academy of Sciences of the United States of America*, *114*, 10642–10647.

Lin, P. H., Lin, H. Y., Kuo, C. C., & Yang, L. T. (2015). N-terminal functional domain of Gasdermin A3 regulates mitochondrial homeostasis via mitochondrial targeting. *Journal of Biomedical Science*, *22*, 44.

Liu, Z., Wang, C., Rathkey, J. K., Yang, J., Dubyak, G. R., Abbott, D. W., et al. (2018). Structures of the Gasdermin D C-terminal domains reveal mechanisms of autoinhibition. *Structure*, *26*, 778–84 e3.

Liu, X., Zhang, Z., Ruan, J., Pan, Y., Magupalli, V. G., Wu, H., et al. (2016). Inflammasome-activated gasdermin D causes pyroptosis by forming membrane pores. *Nature*, *535*, 153–158.

Lu, A., Magupalli, V. G., Ruan, J., Yin, Q., Atianand, M. K., Vos, M. R., et al. (2014). Unified polymerization mechanism for the assembly of ASC-dependent inflammasomes. *Cell*, *156*, 1193–1206.

Maloy, K. J., & Powrie, F. (2011). Intestinal homeostasis and its breakdown in inflammatory bowel disease. *Nature*, *474*, 298–306.

Martinon, F., Burns, K., & Tschopp, J. (2002). The inflammasome: A molecular platform triggering activation of inflammatory caspases and processing of proIL-beta. *Molecular Cell*, *10*, 417–426.

Martinon, F., Petrilli, V., Mayor, A., Tardivel, A., & Tschopp, J. (2006). Gout-associated uric acid crystals activate the NALP3 inflammasome. *Nature*, *440*, 237–241.

Orning, P., Weng, D., Starheim, K., Ratner, D., Best, Z., Lee, B., et al. (2018). Pathogen blockade of TAK1 triggers caspase-8-dependent cleavage of gasdermin D and cell death. *Science*, *362*, 1064.

Petrilli, V., Papin, S., Dostert, C., Mayor, A., Martinon, F., & Tschopp, J. (2007). Activation of the NALP3 inflammasome is triggered by low intracellular potassium concentration. *Cell Death and Differentiation*, *14*, 1583–1589.

Posson, D. J., Rusinova, R., Andersen, O. S., & Nimigean, C. M. (2018). Stopped-flow fluorometric ion flux assay for ligand-gated Ion Channel studies. *Methods in Molecular Biology*, *1684*, 223–235.

Rogers, C., Fernandes-Alnemri, T., Mayes, L., Alnemri, D., Cingolani, G., & Alnemri, E. S. (2017). Cleavage of DFNA5 by caspase-3 during apoptosis mediates progression to secondary necrotic/pyroptotic cell death. *Nature Communications*, *8*, 14128.

Ruan, J., Xia, S., Liu, X., Lieberman, J., & Wu, H. (2018). Cryo-EM structure of the gasdermin A3 membrane pore. *Nature*, *557*, 62–67.

Ruhl, S., Shkarina, K., Demarco, B., Heilig, R., Santos, J. C., & Broz, P. (2018). ESCRT-dependent membrane repair negatively regulates pyroptosis downstream of GSDMD activation. *Science*, *362*, 956–960.

Sarhan, J., Liu, B. C., Muendlein, H. I., Li, P., Nilson, R., Tang, A. Y., et al. (2018). Caspase-8 induces cleavage of gasdermin D to elicit pyroptosis during Yersinia infection. *Proceedings of the National Academy of Sciences of the United States of America*, *115*, E10888–E10897.

Schroder, K., & Tschopp, J. (2010). The inflammasomes. *Cell*, *140*, 821–832.

Shi, J., Zhao, Y., Wang, Y., Gao, W., Ding, J., Li, P., et al. (2014). Inflammatory caspases are innate immune receptors for intracellular LPS. *Nature*, *514*, 187–192.

Shi, J., Zhao, Y., Wang, K., Shi, X., Wang, Y., Huang, H., et al. (2015). Cleavage of GSDMD by inflammatory caspases determines pyroptotic cell death. *Nature*, *526*, 660–665.

Sollberger, G., Choidas, A., Burn, G. L., Habenberger, P., Di Lucrezia, R., Kordes, S., et al. (2018). Gasdermin D plays a vital role in the generation of neutrophil extracellular traps. *Science Immunology*, *3*, aar6689.

Taabazuing, C. Y., Okondo, M. C., & Bachovchin, D. A. (2017). Pyroptosis and apoptosis pathways engage in bidirectional crosstalk in monocytes and macrophages. *Cell Chemical Biology*, *24*, 507–14 e4.

Wang, Y. P., Gao, W. Q., Shi, X. Y., Ding, J. J., Liu, W., He, H. B., et al. (2017). Chemotherapy drugs induce pyroptosis through caspase-3 cleavage of a gasdermin. *Nature*, *547*, 99.

Wang, H., Sun, L., Su, L., Rizo, J., Liu, L., Wang, L. F., et al. (2014). Mixed lineage kinase domain-like protein MLKL causes necrotic membrane disruption upon phosphorylation by RIP3. *Molecular Cell*, *54*, 133–146.

Yang, X., Chang, H. Y., & Baltimore, D. (1998). Autoproteolytic activation of pro-caspases by oligomerization. *Molecular Cell*, *1*, 319–325.

Zanoni, I., Tan, Y., Di Gioia, M., Broggi, A., Ruan, J., Shi, J., et al. (2016). An endogenous caspase-11 ligand elicits interleukin-1 release from living dendritic cells. *Science*, *352*, 1232–1236.

Zhang, Y., Chen, X., Gueydan, C., & Han, J. (2018). Plasma membrane changes during programmed cell deaths. *Cell Research*, *28*, 9–21.

Zhang, J. H., Chung, T. D., & Oldenburg, K. R. (1999). A simple statistical parameter for use in evaluation and validation of high throughput screening assays. *Journal of Biomolecular Screening*, *4*, 67–73.

Zhao, Y., Yang, J., Shi, J., Gong, Y. N., Lu, Q., Xu, H., et al. (2011). The NLRC4 inflammasome receptors for bacterial flagellin and type III secretion apparatus. *Nature*, *477*, 596–600.

CHAPTER EIGHT

Measuring TREX1 and TREX2 exonuclease activities

Wayne O. Hemphill, Fred W. Perrino*
Department of Biochemistry, Wake Forest School of Medicine, Winston-Salem, NC, United States
*Corresponding author: e-mail address: fperrino@wakehealth.edu

Contents

Methods in Enzymology, Volume 625
ISSN 0076-6879
https://doi.org/10.1016/bs.mie.2019.05.004

Abstract

Three-prime Repair Exonuclease (TREX1) degrades ssDNA and dsDNA. TREX1 localizes to the perinuclear space in cells and degrades cytosolic DNA to prevent aberrant nucleic acid sensing and immune activation in humans and mice. Mutations in the *TREX1* gene cause a spectrum of human autoimmune diseases including Aicardi-Goutières syndrome, familial chilblain lupus, retinal vasculopathy with cerebral leukodystrophy, and are associated with systemic lupus erythematosus. More than 60 disease-causing *TREX1* variants have been identified including dominant and recessive, missense, and frameshift mutations that map to the catalytic core region and to the C-terminal cell localization region. The TREX1-disease causing mutations affect exonuclease activity at varied levels. In this chapter, we describe methods to purify variant recombinant TREX1 enzymes and measure the exonuclease activity using ssDNA and dsDNA substrates. The relationships between TREX1 activities, types of TREX1 mutations, and TREX1-associated autoimmune diseases are considered.

1. Introduction

The closely related *TREX1* and *TREX2* single-open-reading-frame genes encode the 314-amino-acid TREX1 and the 236-amino-acid TREX2 deoxyribonucleases located on chromosomes 3p21.2–21.3 and Xq28, respectively (Mazur & Perrino, 1999). The TREX1 catalytic core (amino acids 1–242) is about 40% identical to TREX2 (amino acids 1–236) and TREX1 contains an extended C-terminal region of 76 amino acids, not present in TREX2. These closely related nonprocessive $3' \rightarrow 5'$ exonucleases degrade ssDNA and dsDNA, removing mononucleotides from DNA 3′termini (Mazur & Perrino, 2001b; Perrino, Miller, & Ealey, 1994), and modified DNA 3′ termini are poor substrates for these enzymes (Harrigan, Fan, Momand, et al., 2007; Perrino et al., 1994). TREX1 and TREX2 are Mg^{2+} ion-dependent nucleases belonging to the DnaQ-like subfamily (DEDD-h) characterized by the presence of four conserved carboxylate residues and a histidine positioned in the active site. TREX1 and TREX2 are the only mammalian deoxyribonuclease members of the DEDD-h subfamily, and degrade deoxynucleotide polymers at rates >1000-fold higher than ribonucleotide polymers (Mazur & Perrino, 1999).

TREX1 and TREX2 are functional dimers, indicating that the dimeric architecture is critical in their biological functions. Structural studies reveal the homodimeric nature of TREX1 and TREX2, and distinct structural elements, that point to different biological roles for these proteins (de Silva, Choudhury, Bailey, et al., 2007; Perrino, Harvey,

McMillin, & Hollis, 2005). Between the TREX homodimers' catalytic domains (TREX1 amino acids 1–242; TREX2 amino acids 1–236) exists a twofold axis of symmetry perpendicular to the central β-strands that connect the individual β-sheets of each protomer at the interface to form an extended β-sheet through the dimer core. Due to the dimers' symmetries, active sites are positioned at opposite outer edges providing open access for the DNA. The side chains of the four conserved carboxylate residues in the active sites coordinate two Mg^{2+} ions, and the histidine is positioned to deprotonate a water producing a nucleophile to attack the scissile phosphodiester bond and effect hydrolysis. TREX1 and TREX2 contain a flexible loop of unique structure adjacent to the active site. TREX2 contains three conserved arginine residues on this loop that contribute to tight DNA binding (Perrino et al., 2005). The corresponding loop in TREX1 has only a single conserved arginine residue with additional residues outside the loop region contributing to DNA binding. TREX1 contains a unique polyproline II helix (PPII) formed by a non-repetitive proline rich region that contains six prolines within an 8-amino acid stretch. The corresponding region in TREX2 is a β-hairpin. TREX1 contains a unique C-terminal region of amino acids 243–314 that is responsible for its perinuclear ER localization and oligosaccharyltransferase (OST) complex interaction (de Silva et al., 2007; Hasan, Fermaintt, Gao, et al., 2015; Kucej, Fermaintt, Yang, Irizarry-Caro, & Yan, 2017; Mazur & Perrino, 1999, 2001a, 2001b; Orebaugh, Fye, Harvey, et al., 2013).

TREX1 mutations cause a spectrum of autoimmune diseases including Aicardi-Goutières syndrome (AGS), familial chilblain lupus (FCL), retinal vasculopathy with cerebral leukodystrophy (RVCL), and are associated with systemic lupus erythematosus (SLE). The TREX1-mediated disease spectrum is recapitulated in the mouse models with specific TREX1 mutations. The aggressive immune-activating phenotype of the TREX1 null mouse most closely mimics AGS (Morita, Stamp, Robins, et al., 2004), the spontaneous lupus-like inflammation in the TREX1 D18N mouse is most similar to FCL (Grieves, Fye, Harvey, et al., 2015), and the knock-in mouse carrying human TREX1 frame-shift mutations truncating the TREX1 C-terminus reflect the unique serologic autoimmune-like phenotypes caused by dysregulation of the ER oligosaccharyltransferase (OST) complex (Sakai, Miyazaki, Shin, et al., 2017). TREX1 is ubiquitously expressed with higher levels measured in some lymphoid tissues (Grieves et al., 2015; Mazur & Perrino, 2001a), and degrades cytosolic damaged DNA (Yang, Lindahl, & Barnes, 2007), retroelements (Stetson, Ko, Heidmann, & Medzhitov, 2008), and unprocessed erythroblast DNA (Rego, Harvey, Simpson, et al., 2018) to prevent inappropriate immune activation. TREX2 mutations have not been identified in humans, but TREX2 knockout mice exhibit increased chemically- (Parra, Manils, Castellana, et al., 2009) and UVB-induced skin tumorigenesis phenotypes consistent with the high level TREX2 expression detected specifically in keratinocytes (Ablasser, Hemmerling, Schmid-Burgk, et al., 2014; Manils, Gómez, Salla-Martret, et al., 2015; Rice, Rodero, & Crow, 2015).

The diverse immune-activating phenotypes caused by TREX1 mutations reflect the various types and positions of mutations within the catalytic and the C-terminal regions (Rice et al., 2015). In attempts to understand the link between TREX1 mutation and human disease we have developed strategies to measure TREX1 exonuclease activity focusing on the N-terminal catalytic domain. In this chapter, we describe assays to measure exonuclease activity of recombinant TREX1 using ssDNA and dsDNA substrates *in vitro*. These assays can be used with recombinant TREX1 dimeric enzymes (AA 1–242) of wild-type and mutant variant homo- and heterodimers providing insight into TREX1 catalytic mechanism. While the focus here is the TREX1 catalytic core, the high structural and catalytic similarities to TREX2 make these assays relevant and applicable to both enzymes.

2. Considerations when purifying TREX1 enzymes

2.1 Preparation of truncated enzyme

Attempts to overexpress in bacteria and purify the complete TREX1 1–314 amino acid enzyme have not been successful, presumably due to the hydrophobic properties of residues 287–314. Thus, in order to test exonuclease activities we overexpress the truncated TREX1 N-terminal 1–242 amino acids, which contains the TREX1 catalytic domain and displays potent exonuclease activity. In contrast, the full-length TREX2 1–236 amino acid enzyme is readily overexpressed in bacteria and purified using column chromatography. Hereafter, mention of TREX1 is about the truncated form of the enzyme.

2.2 Preparation of heterodimers

The dimeric structure of TREX1 is unusual for $3' \rightarrow 5'$ exonucleases. The recombinant TREX1 enzyme is a kinetically stable dimer that does not dissociate and re-equilibrate at measurable rates (Orebaugh, Fye, Harvey, Hollis, & Perrino, 2011). This unique property of TREX1 dimers provides opportunity to study the activities of purified heterodimers or compound heterodimers of variant TREX1 forms.

3. Purifying recombinant TREX1 enzymes

The TREX1 expression and purification procedure used in our lab generates large quantities of highly purified TREX1 enzyme for the assays described here. The materials and protocol described are applicable to TREX1 homodimer and heterodimer purification. The materials or procedural steps required specifically for heterodimer purification are indicated in the materials and protocol sections of this chapter. Two different *TREX1* gene-encoding plasmids are listed, each of which should encode a different TREX1 protomer. For homodimers, the TREX1 gene is expressed as a fusion with maltose binding protein (MBP) in pLM303x (Lee-Kirsch, Chowdhury, Harvey, et al., 2007), with a rhinovirus 3C protease (PreScission Protease) recognition site between the *MBP* and *TREX1* genes. For heterodimers, the pLM303x plasmid is co-expressed with a second plasmid containing a distinct *TREX1* gene construct prepared as a fusion of the His-tagged NusA with TREX1 in pCDFDuet-1 (Novagen). Plasmids used for heterodimer expression and purification can encode either a wild-type and mutant *TREX1* gene, or two different mutant *TREX1* genes. Proteins are then purified *via* affinity chromatography. Common techniques such as gel preparation, electrophoresis, vector cloning, are mentioned, but details not provided. Methods of preparation for some materials not purchased directly from commercial sources will be discussed in more detail to ensure replication of our procedures. Enzyme concentration methods are not necessary for assays and will not be discussed here. The following equipment and materials are used in this assay.

3.1 Equipment

1. FPLC or low-pressure chromatography system at 4°C with UV detection
2. Six 1 L culture flasks
3. Temperature-controlled shaking incubator
4. SDS-PAGE system
5. Spectrophotometer UV–vis

6. Temperature-controlled centrifuge(s) and adapters:
   a. Six 1 L samples at 4000 rpm/4°C for 20 min
   b. Four 40 mL samples at 15,000 rpm/4°C for 20 min
   c. 1 mL samples at >16,000 G/20°C for 1 min
7. French Press or equivalent pressurized lysis instrument
   a. Ex. Emulsiflex (Avestin)
8. Thermometer
9. UV–vis gel imager
10. Magnetic stirrer

3.2 Materials

1. pLM303x-TREX1 plasmid (Lee-Kirsch et al., 2007)
2. * pCDFDuet-1-TREX1 (Novagen)
   a. ***Only needed for heterodimer purification.**
3. *Escherichia coli* BL21(DE3) Rosetta 2 cells (Novagen)
4. LB media
5. PreScission Protease (GE Biosciences)
6. Kanamycin
7. Chloramphenicol
8. *Streptomycin
   a. ***Only needed for heterodimer purification.**
9. Ice bath for five 1 L flasks
10. Isopropyl β-D-1-thiogalactopyranoside (IPTG)
11. Complete Protease Inhibitor Tablets (Roche; Sigma-Aldrich 50564890001)
12. Lysozyme
13. Amylose A Buffer
    a. 20 mM Tris base-HCl (pH 7.5)
    b. 200 mM NaCl
    c. 1 mM EDTA
    d. **Filter through 0.2 μm nitrocellulose filter.**
14. Amylose B Buffer
    a. Amylose A Buffer
    b. 10 mM maltose
15. Bacterial transformation materials and reagents
16. SDS-PAGE materials and reagents (ex. Molecular Weight Standard, SDS loading buffer, gels, gel stain)
17. Dialysis tubing of 6–8 kDa MWCO

18. *Ni-NTA Agarose resin (Qiagen)
    a. ***Only needed for heterodimer purification.**
19. *Wash Buffer
    a. ***Only needed for heterodimer purification**
    b. 50 mM NaH_2PO_4-HCl (pH 7.5)
    c. 300 mM NaCl
    d. 10 mM imidazole
    e. **Filter with 0.2-μm filter.**
20. *Elution Buffer
    a. ***Only needed for heterodimer purification.**
    b. 50 mM NaH_2PO_4-HCl (pH 7.5)
    c. 300 mM NaCl
    d. 250 mM imidazole
    e. **Filter with 0.2-μm filter.**
21. Phosphate Cellulose (P-cell) Resin (de Silva et al., 2007)
    a. Cellulose
    b. Dibasic sodium phosphate
    c. Sodium azide.
22. P-cell A Buffer
    a. 50 mM Tris base-HCl (pH 7.5)
    b. 1 mM EDTA
    c. 10% (v/v) glycerol
    d. **Filter through 0.2-μm filter.**
23. P-cell B Buffer
    a. 50 mM Tris base-HCl (pH 7.5)
    b. 1 mM EDTA
    c. 10% (v/v) glycerol
    d. 1 M NaCl
    e. **Filter through 0.2-μm filter.**

4. Measuring TREX1 activity on ssDNA

TREX1 ssDNA exonuclease activity is measured by quantifying the removal of mononucleotides from the 3′ termini of mixed or homogeneous sequence ssDNA oligonucleotides (Mazur & Perrino, 2001b). Reaction products are separated on urea-polyacrylamide gels that provide single nucleotide resolution in order to quantify the individual DNA band intensities and calculate activities. A modification of the standard assay has been used to measure Michaelis-Menten kinetics for precise K_{cat} and K_m value determinations. Necessary deviations from the standard protocol are indicated in the corresponding protocol section of this chapter. In our standard ssDNA assay we measure the degradation of a FAM-labeled 30-mer oligonucleotide visualized on a denaturing urea-polyacrylamide sequencing-type gel. The amount of dNMP excised in a reaction is quantified by scanning the gel using a large gel imager to capture the FAM fluorescence present within each oligomer at each band position within a lane. The fraction of oligomer at each position is multiplied by the number of dNMPs excised from the 30-mer to generate the oligomer, and by the total fmol of 30-mer in the starting reaction. The sum of these values yields the total fmol of dNMP excised, which is used to calculate the rate of the enzyme (Fye, Orebaugh, Coffin, Hollis, & Perrino, 2011; Orebaugh et al., 2011; Perrino et al., 2005; Perrino, Krol, Harvey, et al., 2004). The following equipment and materials are used in this assay.

4.1 Equipment

1. Large sequencing-style polyacrylamide gel electrophoresis system
2. System for drying samples *in vacuo*
   a. Ex. SpeedVac Plus SC110 A and Refrigerated Vapor Trap RVT400 (Savant)
3. Large gel imager capable of imaging at excitation/emission of 495/520 nm
   a. Ex. Typhoon FLA 9500 (GE Healthcare Life Sciences)

4.2 Materials

1. 23% urea polyacrylamide gel (large, sequencing-style)
2. Reagents/buffers for polyacrylamide gel electrophoresis
3. "Rxn mix"
   a. 50 mM Tris base-HCl (pH 7.5)
   b. 12.5 mM $MgCl_2$
   c. 5 mM DTT
   d. **We recommend pre-making aliquots of this mix and storing at −20°C until use, or making the mix fresh immediately preceding the assay. Aliquots stored up to 6 months at −20°C have no detectable effect on enzyme activity compared to fresh mix.**
4. 5′-FAM-labeled 30-mer oligonucleotide
5. Bovine serum albumin

6. TREX1 enzyme(s)
7. Sample Loading Buffer (SLB)
   a. 4% (v/v) 500 mM EDTA
   b. 10% (v/v) Bromophenol Blue
   c. 10% (v/v) xylene cyanol
   d. 76% (v/v) deionized formamide
8. Cold, 100% ethanol

5. Measuring TREX1 activity on dsDNA with fluorescence

A fluorescence-based assay has been developed to measure TREX1 activities using dsDNA. A single nick is first introduced into the supercoiled plasmid pMYC using the restriction endonuclease nicking enzyme Nt. *Bbv*CI. Reactions are prepared containing TREX1 and the nicked dsDNA plasmid, incubated, and samples are removed and quenched at various times in 15 × SYBR green dye solution. Samples are then excited, and emission intensity measured on a microplate reader. The fluorescent signal is directly related to the amount of remaining dsDNA. The rate of decreased fluorescent signal is used to quantify TREX1 activity. Some TREX1 mutant enzymes competitively bind to the nicked-dsDNA substrate, inhibiting TREX1 wild-type enzyme activity that is added in the same reaction. This competitive binding is not observed using ssDNA substrates (Lehtinen, Harvey, Mulcahy, Hollis, & Perrino, 2008). Thus, a variation of this dsDNA assay can be used with TREX1 mutants and wild-type enzyme to assess competitive DNA binding activity. The necessary deviations from the protocol are indicated in the protocols section of the chapter. Common techniques are mentioned, but not discussed in detail. The following equipment and materials are used for the assay.

5.1 Equipment

1. Microplate reader with 495/520 nm excitation/emission

5.2 Materials

1. "Rxn mix"
   a. **Defined in Section 4.2.**
2. Bovine serum albumin
3. TREX1 enzyme(s)

4. Black, 384-well plate
   a. Ex. Corning #3573.
5. SYBR green dye (ThermoFisher), 10,000 × concentrate
6. Nicked plasmid of about 10 kb
   a. Ex. pMYC nicked with Nt.*Bbv*CI (NEB)

6. Visualizing TREX1 activity on dsDNA with agarose gel electrophoresis

TREX1 degradation of dsDNA can be visualized using agarose gels. Reactions are prepared and performed as described for the fluorescence assay, then quenched by addition of ethanol, dried *in vacuo*, and resuspended in gel loading solution. Products are separated and visualized by electrophoresis through agarose gels prepared with GelRed dsDNA stain. All variations of the fluorescence-based assay can be applied to the agarose gel-based assay. Common techniques used for this assay are mentioned but not detailed. The following equipment and materials are used in this assay.

6.1 Equipment

1. Agarose gel electrophoresis system
2. System for drying samples *in vacuo*
   a. Ex. SpeedVac Plus SC110 A and Refrigerated Vapor Trap RVT400 (Savant)
3. UV–vis gel imaging system
   a. Ex. ChemiDoc (Bio-Rad)

6.2 Materials

1. "Rx mix"
   a. **Defined in Section 4.2.**
2. Bovine serum albumin
3. TREX1 enzyme(s)
4. Plasmid (~10 kb)
5. Nicked plasmid (~10 kb)
   a. Ex. pMYC nicked with Nt.*Bbv*CI (NEB)
6. Agarose gel electrophoresis buffers/reagents
7. GelRed dsDNA stain (Biotium #41003)
8. Loading dye
   a. Ex. 6 × Gel Loading Dye Blue (NEB)
9. Cold, 100% ethanol

7. Protocols

7.1 Preparation of phosphate cellulose resin

1. Add 100 g cellulose to a 2 L beaker.
2. Suspend in 0.5 N NaOH to 1.5 L volume, stir, and let sit at room temperature for 5 min.
3. Decant the top layer of fluid, add deionized water to 1.5 L volume, and let cellulose sit again until it settles. Repeat this wash phase 2 more times.
4. Suspend washed cellulose in 0.6 N HCl to 1.5 L volume, then let sit at room temperature for 5 min.
5. Repeat step 3.
6. Resuspend slurry to 1.5 L volume with deionized H_2O.
7. Adjust pH of slurry to 7.0 using dibasic sodium phosphate.
8. Let the slurry sit until it packs, then decant to a remaining 1 L volume.
9. Resuspend the slurry, add 0.1% sodium azide, and transfer to bottle for long term storage. **Final product should be a ~50% slurry.**

7.2 Overexpression and affinity chromatography purification of TREX1

1. Protein Induction
   a. Transform Rosetta II cells with pLM303x-TREX1 plasmid, and plate on selective media using kanamycin (Kan) and chloramphenicol (CP).
      i. ***If performing a heterodimer purification, co-transform cells with pCDFDuet-1-TREX1 and pLM303x-TREX1 plasmids, and include Kan, CP, and streptomycin (SP) in selective media and subsequent culturing broths.**
   b. Using a colony from the transformation plate, inoculate a ~200 mL LB/Kan/CP broth. Incubate this starter culture overnight at 37°C/180 rpm.
   c. Determine the A600 of the starter culture. Then, dilute the starter culture into five 1 L broths such that their A600 ~ 0.1.
   d. Let the new cultures incubate at 37°C/180 rpm, periodically measuring A600 of the cultures, until A600 = 0.5–0.8. This should take ~2 h.
   e. Induce each 1 L culture with 0.5 mM IPTG (final concentration).
   f. Place the cultures back in incubation at 37°C/180 rpm for 10 min. Then, retrieve the cultures and cool them in ice baths until their

temperature reaches 20°C as measured using a thermometer. After cooling, place the cultures into incubation at 16°C/180rpm to incubate overnight (15–20h).

2. Cell Lysis
   a. Retrieve induced cultures and centrifuge them at 4000rpm/4°C for 20min (JLA-8.1000 rotor).
   b. Decant the supernatants, and transfer the pellets by spatula to a beaker containing 120mL Amylose A Buffer, three protease inhibitor tablets, and 120mg lysozyme. Cover the beaker and place it a on a magnetic stirrer at 4°C until the pellet is resuspended. **This usually takes about 20** min.
   c. Keeping sample and lysate on ice at all times, lyse the cell suspension by pressure homogenization at 15,000–18,000psi. Two rounds of lysis are recommended. **Do not cycle the sample through the homogenizer continuously. Allow the outgoing lysate to cool on ice until all sample is finished, then reapply lysate for second lysis.**
   d. Centrifuge the lysate for 20min at 15,000rpm/4°C (JA-25.50 rotor).
   e. Retrieve the supernatant by decanting, and place it at 4°C until further use.
3. Amylose Column Purification
   a. Prepare a 15mL amylose column in a 4°C chromatography system. Retrieve the supernatant sample from 4°C and place it on ice. Then, using a flow rate of 1.5mL/min, perform the purification as follows making sure to record A280 throughout:
      i. Equilibrate column with 75mL Amylose A Buffer.
      ii. Apply the sample to the column.
      iii. Wash column with 120mL Amylose A Buffer, or until A280 returns to baseline.
      iv. Elute from column with Amylose B Buffer, and collect protein-containing eluent as a single sample.
   b. Add glycerol to eluted sample to a final concentration of 10% (v/v).
      i. ***If performing a heterodimer purification, skip the glycerol addition. Glycerol will be added after elution from the Ni-NTA column.**
   c. Add PreScission Protease (final concentration 7.5μg/mL) to the eluted sample, and mix gently.
      i. ***If performing a heterodimer purification, skip the protease addition. Protease will be added after the Ni-NTA column purification.**

   - **d.** Dialyze the protein at 4°C overnight in 1 L of P-cell A Buffer containing 50 mM NaCl, using 6–8 kDa MWCO dialysis tubing.
      - **i. *If purifying heterodimer, protein should instead be dialyzed in 1 L of Wash Buffer.**
4. **Ni-NTA Column Purification **(**Skip step 4 procedures contained here if purifying homodimer.)**
   - **a.** Prepare a 15 mL Ni-NTA Agarose resin column in a 4° C chromatography system. Retrieve the dialyzed sample from 4°C and place it on ice. Then, using a flow rate of 1.5 mL/min, perform the purification as follows making sure to record A280 throughout:
      - **i.** Equilibrate column with 75 mL Wash Buffer.
      - **ii.** Apply the sample to the column.
      - **iii.** Wash column with 120 mL Wash Buffer, or until A280 reaches baseline.
      - **iv.** Elute column with Elution Buffer, and collect protein-containing eluent in a single sample.
   - **b.** Add glycerol to the eluted sample to a final concentration of 10% (v/v).
   - **c.** Add PreScission Protease (final concentration 7.5 μg/mL) to the eluted sample, and mix gently.
   - **d.** Dialyze the protein at 4°C overnight in 1 L of P-cell A Buffer containing 50 mM NaCl, using 6–8 kDa MWCO dialysis tubing.
5. Phosphocellulose Column Purification
   - **a.** Prepare a 15 mL phosphocellulose column in a 4°C chromatography system. Retrieve the dialyzed sample from 4°C and place it on ice. Then, using a flow rate of 1.5 mL/min, perform the purification as follows making sure to record A280 throughout:
      - **i.** Equilibrate column with 75 mL P-cell A Buffer.
      - **ii.** Apply sample to column.
      - **iii.** Wash column with 75 mL P-cell A Buffer.
      - **iv.** Elute column with 200 mL gradient of 0 ≥ 100% P-cell B Buffer. Collect eluent in 2 mL fractions. **The elution peak should appear at ~15% P-cell B Buffer.**
   - **b.** Combine the eluent fractions containing the TREX1 enzyme. **This should be indicated on the chromatogram by a single large A280 peak during the elution phase. We recommend confirming the protein content of the fractions with SDS-PAGE.**
6. Fig. 1A illustrates the expected protein content during various stages of purification.
7. The concentration of purified enzyme is determined by absorbance (A280) and confirmed by subjecting equivalent-mass samples of multiple

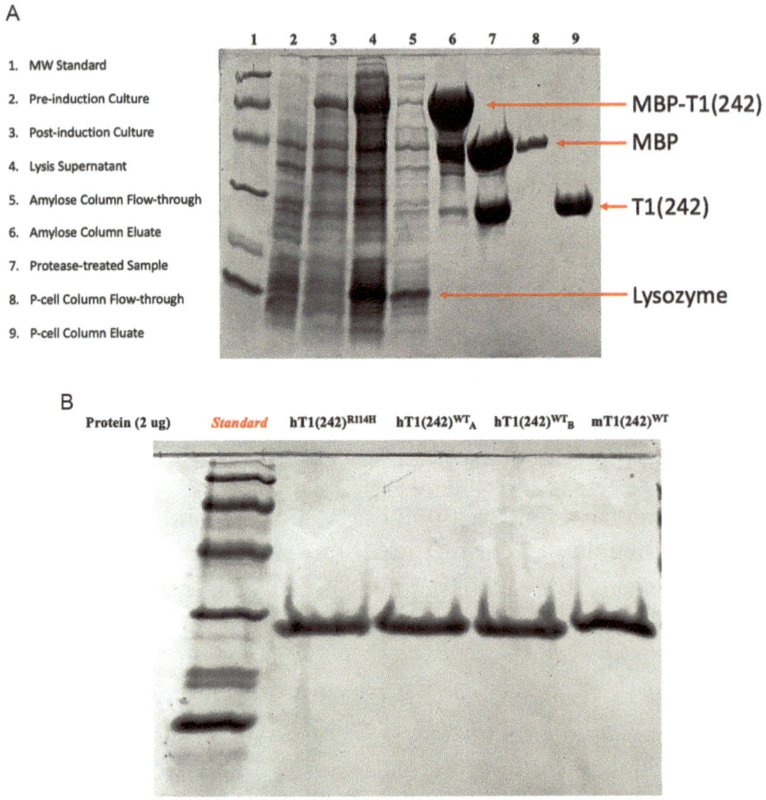

Fig. 1 Purification of hTREX1$^{WT/WT}$. (A) SDS-PAGE samples from critical phases in the enzyme purification process. Bands corresponding to various key proteins are identified using 12% polyacrylamide gels. (B) SDS-PAGE gel results demonstrating equal-mass quantities of TREX1 enzymes using 12% polyacrylamide gels. All non-standard lanes contain ~2 µg of protein. Two lanes contain the same human TREX1 wild-type enzyme from different preparations (labeled subscripts "A" and "B"). Note, the TREX1 murine enzyme migrates slightly more slowly relative to the TREX1 human enzyme.

different TREX1 preparations to SDS-PAGE and visualizing polypeptides by Coomassie blue dye staining. Gels are imaged, and ImageJ software is used to determine the densitometric profile of the TREX1 electropherogram. The intensity profile for each lane is baseline corrected and the area under the TREX1 peak determined to calculate the percent purity and final concentration of each preparation. Example data of this step using various enzymes is illustrated in Fig. 1B.

8. Enzyme activity is confirmed using one of the assays described in this chapter.
9. Store the enzyme at −80°C. **We have found that the enzyme can be thawed up to three times without detectable loss of activity.**

7.3 Measuring TREX1 activity on ssDNA

1. This procedure is performed in two phases described below. **All steps should be performed on ice unless otherwise stated.**
   a. Phase I: Identify the concentration range for each TREX1 enzyme that is within the linear range of the assay. This is determined by the generation of DNA bands migrating to positions on the gel that are visibly degraded, but do not migrate to positions near the bottom of the gel where background bands are evident as indicated in the no-enzyme control lanes. Multiple experiments might be required to identify appropriate TREX1 concentrations in the linear range of the assay. It is recommended to begin with broad concentration ranges and to narrow the selected concentrations identifying three different enzyme concentrations within the linear range of the assay. Data demonstrating the identification of linear range concentrations for hTREX1$^{WT/WT}$ and several hTREX1 mutants is shown in Fig. 2.
   b. Phase II: TREX1 activity is determined using the identified three enzyme concentrations in the assay linear range with triplicate reactions for each concentration. Band intensities for each reaction are quantified by densitometry and used to calculate the initial velocities of each reaction. Initial velocities of reaction triplicates are averaged to calculate activity rates for each enzyme tested, and rates are compared among enzyme variants (if applicable).

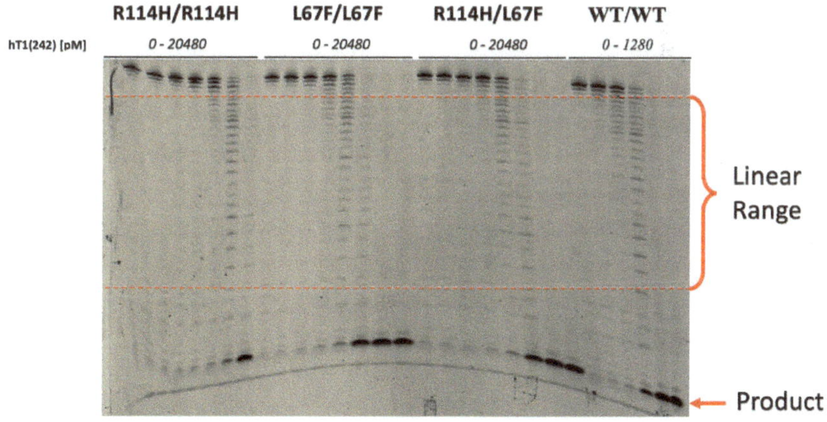

Fig. 2 Phase I gel results for TREX1 ssDNA exonuclease assay. Degradation of a FAM-labeled 30-mer oligonucleotide substrate [50 nM] by varying concentrations of hTREX1 enzymes visualized on a 23% urea polyacrylamide gel. Banding consistent with the linear range of the assay is identified as "Linear Range." Banding that corresponds to accumulating product is labeled as "Product."

2. Prepare a volume of the following "Master Mix" equal to the product of 30 μL and the number of reactions to be performed. Make sure to account for an extra no-enzyme control reaction.
   a. 44.4% Rxn mix
   b. 22.2% 250 nM FAM-labeled 30-mer oligonucleotide
   c. 33.3% H_2O
   d. **For an experiment consisting of 10 reactions you would need 300 μL of Master Mix composed of 133.3 μL Rx mix, 66.7 μL of 30-mer, and 100 μL of deionized water.**
3. Obtain a microcentrifuge tube for each reaction to be performed, and add 27 μL of Master Mix to each tube. Leave the tubes at room temperature.
4. For each enzyme, prepare aliquots of each concentration to be tested at 10 × the desired reaction concentration. This should be done by dilution of stock enzyme into 1 mg/mL BSA. **If the inhibitive properties of a mutant enzyme on wild-type enzyme are being tested, include wild-type enzyme in the prepared mutant enzyme aliquots at 10 × the desired reaction concentration.**
5. Add 3 μL of each enzyme aliquot to its respective reaction tube to initiate the reactions. Vortex quickly after enzyme addition to mix. For the no-enzyme control reaction, add 3 μL of 1 mg/mL BSA. Incubate reactions at room temperature.
6. After a 20 min incubation at room temperature, quench each reaction with 90 μL of cold, 100% ethanol. **Quenched samples may be stored at −20°C overnight or until the protocol can be continued.**
7. Evaporate the quenched reactions *in vacuo* at low/no heat setting.
8. Resuspend the evaporated samples in 8 μL each of SLB.
9. **Make sure at this point you have a 23% urea polyacrylamide gel prepared with VERY clean casting plates. The cleanliness of the gel image will be critical for quantification later.**
10. Load 8 μL of each sample onto a 23% urea polyacrylamide gel, and electrophorese the gel with constant power at 55 W/3000 V for ~3.5 h. **These electrophoresis conditions may vary depending on the gel apparatus and electrophoresis system. In our gel system, optimal band separation with minimal "smiling" of the bands occurs at ~60% progression of the dye front down our gels.**
11. After electrophoresis, separate the two plates of the gel cast, making certain the gel completely adheres to the bottom plate. Cover the gel in clear plastic wrap (ex. all-purpose polyvinyl-chloride wrap),

making sure that there are no wrinkles or bubbles in the wrap over the gel or excess wrap accumulating on the back of the casting plate. Be wary of touching the gel, wrap, or plate in areas other than the edges, as this can affect image quality. We recommend wiping the front and back of the finished gel/plate with 70% ethanol and letting it dry in a vertical position for a few minutes before imaging to obtain cleaner gels.

12. Image the prepared gel at excitation/emission of 495/520 nm.
13. Quantify and record band intensities for every reaction (lane) on the gel image by densitometry. These values will be used for activity quantification, to be discussed later in this chapter.

7.4 Measuring TREX1 activity on dsDNA with fluorescence

1. Prepare a volume of Master Mix equal to the product of the number of reactions and 140 μL. When determining the number of reactions, we recommend performing reactions in triplicate for each concentration of each enzyme tested. The Master Mix is prepared as follows:
   a. 44.4% Rx mix
   b. 55.6% 20 ng/μL nicked 10 kb plasmid
   c. **For 10 reactions, you would need 1.4 mL of Master mix composed of 622 μL of Rx mix and 778 μL of nicked plasmid at 20 ng/μL.**
2. Obtain microcentrifuge tubes for every reaction to be performed, and place 135 μL of Master Mix into each reaction tube.
3. Set up the 384-well microplate and add 20 μL of 15 × SYBR green (stock dissolved in H_2O) to every well to be used. **You will need seven wells per reaction, corresponding to the seven samples that will be taken for each of the time points used. Prepare this solution no >1 h before reaction initiation, and make sure to cover the microplate throughout the experiment whenever possible to protect photosensitive contents.**
4. For each enzyme to be tested, prepare aliquots for each of the intended enzyme concentrations at 10 × the desired reaction concentration. This should be done *via* dilution of stock enzyme into 1 mg/mL BSA. **If the inhibitive properties of a mutant enzyme on wild-type enzyme are being tested, include wild-type enzyme in the prepared mutant enzyme aliquots at 10× the desired reaction concentration.**
5. Take an 18 μL sample of each reaction tube and quench it in its corresponding well in the microplate, then add 2 μL of 1 mg/mL

BSA. This will act as the 0 min time point for the assay. **Here, and at every sample quenching after, make sure to pipette-mix the well contents thoroughly to completely inactivate the TREX1. This is a major source of result variation if not done adequately.**

6. Add 13 μL of the corresponding enzyme aliquot to each reaction tube, and vortex thoroughly to initiate the reactions.
7. Allow the reactions to incubate for 1 h at room temperature. Take 20 μL samples from each reaction tube at the 5, 10, 20, 30, 45, and 60-min time points and quench the samples in their intended wells on the microplate.
8. After reaction completion, read the fluorescence intensity of the plate wells at excitation/emission of 495/520 nm and record the values. These values will be used for activity quantification, to be discussed later in the chapter.

7.5 Visualizing TREX1 activity on dsDNA with agarose gel electrophoresis

1. Prepare a volume of Master Mix equal to the product of the number of reactions and 140 μL. The Master Mix is prepared as follows:
   a. 44.4% Rx mix
   b. 55.6% 20 ng/μL nicked 10 kb plasmid
   c. **For 10 reactions, you would need 1.4 mL of Master mix composed of 622 μL of Rx mix and 778 μL of nicked plasmid at 20 ng/μL.**
2. Obtain microcentrifuge tubes for every reaction to be performed, and place 135 μL of Master Mix into each reaction tube.
3. Obtain five microcentrifuge tubes per planned reaction, and add 80 μL of cold 100% ethanol to each. These will be used to quench the samples from each reaction.
4. For each enzyme to be tested, prepare aliquots for each of the intended enzyme concentrations at 10 × the desired reaction concentration. This should be done *via* dilution of stock enzyme into 1 mg/mL BSA. **If the inhibitive properties of a mutant enzyme on wild-type enzyme are being tested, include wild-type enzyme in the prepared mutant enzyme aliquots at 10× the desired reaction concentration.**
5. Take an 18 μL sample from each reaction tube and quench it in its corresponding ethanol tube, then add 2 μL of 1 mg/mL BSA. This will act as the 0 min time point for the assay.

6. Add 13 μL of the corresponding enzyme aliquot to each reaction tube, and vortex well to initiate the reactions.
7. Allow the reactions to incubate at room temperature. Take 20 μL samples from each reaction tube at the 5, 10, 20, and 40-min time points and quench the samples in their intended tubes.
8. Evaporate the quenched reactions *in vacuo* at low/no heat setting. This should take ~20 min.
9. Resuspend the evaporated samples in 11 μL of solution containing 80% 1 × TAE with 20% 5 × loading dye.
10. Load the samples onto a 0.8% agarose gel prepared with a GelRed dsDNA stain, and electrophorese the gel for 50 min at 110 V.
11. After electrophoresis, image the gel. Fig. 3A shows example data from this assay, and identifies the banding present.

8. Calculating TREX1 exonuclease activities

8.1 Quantifying TREX1 activity on ssDNA

All band intensities are determined in each lane (reaction) and summed (referred to hereafter as "lane intensity"). The lane intensity represents the total FAM-labeled substrate used in the reaction (for 50 nM in 30 μL, this would be 1.5 pmol of FAM-labeled oligonucleotide). The ratio of band intensity to lane intensity multiplied by the moles of FAM-labeled oligonucleotide yields the moles of *n*-mer oligonucleotide represented by each band in the lane. The highest molecular weight band is the starting 30-mer, and each band below represents oligonucleotide products of decreasing length by a single nucleotide. The amount of dNMPs excised in each reaction (lane) is determined by multiplying the moles of each length oligonucleotide by the number of nucleotides excised to produce the *n*-mer oligonucleotide associated with each band, and summing these values for all bands in the lane. The equation and variable definitions for calculating dNMPs excised in each reaction (lane) are detailed below.

If:

N = the band number of a given band in the chosen lane, such that $N=1$ is the highest apparent molecular weight band in the chosen lane, $N=2$ is the next highest, *etc.*

B = the band intensity value associated with the band identified by N.

L = the lane intensity of the chosen lane (sum of B for all N in chosen lane).

T = the total number of bands present in the chosen lane.

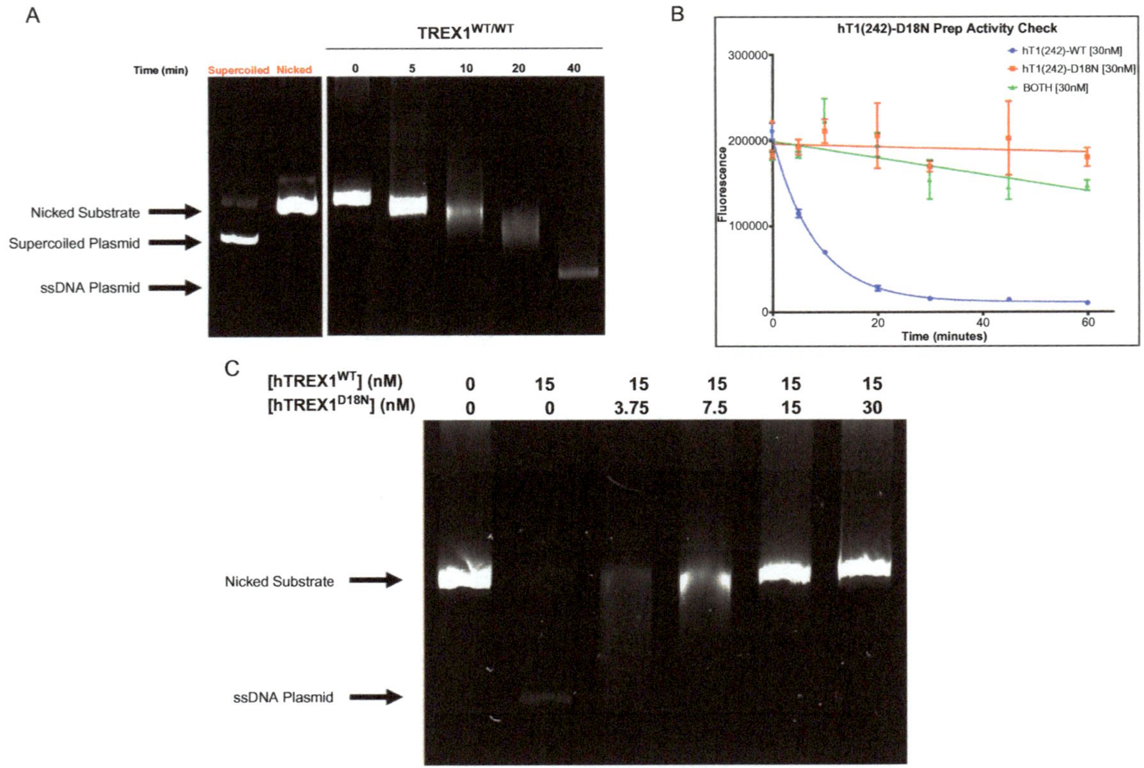

Fig. 3 Results of TREX1 dsDNA fluorescence and agarose gel assays. (A) Degradation of 10 ng/μL Nt.*Bbv*CI-nicked pMYC substrate by 15 nM hTREX1$^{WT/WT}$ visualized on a 0.8% agarose gel. Bands corresponding to supercoiled and nicked plasmid are indicated. TREX1 excises dNMPs from the nicked polynucleotide of the dsDNA plasmid generating a single-stranded DNA polymer. The position of migration of the ssDNA polymer is identified as "ssDNA." The gel lanes from left to right are: supercoiled (un-nicked) plasmid, nicked plasmid, reaction samples at the indicated times (min). (B) Degradation of 10 ng/μL Nt.BbvCI-nicked pMYC dsDNA by 30 nM hTREX1$^{WT/WT}$ and/or 30 nM hTREX1$^{D18N/D18N}$, visualized *via* fluorescence. Fluorescence values for the indicated time points for each reaction indicated were plotted as a function of fluorescence *vs* time in Prism7 (GraphPad), and fitted with one-phase decay nonlinear regression. (C) Degradation of 5 ng/μL Nt.BbvCI-nicked pMYC substrate by 15 nM hTREX1$^{WT/WT}$ in the presence of increasing concentrations of hTREX1^{D18N}, visualized on a 0.8% agarose gel. Bands corresponding to undegraded substrate are indicated.

Then:

$$dNMPs\ Excised = \sum_{N=1}^{T} \frac{(1.5)*B*(N-1)}{L}, \text{(pmol)}$$

For the purposes of comparing TREX1 variant activities, we calculate TREX1 enzyme activity as the picomoles of dNMPs excised divided by seconds of incubation time divided by picomoles of enzyme. This is shown in the equation:

$$Trex1\ Activity = \frac{(dNMPsExcised)}{(Time)*(Enzyme)}, \left(\text{s}^{-1}\right)$$

8.2 Quantifying TREX1 activity on dsDNA with fluorescence

Fluorescence values are obtained for each reaction sample and plotted as a function of time. The time-course reaction curve is generated and fit using single-phase decay nonlinear regression with Prism (GraphPad) as illustrated in Fig. 3B. Reaction rates are calculated using initial slopes of the fitted line and expressed as fluorescence units per second. A standard curve equating fluorescence units to DNA concentration can be generated if units of dNMPs per second are desired by including no-enzyme reactions and a range of dsDNA substrate concentrations in the experiment. In either case, the calculated reaction rates can be normalized to determine relative activities of TREX1 variant enzymes.

9. Exonuclease activities of TREX1 mutant enzymes

TREX1 mutations have been identified at positions throughout the gene with varying effects on ssDNA and dsDNA exonuclease activities and cell localization. Four nucleotides of ssDNA bind at the TREX1 active site through sequence independent hydrogen bonding and hydrophobic interactions (de Silva et al., 2007). The phosphate oxygens of the DNA 3′ nucleotide, along with the carboxylate oxygens of Asp18, Glu20, Asp130, and Asp200 and several water molecules contribute to the coordination of two divalent metal ions at the active site. The D18N, D18H, D200N, and D200H disease causing mutations at these active site residues abolish DNA exonuclease activity in TREX1 homodimers resulting from dysfunctional metal ion coordination (Lehtinen et al., 2008).

However, heterodimers containing one of these active site mutant protomers with the wild-type protomer exhibit ~50% wild-type activity using the ssDNA assay and undetectable activity using the dsDNA assay. In addition, the heterodimers competitively inhibit TREX1 wild-type activity in the dsDNA assay (Lehtinen et al., 2008). DNA exonuclease assays support the idea that heterodimers bind DNA and position the 3′ nucleotide into the active site. The inability to perform chemistry of phosphodiester bond cleavage resulting from dysfunctional metal-ion coordination traps the TREX1 mutant enzyme onto the dsDNA in a nonproductive enzyme-DNA complex at the site of the nick and accounts for the inhibition of TREX1 WT DNA degradation activity detected in the presence of the Asp18 and Asp200 TREX1 mutant enzymes. Thus, the combination of TREX1 catalytic deficiency and DNA binding proficiency generates TREX1 mutant enzymes that fail to process nicks in dsDNA and block access to these sites by TREX1 WT enzyme helping to explain the dominant genetics exhibited by the *TREX1 D18N, D18H,* and *D200N* alleles (Rice et al., 2015) (see Fig. 3B and C) The TREX1 residues Arg-174 and Lys-175 positioned adjacent to the active sites act with the Arg-128 residues positioned in the catalytic core facilitate dsDNA melting and generate ssDNA for entry into the active sites. TREX1 mutations at R174 and K175 located in the flexible loop have varied effects on ssDNA and dsDNA exonuclease activities by altering TREX1 DNA binding (Fye et al., 2011).

The TREX1 R114H mutation is one of the most common autoimmune disease-causing mutations. DNA degradation properties of R114H have provided direct evidence that TREX1 residues in one protomer contribute to DNA degradation catalyzed in the opposing protomer and help to explain the dimeric TREX1 structure required for full catalytic competency (Orebaugh et al., 2011). The TREX1 Arg-114 residue is about 15 Å away from the active site positioned at the dimer interface where it hydrogen bonds with two backbone carbonyl oxygens of the opposing protomer where it contributes to the activity in the opposing protomer of the TREX1 dimer. The TREX1 R114H homodimer exhibits severely reduced enzymatic activity on ss- and dsDNA. The R114 H mutation causes recessive TREX1-associated autoimmune disorders (Rice et al., 2015).

The TREX1 flexible regions contribute to a nucleic acid-kinking mechanism used to unwind, melt, and reshape dsDNA and position the ssDNA 3′termini required for catalysis into the active site (de Silva et al., 2007; Grieves et al., 2015). This DNA binding mechanism likely explains the varied effects on exonuclease activities using ssDNA and dsDNA assays exhibited by some TREX1 mutations such as R174 and K175 positioned on the flexible DNA-binding loop. Additionally, residues like Arg-62 in one TREX1 protomer provide necessary structural elements for full catalytic activity in the opposing TREX1 protomer (Fye, Coffin, Orebaugh, Hollis, & Perrino, 2014). These data provide compelling evidence for the required TREX1 dimeric structure for full catalytic function.

10. Summary

The nonprocessive $3' \rightarrow 5'$ exonuclease TREX1 degrades cytosolic DNA to prevent aberrant nucleic acid sensing. When TREX1 exonuclease activity is attenuated by mutation or altered cellular location autoimmune disease can result. In this chapter, we detailed methods to overexpress and purify recombinant TREX1 enzyme and to measure ssDNA and dsDNA exonuclease activities *in vitro*. The biochemical activities of TREX1 mutants determined using these assays have contributed to our understanding of TREX1-associated autoimmune disease. In addition, these methods help elucidate the mechanisms by which TREX1 mutation affects exonuclease activity and cytosolic nucleic acid sensing.

Acknowledgments

This work was supported by the National Institute of Allergy and Infectious Diseases (AI116725), National Institute of General Medical Sciences (GM110734), Lupus Research Alliance, Wake Forest Innovations, Comprehensive Cancer Center of Wake Forest University National Cancer Institute (Center Support Grant, P30CA012197), and a NIH NRSA Predoctoral Fellowship (T32-GM095440).

References

Ablasser, A., Hemmerling, I., Schmid-Burgk, J. L., et al. (2014). TREX1 deficiency triggers cell-autonomous immunity in a CGAS-dependent manner. *The Journal of Immunology*, *192*(12), 5993–5997.

Silva de, U., Choudhury, S., Bailey, S. L., et al. (2007). The crystal structure of TREX1 explains the 3′ nucleotide specificity and reveals a polyproline II helix for protein partnering. *Journal of Biological Chemistry*, *282*(14), 10537–10543.

Fye, J. M., Coffin, S. R., Orebaugh, C. D., Hollis, T., & Perrino, F. W. (2014). The Arg-62 residues of the TREX1 exonuclease act across the dimer interface contributing to catalysis in the opposing protomers. *The Journal of Biological Chemistry*, *289*(16), 11556–11565.

Fye, J. M., Orebaugh, C. D., Coffin, S. R., Hollis, T., & Perrino, F. W. (2011). Dominant mutations of the TREX1 exonuclease gene in lupus and aicardi-goutières syndrome. *The Journal of Biological Chemistry*, *286*(37), 32373–32382.

Grieves, J. L., Fye, J. M., Harvey, S., et al. (2015). Exonuclease TREX1 degrades double-stranded DNA to prevent spontaneous lupus-like inflammatory disease. *Proceedings of the National Academy of Sciences of the United States of America, 112*(16), 5117–5122.

Harrigan, J. A., Fan, J., Momand, J., et al. (2007). WRN exonuclease activity is blocked by DNA termini harboring 3' obstructive groups. *Mechanisms of Ageing and Development, 128*(3), 259–266.

Hasan, M., Fermaintt, C. S., Gao, N., et al. (2015). Cytosolic nuclease TREX1 regulates oligosaccharyltransferase activity independent of nuclease activity to suppress immune activation. *Immunity, 43*(3), 463–474.

Kucej, M., Fermaintt, C. S., Yang, K., Irizarry-Caro, R. A., & Yan, N. (2017). Mitotic phosphorylation of TREX1 C terminus disrupts TREX1 regulation of the oligosaccharyltransferase complex. *Cell Reports, 18*(11), 2600–2607.

Lee-Kirsch, M. A., Chowdhury, D., Harvey, S., et al. (2007). A mutation in TREX1 that impairs susceptibility to granzyme A-mediated cell death underlies familial chilblain lupus. *Journal of Molecular Medicine, 85*(5), 531–537.

Lehtinen, D. A., Harvey, S., Mulcahy, M. J., Hollis, T., & Perrino, F. W. (2008). The TREX1 double-stranded DNA degradation activity is defective in dominant mutations associated with autoimmune disease. *The Journal of Biological Chemistry, 283*(46), 31649–31656.

Manils, J., Gómez, D., Salla-Martret, M., et al. (2015). Multifaceted role of TREX2 in the skin defense against UV-induced skin carcinogenesis. *Oncotarget, 6*(26), 22375–22396.

Mazur, D. J., & Perrino, F. W. (1999). Identification and expression of the TREX1 and TREX2 CDNA sequences encoding mammalian $3' \rightarrow 5'$ exonucleases. *Journal of Biological Chemistry, 274*(28), 19655–19660.

Mazur, D. J., & Perrino, F. W. (2001a). Structure and expression of the TREX1 and TREX2 $3' \rightarrow 5'$ exonuclease genes. *Journal of Biological Chemistry, 276*(18), 14718–14727.

Mazur, D. J., & Perrino, F. W. (2001b). Excision of $3'$ Termini by the Trex1 and TREX2 $3' \rightarrow 5'$ Exonucleases characterization of the recombinant proteins. *Journal of Biological Chemistry, 276*(20), 17022–17029.

Morita, M., Stamp, G., Robins, P., et al. (2004). Gene-targeted mice lacking the trex1 (DNase III) $3' \rightarrow 5'$ DNA exonuclease develop inflammatory myocarditis. *Molecular and Cellular Biology, 24*(15), 6719–6727.

Orebaugh, C. D., Fye, J. M., Harvey, S., Hollis, T., & Perrino, F. W. (2011). The TREX1 exonuclease R114H mutation in aicardi-goutières syndrome and lupus reveals dimeric structure requirements for DNA degradation activity. *The Journal of Biological Chemistry, 286*(46), 40246–40254.

Orebaugh, C. D., Fye, J. M., Harvey, S., et al. (2013). The TREX1 C-terminal region controls cellular localization through ubiquitination. *The Journal of Biological Chemistry, 288*(40), 28881–28892.

Parra, D., Manils, J., Castellana, B., et al. (2009). Increased susceptibility to skin carcinogenesis in TREX2 knockout mice. *Cancer Research, 69*(16), 6676–6684.

Perrino, F. W., Harvey, S., McMillin, S., & Hollis, T. (2005). The human TREX2 $3' \rightarrow 5'$-exonuclease structure suggests a mechanism for efficient nonprocessive DNA catalysis. *Journal of Biological Chemistry, 280*(15), 15212–15218.

Perrino, F. W., Krol, A., Harvey, S., et al. (2004). Sequence variants in the $3' \rightarrow 5'$ deoxyribonuclease TREX2: identification in a genetic screen and effects on catalysis by the recombinant proteins. *Advances in Enzyme Regulation, 44*(1), 37–49.

Perrino, F. W., Miller, H., & Ealey, K. A. (1994). Identification of a $3' \rightarrow 5'$-exonuclease that removes cytosine arabinoside monophosphate from 3' termini of DNA. *Journal of Biological Chemistry, 269*(23), 16357–16363.

Rego, S. L., Harvey, S., Simpson, S. R., et al. (2018). TREX1 D18N mice fail to process erythroblast DNA resulting in inflammation and dysfunctional erythropoiesis. *Autoimmunity*, *51*(7), 333–344.
Rice, G. I., Rodero, M. P., & Crow, Y. J. (2015). Human disease phenotypes associated with mutations in TREX1. *Journal of Clinical Immunology*, *35*(3), 235–243.
Sakai, T., Miyazaki, T., Shin, D.-M., et al. (2017). DNase-active TREX1 frame-shift mutants induce serologic autoimmunity in mice. *Journal of Autoimmunity*, *81*, 13–23.
Stetson, D. B., Ko, J. S., Heidmann, T., & Medzhitov, R. (2008). Trex1 prevents cell-intrinsic initiation of autoimmunity. *Cell*, *134*(4), 587–598.
Yang, Y.-G., Lindahl, T., & Barnes, D. E. (2007). Trex1 exonuclease degrades SsDNA to prevent chronic checkpoint activation and autoimmune disease. *Cell*, *131*(5), 873–886.

CHAPTER NINE

Kinetics of DNA–protein association and dissociation by stopped-flow spectroscopy

Johannes Rudolph[a,*], Karolin Luger[a,b]
[a]Department of Biochemistry, University of Colorado Boulder, Boulder, CO, United States
[b]Howard Hughes Medical Institute, University of Colorado Boulder, Boulder, CO, United States
*Corresponding author: e-mail address: johannes.rudolph@colorado.edu

Contents

Abstract

The ability of proteins to recognize a specific nucleotide sequence or structural feature in the genome is central to DNA replication and repair as well as gene transcription. To understand how DNA-binding proteins accomplish this feat it is important to characterize how they efficiently recognize specific DNA sequences or various structural features in the milieu of a crowded nucleus containing 50 mM basepairs of DNA. This minimally entails the determination of thermodynamic (equilibrium) parameters that characterize the relative affinities of DNA-binding proteins with their cognate and non-cognate DNA sequences of interest. Additionally, elucidating the kinetic parameters that govern DNA–protein association and dissociation can provide essential mechanistic insights into the underlying mechanisms, and can inform on how these processes can be controlled under varying cellular conditions. Unfortunately, the acquisition of kinetic data can appear intimidating and are usually not included in the characterization of most DNA binding proteins. Here we provide detailed theoretical considerations and practical, easy-to-follow protocols for using stopped-flow spectroscopy to study the kinetics of DNA–protein association and dissociation, using poly-(ADP ribose) polymerase 1 as a practical example.

Methods in Enzymology, Volume 625
ISSN 0076-6879
https://doi.org/10.1016/bs.mie.2019.04.025

1. Introduction

Poly-(ADP-ribose) polymerase 1 (PARP1) plays an important role in the DNA damage response (Beck, Robert, Reina-San-Martin, Schreiber, & Dantzer, 2014; Crawford, Bonfiglio, Mikoč, Matic, & Ahel, 2018; Gibson & Kraus, 2012; Morales et al., 2014). When PARP1 binds to a DNA lesion caused by endogenous or exogenous insults from free radicals, ionizing radiation, or DNA-modifying chemicals, it becomes activated to polymerize long chains of PAR onto itself and other nuclear acceptor proteins such as histones and DNA repair proteins. These PAR chains recruit a variety of repair proteins that contain PAR-binding motifs (Rack, Perina, & Ahel, 2016). Knock-out experiments in mice show that PARP1 directly activates base excision repair, homologous recombination, nucleotide excision repair, and alternative non-homologous end-joining (de Murcia et al., 1997). PARP1 is implicated in nuclear processes other than DNA repair including chromatin compaction, regulation of transcription, DNA methylation, and regulation of RNA metabolism (Bai, 2015; Bock & Chang, 2016; Gibson & Kraus, 2012; Thomas & Tulin, 2013). Recently, PARP1 has been implicated in the innate immune response to foreign DNA via its interaction with cyclic GMP–AMP synthase (cGAS) (Liu et al., 2018) and its activation of IFI16 (Dunphy et al., 2018). PARP1 is also a validated target for cancer therapy (Franzese et al., 2019). Most notably, olaparib, niraparib, and rucaparib are in clinical use for treatment of ovarian and/or breast cancer in BRCA1/2 negative patients and 100+ clinical trials for inhibitors of PARPs are currently ongoing to treat breast and ovarian cancer, non-small cell lung cancer, and glioblastoma, either as monotherapy or in combination with chemo- or radiotherapy.

2. Motivation

The interaction between PARP1 and DNA is critical to its many nuclear functions as well as its inhibition by clinically-relevant inhibitors. Thus, it is important to understand the structural and functional details of the DNA–PARP1 interaction, or for that matter, **any** DNA–protein interaction with important biological implications. PARP1 has a significant affinity for both damaged and undamaged DNA. Yet under normal growth conditions, the basepairs of intact DNA outnumber those of damaged DNA by many orders of magnitude ($>10^8$). Thus, it is hard to envision how PARP1 manages to find the few sites of damage in the presence of

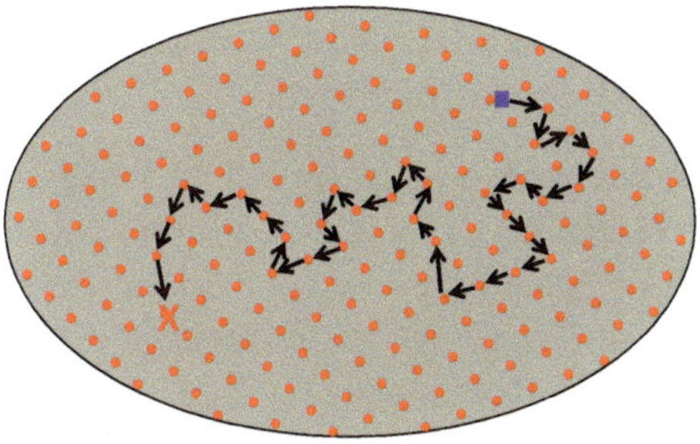

Fig. 1 Finding a specific DNA-binding site in the presence of a high concentration of non-specific DNA-binding sites is a "hard" problem that must be overcome by proteins such as PARP1 and transcription factors. The upper limit on the number of thermodynamically available binding sites in the human nucleus is 10^9, the number of base pairs in the genome. Many of these potential binding sites may be occupied or obscured by other DNA binding proteins, reducing the number of sites to 10^7–10^8. If the binding preference for specific sites over non-specific sites is 100–1000-fold, a DNA-binding protein will spend the majority of its time bound to non-specific DNA, even while searching for its specific binding site. Performing kinetic experiments that characterize DNA binding and release in vitro can help elucidate the mechanism by which DNA-binding proteins can execute this search with efficiency.

such an overwhelming excess of undamaged DNA (Fig. 1). In this way, PARP1 faces the same dilemma as that of transcription factors: how to find a specific DNA-binding site that is outnumbered by non-specific binding sites by many orders of magnitude (Berg, Winter, & von Hippel, 1981; Ferreiro & De Prat-Gay, 2003; Halford & Marko, 2004; Ruusala & Crothers, 1992). Understanding the kinetics of protein binding and release from DNA, as elucidated by stopped-flow spectroscopy, can provide much-needed mechanistic insight into how this real biological dilemma is solved in cells. The methods we describe here are therefore of interest not only to the research community studying PARP1 but are applicable to studying the behavior of any protein that binds DNA.

In this methods chapter, we provide details on how to collect "shots" on a stopped-flow instrument in order to measure the kinetics of DNA–protein association and dissociation, using PARP1 and an 18-mer oligomer as an example (Rudolph, Mahadevan, Dyer, & Luger, 2018). We provide theoretical and practical considerations, details about sample preparation and data collection, troubleshooting advice, and an introduction to analysis of the stopped-flow data. Importantly, this methods article does not intend to replace the user manual of your stopped-flow instrument, which you are highly encouraged to study frequently and intensely before and during any stopped-flow experiment.

3. Theoretical considerations

In order to monitor the association of DNA with a protein, one must be able to measure formation of the DNA–protein complex. While simple absorbance or intrinsic protein fluorescence can at times lead to detectable signals indicative of complex formation, fluorescence provides much greater sensitivity. Fluorescein and yellow/green Alexa analogs are particularly good choices because of their strong signal and their compatibility with the excitation and detection range of all stopped-flow instruments. Fluorescently labeled DNA oligomers can be purchased commercially and thus are a convenient and economical way to obtain a specifically and quantitatively labeled binding partner. It is important to attach the fluorescent label in a location that does not interfere with binding to protein. As labeling can be done at the 5′- or 3′-ends, as well as internally, and as oligomers can be made of varying lengths, the experimentalist has a lot of flexibility. If there is insufficient structural information to rationally choose a labeling location, it is simple enough to try a few different ones. While labeling the protein is also feasible, typical labeling reactions that modify cysteine or lysine residues using maleimide or succinimidyl chemistry, respectively, often yield heterogeneous samples. This heterogeneity is due to incomplete or non-specific labeling where some proteins may be labeled at different positions, some proteins remain unlabeled, and some proteins have multiple labels at different locations. For DNA-binding proteins containing Zn-fingers, like PARP1, maleimide-based modifications are particularly problematic as the Zn^{2+}-ion in the motif is coordinated by two cysteine residues. If these cysteines are modified, the Zn^{2+} presumably dissociates and the Zn-finger fold may unfold, disrupting DNA binding. Thus, fluorescently-labeled DNA is a much better choice.

Once one has obtained a fluorescently-labeled DNA oligomer, there are two possible detection methods for monitoring DNA–protein association (Fig. 2A). First, one can monitor the fluorescence of the oligomer, which may increase or decrease upon binding to protein. As it is also possible that the fluorescence signal remains unchanged upon binding to protein, this approach might not work. Also, fluorescence signals are prone to artifacts introduced by mixing unmatched buffers. Thus, fluorescence anisotropy is a powerful technique that will more reliably yield a detectable signal. DNA oligomers are typically sufficiently small (<40 nucleotides) and proteins are usually sufficiently large (>40 kDa) to significantly increase the

Fig. 2 Experimental set-up for monitoring (A) association of DNA with protein and (B) dissociation of DNA from a DNA–protein complex. The simplest kinetic schemes for DNA–protein association and dissociation are shown. For an association experiment, one syringe is filled with protein, the other with labeled DNA. For a dissociation experiment, one syringe is filled with the labeled DNA–protein complex, the other syringe is filled with competitor DNA.

anisotropy of the labeled oligomer upon binding to protein. Fluorescence anisotropy is exquisitely sensitive, yielding good signals in the low nanomolar concentration range. When choosing a fluorophore for anisotropy experiments, it is important to take into account the relationship between rotational diffusion and fluorophore lifetime for anisotropy (Jameson & Ross, 2010). That is, polarization may be harder to observe for fluorophores with a short lifetime (i.e. 0.5 ns for Alexa 555) and a long (20 ns) rotation correlation time. It is very simple to test the feasibility of both fluorescence and fluorescence anisotropy using a plate assay, as described below.

While association of protein with DNA can be studied by simply mixing these two reagents (with the appropriate signal change upon binding, Fig. 2A), studying the dissociation typically requires the use of an unlabeled competitor (Fig. 2B). Theoretically one could study dissociation by pre-forming a DNA–protein complex and then diluting this complex sufficiently to promote dissociation. Although most stopped-flow instruments can be used in a variable volume mode (e.g. mixing 1 part complex with 5 parts buffer), setting up the right conditions to start with sufficient complex and achieve sufficient dissociation can be tricky. Given the simplicity of using competitor DNA, and the additional advantage that one can deconvolute the kinetic mechanism of dissociation (see below), using varying concentrations of competitor DNA is the preferred approach to studying dissociation.

4. Practical considerations

Before embarking on kinetic experiments that are relatively difficult and sample-consuming, it is wise to perform preliminary tests using a simple plate assay. Assays under these equilibrium conditions will allow you to determine if you have a detectable signal for measuring binding of DNA to protein and derive a dissociation constant (K_d). Also, these simple assays allow you to explore buffer conditions that may affect the K_d, such as pH, ionic strength, and the presence of divalent cations. We provide here a typical protocol:

1. Prepare a stock solution of protein at 1 μM (25 μL) and place in column 1 of a 384-well plate (e.g. Costar 9017). We recommend performing all reactions in at least duplicate, so repeat this procedure for the next row of the plate. If your binding constant is very weak (>0.2 μM), a higher concentration of protein may be needed.
2. Add 10 μL of binding buffer to all other columns of that row in the plate. To speed up your protocol you can use a multichannel pipettor for these additions, sequentially filling all the even numbered columns and then the odd numbered columns due to the tighter spacing of a 384-well plate than a typical multichannel pipettor. Binding buffer should include a buffer to control pH (e.g. 50 mM Tris–HCl, pH 8, or HEPES, pH 7), some ions (e.g. 100 mM NaCl or KCl), potentially some divalent cations (1–10 mM Mg^{2+} or 0.1–1 mM Mn^{2+}), and other protein additives (e.g. DTT). Also, we recommend the use of a low concentration of detergent (e.g. 0.01% IGEPAL) to prevent protein from adhering to the plastic of the 384-well plate.
3. Perform a serial dilution of the protein down the columns of the row of the 384-well plate, pipetting up 15 μL, and then dispensing those 15 μL into the next column, with adequate mixing by up-and-down pipetting in each well. Leave the last four columns with buffer only. These will serve as controls. We recommend you do not use a multichannel pipettor for these serial dilutions as it is essentially impossible to achieve even bubble-free pipetting in multiple wells at the same time.[1]

[1] Reproducible pipetting by hand in 384-well plates takes some practice. We recommend starting with trial dilutions of the labeled DNA. Use 250 nM as the highest concentration and perform the same serial dilution as described in the text. Using the fluorescence mode on your plate reader, set the gain for the first well (highest concentration) and then read the entire row. Good pipetting will yield a straight line with $R^2 \geq 0.99$ upon plotting observed fluorescence vs concentration of DNA.

4. Prepare a stock solution of fluorescently labeled DNA at 5 nM (110 μL/row) and add 10 μL of the diluted DNA to every well in the row, including the last four.
5. Mix by gentle agitation for a few minutes and then spin the plate in a microplate centrifuge or "salad spinner" (e.g. Handyfuge) to ensure all the liquid is at the bottom of the well.
6. Read the plate in a fluorescent plate reader. (There are many excellent brands of plate readers, including BMG LabTech, Tecan, and Molecular Devices, and the capability of reading anisotropy should be a significant factor when deciding which one to purchase.) Use any well in columns 20–24 to set the gain (50% of max) and the mP (~100), if reading anisotropy.
7. For fluorescence readings, units are arbitrary (in the thousands), and even relatively small changes upon protein binding can yield reproducible binding curves. For anisotropy, units are in milli-polarization units (mP), and overall changes from unbound to bound DNA of 20–200 mP can yield reproducible binding curves.
8. The observed data (*F*) for binding of DNA (*D*) to protein (*P*) should be fitted using the standard binding Eq. (1) to derive the dissociation constant (K_d, in units of M). If the K_d is sufficiently tight to approach the concentration of the probe (2.5 nM), the quadratic binding equation should be used instead (2). Remember that all your final concentrations are 50% of the concentration in which they were added.

$$F = \frac{[P]}{K_d + [P]} + F_0 \tag{1}$$

$$F = \frac{1}{2}\left\{ (F_{max} - F_0)([D] + [P] + K_d) - \sqrt{([D] + [P] + K_d)^2 - 4[D][P]} \right\} + F_0 \tag{2}$$

The experimentalist's choice of signal for monitoring time-dependent DNA–protein association and dissociation (fluorescence or fluorescence anisotropy) will depend not only on the outcome of the plate experiment described above, but also on the availability of a suitable instrument. Stopped-flow instruments are expensive and are often found as part of a core or shared equipment facility. The four major manufacturers of these instruments are Applied Photophysics, TgK Scientific, KinTek Corporation, and Bio-Logic. For all of these instruments, a fluorescence polarization module is available as an add-on beyond the basic set-up.

A crucial question asked by most experimentalists, especially those who prepare their own protein, is "How much protein will I need to generate publication-worthy data?" Not unexpectedly, the answer is "that depends." In the following section we provide some guidelines.

1. The amount of protein needed depends on its affinity for DNA. A stopped-flow experiment to measure the binding to, or dissociation from protein must be performed at concentrations above the apparent K_d, as determined in the plate assay above. For purposes of this calculation, let us assume a MW of the protein of 50 kDa and K_d of 50 nM, which is on the high end for many DNA–protein interactions.
2. Volumes (and thus amounts of protein) depend on which stopped-flow apparatus you are using. For purposes of this calculation, we assume use of the instrument from Applied Photophysics, which will use similar volumes as the instruments from TgK Scientific and KinTek Corporation, but larger volumes than the instrument manufactured by Bio-Logic.
3. A series of mixing experiments (with 10 + shots for averaging of signals, see below) can be readily performed using 1.5 mL of protein and 1.5 mL of DNA. If this experiment is performed at 100–1000 nM protein, one would need 7.5–75 μg of protein.
4. If one is measuring both association and dissociation of DNA with protein, one needs to multiply the above value by two, thus requiring 15–150 μg of protein.
5. Typically one will perform a concentration series (varying DNA or protein) in order to determine mechanisms (see below). If one uses five different concentrations, one would need to multiply the numbers from above by a factor of 5, thus requiring 75–750 μg of protein.
6. Finally, if one desires three biological replicates of the above experiments, one would need a total of 0.2–2.2 mg. Considering how much useful mechanistic information one can derive from stopped-flow experiments, this minimal investment in protein is more than worthwhile.

The amount of fluorescent DNA you will require is small and a minimal order of 100 nmol should be sufficient for all association and dissociation experiments. Because dissociation experiments are performed with a large excess of competitor DNA, order at least 400 nmol of the unlabeled DNA.

5. Sample preparation

For a single association experiment, prepare one solution of protein at 25–1000 nM (1.5 mL), depending on the affinity measured in the plate assay above (Fig. 2A). Also prepare a solution of labeled DNA (50 nM, 1.5 mL). Particularly if you are detecting binding by fluorescence, it is important that the buffers in the two solutions are well-matched. If you are performing an experiment wherein you are titrating protein, make additional samples containing varying concentrations of protein (25–1000 nM). In this case, prepare a correspondingly larger volume of labeled DNA. Remember that all solutions will be diluted twofold in the stopped-flow instrument, so make all of your solutions twofold higher than your final desired concentration.

For a single dissociation experiment, prepare one solution of DNA complexed to protein at a concentration that is above the K_d (e.g. 50 nM), as measured in the plate assay above (Fig. 2B). It is best to use a slight excess of protein to labeled DNA (1.2–1.5-fold) to ensure that there is no excess labeled DNA remaining in the solution. Also prepare a solution of competing DNA at concentrations that are minimally twofold, preferably >10-fold above the concentration of the complex. If you are performing an experiment wherein you are titrating the competitor DNA, make additional samples containing varying concentrations of unlabeled competitor DNA (e.g. 1–10 μM). In this case, make a correspondingly larger volume of labeled DNA complexed with protein. Again, remember that all solutions will be diluted twofold in the stopped-flow instrument, so make all of your solutions twofold higher than your final desired concentration.

6. Instrument operation and data collection

Although each stopped-flow instrument will have subtle differences in their mode of operation (see user manuals), the steps outlined here for the SX20 instrument from Applied Photophysics are generally applicable.

1. Getting ready
   a. Set-up the optics correctly for the experiment you will be performing (fluorescence vs fluorescence anisotropy; see user manual for your particular instrument).
   b. Ensure that you have the appropriate filters for your fluorescent tag in the photomultiplier tubes (PMTs) and the appropriate excitation wavelength set on the monochromator.
   c. Open the valve on the nitrogen tank that is used to propel the drive syringes.
   d. Since all rate constants are temperature-dependent (about a factor of two for every 10 °C), turn on the water bath if you are using it to regulate the temperature of the drive syringe chamber. That said, we have found that modern climate control keeps room temperature sufficiently constant for collecting reproducible data. The advantages of not using the temperature control bath include better leak detection from one of the drive syringes and prevention of

green sludge in the chamber surrounding the drive syringes from leaving the chamber filled with water.

e. Warm-up the bulb for 30 min prior to collecting any data. In general, turn on the bulb before turning on any other instrument components (i.e. syringe driver, computer) to prevent damage from power surges.

f. Ensure that the instrument and computer are communicating.

g. Use a hypodermic syringe to rinse and then load both drive syringes with water to remove the 20% ethanol used for storage (Fig. 3).

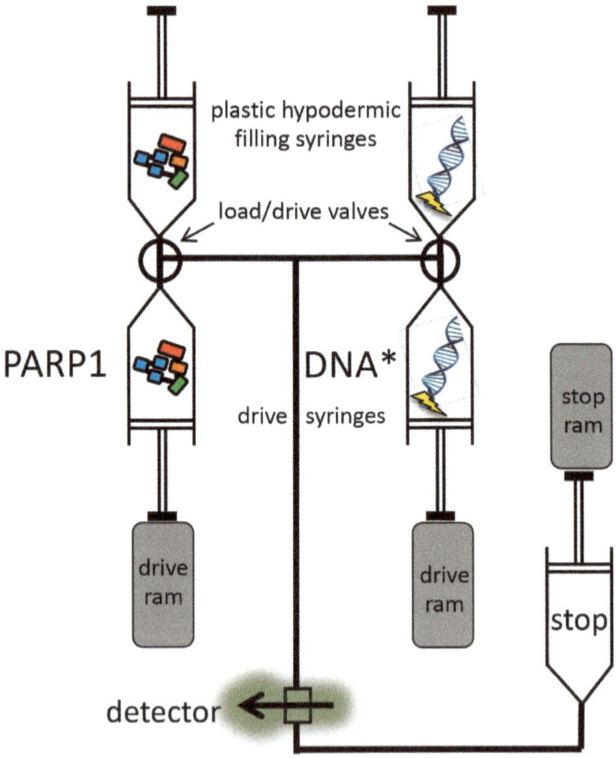

Fig. 3 Schematic of the plumbing of a stopped-flow apparatus. The drive syringes are made of glass and are typically enclosed in a chamber (not shown for clarity) that can be temperature controlled by means of a circulating water bath. The loading syringes are typically disposable hypodermic syringes. We recommend establishing a consistent pattern of loading the fluorescent sample in the left syringe and the non-labeled sample in right syringe to prevent loading mishaps. Pushing the drive and loading syringes back and forth (with manual assistance to the shafts) is important for ensuring there are no air bubbles in the drive syringes. The drive rams (or plate, depending on instrument brand) must always be in good contact with the bottom of the drive syringes prior to each shot. Additionally, the load/drive valves must be in the correct direction prior to each shot.

Then flush the remainder of the plumbing and stopped syringe with water. For the Applied Photophysics instrument, this is done most efficiently by manual emptying of the stop syringe and manual pushing of the drive syringes. (Manual rinsing also avoids potential damage to the instrument by having the valves in the incorrect position or incomplete contact between the drive ram and the syringe plungers.)

h. Rinse and flush both drive syringes with assay buffer. Flush the remainder of the plumbing and stopped syringe with buffer as well. Because buffer is cheap and real samples are not, we generally perform this procedure at least twice to ensure that everything is well equilibrated.

2. Time to collect data

a. Fill two hypodermic syringes, one with protein and one with labeled DNA, using a needle in order to remove all of the valuable protein/DNA reagents from the bottom of your Eppendorf tube. With *both* valves in the **load** position, fill the protein solution into one drive syringe and the DNA solution into the other drive syringe (Fig. 3). We recommend that the two valves are always in equivalent positions (both in **load**, or both in **drive**) to prevent accidental mixing of the protein and DNA during the loading procedure. Ensure that there are no air bubbles in either of the drive syringes by shunting the liquid from the hypodermic syringe and the drive syringe back and forth a few times. Air bubbles are one of the most common causes of artifacts in stopped-flow data. We recommend establishing a habit of always loading the protein into the left syringe and the DNA into the right syringe. A mistake in refilling one of the drive syringes in order to collect more replicates will prematurely end your day of data collection.

b. Switch *both* drive valves to the **drive** position and ensure that both syringe pistons are in good contact with the drive ram (Fig. 3). Checking the drive position and good piston contact before *each and every* shot should be a habit that becomes firmly ingrained into your data-collection process.

c. Flush the plumbing, thus filling the optical cell, by performing 3–4 shots, where you do not collect any data.

d. Adjust the gain (fluorescence and fluorescence anisotropy) and measure the G-factor (fluorescence anisotropy only). For fluorescence, you should know whether to expect an increase or a decrease in signal upon binding based on your preliminary plate assay. Thus, you should adjust the gain appropriately so as not to max out your detector or lose your signal. For fluorescence anisotropy it is safe to adjust the gain to 90%. Remember, you should only adjust the gain after you have loaded fluorescent sample into the detector, and you should re-adjust the gain if you significantly change the concentration of the fluorescent sample in your experiment.

e. Perform real shots where data collection is enabled. As you will most likely not know how fast association or dissociation will occur for your particular DNA-binding protein under your chosen experimental conditions, and as such kinetics can occur on widely different timescales, it is best to start by scanning different time ranges. It is entirely possible to miss your reaction because you "looked at the wrong time." As a starting point we recommend 0–25 ms for really fast reactions, 0–200 ms for fast reactions, 0–2 s for slow reactions, and 0–200 s for really slow reactions. If your reaction takes longer than 200 s, you can perform your kinetics in a plate reader using much less sample. Also, to achieve a more evenly-spaced distribution of data, we recommend that you ensure that you have set the software to collect your data in log mode, since most reactions will exhibit exponential behavior.

f. An important control experiments that you should perform each time is mixing samples containing the fluorophore (labeled DNA or protein complexed with labeled DNA) with buffer alone. This baseline reading is important for a number of reasons.

   i. You can evaluate whether your light source and electronics are stable.

   ii. You can demonstrate that there is no change upon performing a twofold dilution of your fluorophore.

   iii. The baseline value from this experiment will be used in data processing and fitting, and because this value will vary based on the gain setting for the instrument that day and for each dilution of fluorophore, it is important to perform this control for each different fluorophore-containing sample each day.

   iv. The baseline value can be used to help troubleshoot any failure to see expected results (see below).

g. When you wish to load a different protein concentration (for an association experiment, Fig. 2A) or a different competitor DNA concentration (for a dissociation experiment, Fig. 2B) into the drive syringe, it is best to work from lowest to highest concentration (Fig. 3).

i. Turn *both* valves to load, even if you are only exchanging the contents of one of the drive syringes.
ii. Empty the drive syringe you wish to exchange of any remaining unused sample.
iii. Load the drive syringe with the next higher concentration of reagent, ensuring as before that there are no bubbles in the drive syringe.
iv. Reload the other (unchanging) drive syringe with more of the same reagent, again ensuring that you have not introduced any air bubbles.
v. Switch *both* drive valves to the **drive** position and ensure that both syringe pistons are in good contact with the drive ram.
vi. Flush the tubing and stop syringe with the new reagents by performing 3–4 "empty" shots.
vii. Collect more real shots.

3. Cleaning up
   a. After collecting all of your data for the day, first turn off the lamp to conserve lamp hours.
   b. Remove all sample from both drive syringes and flush each syringe three times with water.
   c. Load the drive syringes with water and manually flush the remainder of the plumbing and stop syringe with water. Repeat this process three times. Having a clean instrument for the next time you do an experiment (or for the next user) is very worthwhile.
   d. Load the drive syringes with 20% EtOH and flush the remainder of the plumbing and the stop syringe. This is the appropriate way to store your instrument for short periods. For long term storage, check your instrument manual.

7. Experimental troubleshooting

If despite seeing a good fluorescent or anisotropy signal in the plate reader, you do not see any signal in the stopped-flow experiment, here are some suggestions.

1. The reaction was too fast to measure. This result is much more likely for an association experiment, and extremely unlikely for a dissociation experiment. To diagnose this potential outcome, compare your data for mixing protein + labeled DNA with the control buffer + labeled DNA. In an anisotropy experiment, if binding occurred, the average

signal in the protein sample will be higher than the average signal in the control sample. Changes as small as 0.04 anisotropy units can reflect an observable binding event that was missed. If you did not do so already, use a faster time-range for collection (0–25 ms). Alternatively, since the apparent rate of association is dependent on the concentration of the reactants, use lower concentrations, particularly of protein. We have successfully observed noisy but reproducible binding kinetics ($k_{obs} = 300\,s^{-1}$) at concentrations of 25 nM protein and 25 nM fluorescent DNA.

2. The reaction was too slow to measure. Although stopped-flow instruments allow you to collect data for essentially an infinite amount of time, they are not designed or optimized for slow reactions. Because the reactants are not physically confined to the observation cell, they will slowly diffuse out and/or mix with the preceding/next shot. In this case you should attempt to perform your kinetics in a plate assay, which as a bonus, will consume significantly lower amounts of protein.
3. There is protein aggregation/precipitation that prevents detection of association or dissociation. It may be that the protein is prone to aggregation, which becomes more of a problem during the rather long process of collecting stopped-flow data compared to pipetting the reagents for a plate assay. We have even had the situation where good DNA–protein association kinetics could be observed, however, because of the instability of the DNA-bound protein complex, dissociation experiments yielded erratic, if any, signal. It may be hard to know for certain that aggregation/precipitation is occurring as the concentrations of protein used are typically too low to detect obvious haziness in the solution. When protein aggregation/precipitation is suspected, it is best to perform some (unfortunately tedious) experiments exploring protein stability in different buffers (vary pH), salt concentrations, and detergents.

8. Data analysis

1. Averaging of raw signals. Even with strong signals from fluorescent labeling of DNA, stopped-flow data for DNA–protein association and dissociation experiments are intrinsically noisy. High noise levels are particularly a problem for fast reactions where there is reduced sampling per data point. Assuming the same overall signal strength, collecting 1000 data points between 0 and 25 ms yields much noisier data than collecting the same number of points between 0 and 2.5 ms. Thus, it is important to collect and average multiple individual shots, whereby it is important to

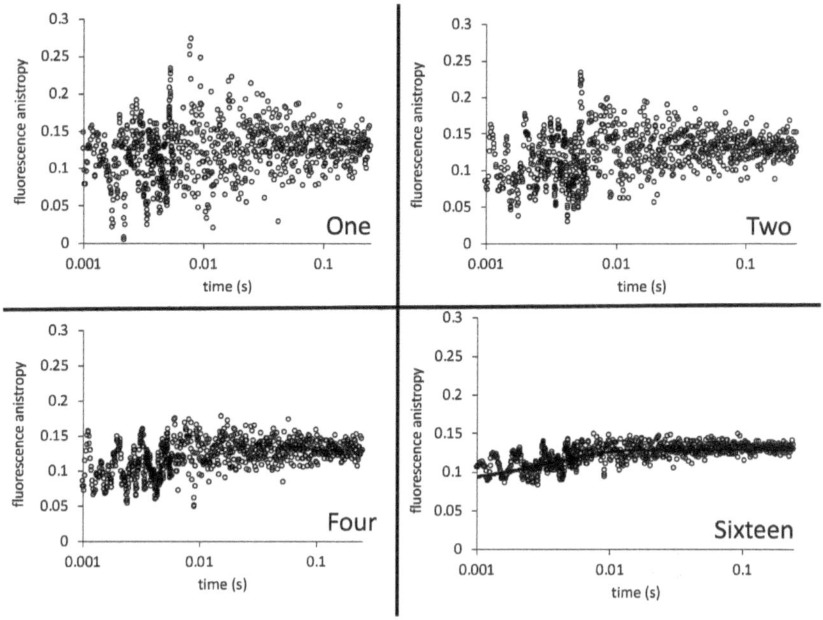

Fig. 4 Averaging multiple shots greatly improves signal-to-noise. Monitoring the association of protein (37 nM) with fluorescently labeled DNA (25 nM), we compare data collected for one shot with averaging of 2, 4, or 16 shots to demonstrate the improvement in signal-to-noise. For the graph with 16 shots, we indicate the fit to a first-order exponential ($k_{obs} = 273 \pm 29\,s^{-1}$) with a line.

remember that the signal-to-noise improves by root(n), where n is the number of shots (Fig. 4). That is, it takes four times as many shots to increase the signal-to-noise by a factor of two. When averaging different shots, it is important to visually inspect the data and remove shots that contain significant anomalies that can arise from air bubbles or inadequately rinsed plumbing after changing reaction conditions.

2. Kinetic fitting to different equations under ideal reaction conditions

a. *Association.* Assuming the simplest mechanism (Fig. 2A) and under ideal reaction conditions ([protein] > [labeled DNA]), the averaged raw data can be fitted using a simple exponential (Eq. (3) to derive the apparent rate k_{obs}, the overall amplitude of the signal (A) and the baseline value (b).

$$[P] = A \times \left(1 - e^{-k_{obs}t}\right) + b \tag{3}$$

Plotting and inspecting the residuals is by far the most powerful tool for evaluating the suitability of your fits (Fig. 5). If the residuals exhibit systematic deviations from an average of zero for a first-order fit, you should

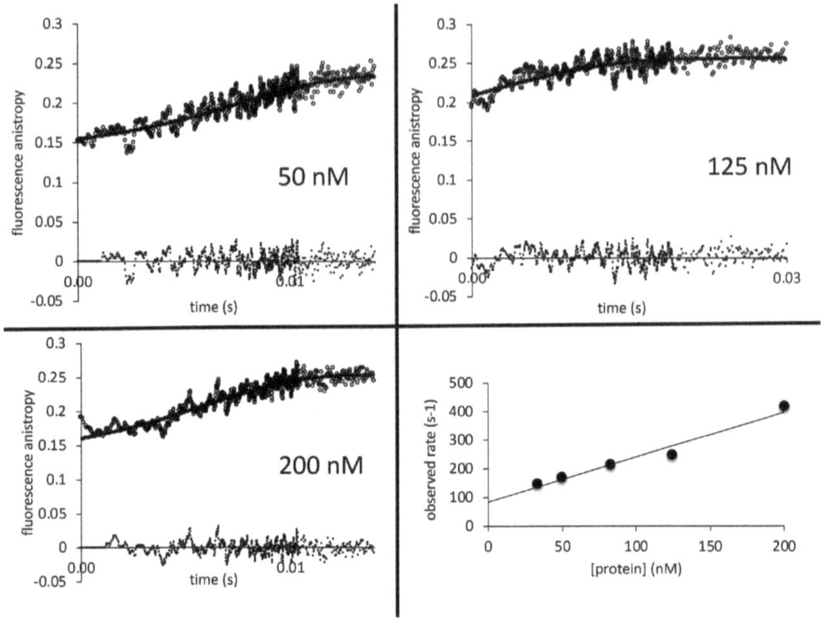

Fig. 5 Representative data from an experiment to measure DNA–protein association. Different concentrations of protein (33–200 nM) were mixed with fluorescently-labeled DNA (37 nM). The data shown for three different concentrations represent the averages of 12 shots each, and the residuals after fitting with a single exponential are shown as small dots centered on the *x*-axis. The replot of k_{obs} vs different concentrations of protein is shown in the bottom right panel, where the slope reflects the second order rate of association ($k_{on} = 1.6\,nM^{-1}\,s^{-1}$) and the intercept the apparent rate of dissociation ($k_{off} = 84\,s^{-1}$).

consider other models, always being wary, however, of overfitting by introduction of unnecessary variables. Textbooks such as Fersht's Structure and Mechanism in Protein Science (Fersht, 1998) can be very useful for guiding the thought process about the intricacies of fitting to more complex models. For the simplest mechanism of association (Fig. 2A), a replot of k_{obs} vs different concentrations of protein at fixed concentrations of labeled DNA should yield a straight line where the slope reflects the second-order rate constant for DNA–protein association (k_{on}) and the intercept reflects the rate of dissociation (k_{off}) (Figs. 2A and 5). Typically observed rates of DNA–protein association vary from 0.002 to $3\,nM^{-1}\,s^{-1}$ (Biro, Zhai, Doucette, & Hingorani, 2010; Johnson & Chester, 1998; Lukina, Koval, Lomzov, Zharkov, & Fedorova, 2017; Rudolph et al., 2018). Often k_{off} will be poorly defined by this experiment, as the rate of association will require short reaction times where dissociation rarely occurs. A poorly determined k_{off} is one reason why it is important to actually perform a separate dissociation experiment.

b. *Dissociation.* Assuming the simplest mechanism (Fig. 2B) and under ideal reaction conditions ([competitor DNA] > [labeled DNA–protein complex], and the rate of association of unlabeled DNA under the reaction conditions is much greater than k_{off}), the averaged raw data can be fitted using a simple exponential (Eq. (4, Fig. 6), with k_{obs}, A, and b defined as for Eq. ((3).

$$[P] = A \times \left(e^{-k_{obs}t}\right) + b \tag{4}$$

Again, inspection of residuals from the fitting is a powerful tool for evaluating the goodness of fit. For this simple mechanism (Fig. 2B), a replot of k_{obs} vs different concentrations of competitor DNA should yield a horizontal

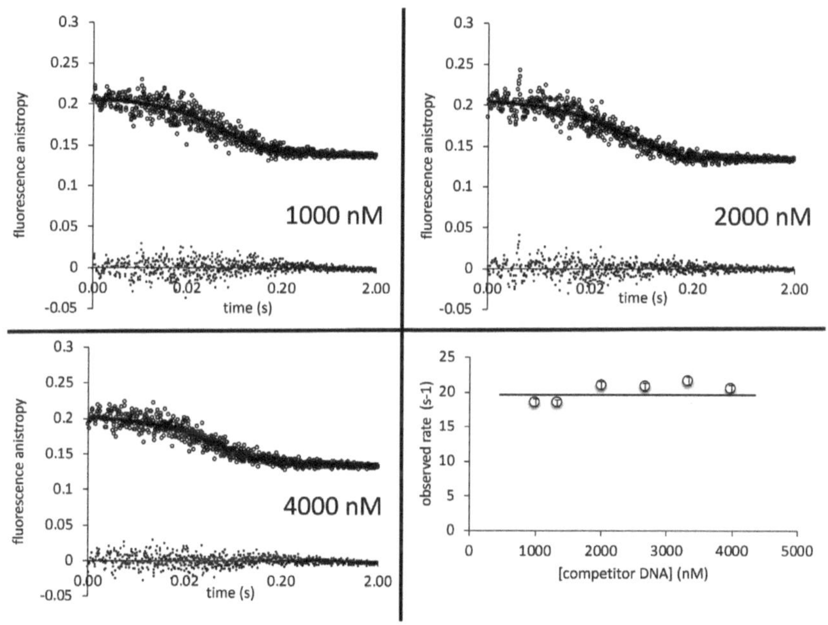

Fig. 6 Representative data from an experiment to measure DNA–protein dissociation. Different concentrations of competitor DNA (1000–4000 nM) were mixed with a complex of fluorescently-labeled DNA (25 nM) mixed with protein (37 nM). The data shown for three different concentrations represent the averages of 12 shots, and the residuals after fitting with a single exponential are shown as small dots centered on the *x*-axis. The replot of k_{obs} vs different concentrations of competitor DNA is shown in the bottom right panel, where the *y*-intercept reflects a more accurate measurement of the rate of dissociation ($k_{off} = 20\,s^{-1}$).

line (Fig. 6). That is, when the concentration of competitor DNA is significantly higher than the concentration of the starting complex, the rate of association of the competitor DNA will be sufficiently fast that the rate of dissociation (k_{off}) will be the rate-determining-step.

3. Global fitting. There are many cases where fitting experimental data using the above methods is inadequate and when global fitting of data may be more suitable. Global fitting allows you to fit multiple data sets (i.e. under different concentrations of protein or DNA) to a single unifying model. We provide two such examples here.

   a. Often it may difficult to perform experiments under ideal limiting conditions. For example, in an experiment to measure association of DNA and protein, if the rate of association is extremely fast ($>10^9$ M^{-1} s^{-1}), high concentrations of protein (i.e. 1 μM) vs fluorescent DNA (20 nM), will yield apparent rates of $>1000\,s^{-1}$, exceeding the dead-time of the stopped-flow instrument. Thus, data will necessarily be collected under conditions where the concentrations of protein are much lower, approaching the concentration of DNA. Under these conditions, and still assuming the simplest model for association (Fig. 2A), the replot of k_{obs} vs [protein] is no longer linear and thus one cannot adequately derive k_{on} or k_{off} from this analysis.

   b. Many reaction mechanisms do not conform to simple limiting cases. For example, in the case of PARP1, the apparent rate of dissociation is not constant as a function of competitor concentration (Rudolph et al., 2018). In other words, the dissociation of DNA is promoted by binding of an additional DNA prior to dissociation of the originally-bound DNA. This mechanism of dissociation is known as intersegment transfer (Doucleff & Clore, 2008; Fried & Crothers, 1984; Iwahara, Zweckstetter, & Clore, 2006; Lieberman & Nordeen, 1997; Ruusala & Crothers, 1992; Zandarashvili et al., 2015), or a monkey-bar mechanism, and helps explain how PARP1 can rapidly scan intact DNA in its search for sites of DNA damage.

   There are several powerful and relatively easy to use software packages that allow for fitting of more complex reaction schemes. This type of software also allows for fitting of data from multiple different types of experiments simultaneously, thus allowing for a robust global analysis. These software packages include the no-cost options of Kinetiscope (http://hinsberg.net/kinetiscope/), and ReKinSim (Gharasoo, Thullner, & Elsner, 2017), as well as the commercially available KinTek Explorer (https://kintekcorp.com/software/). Global fitting allows you to build complex kinetic models by simply writing down the network of reactions you wish to model, and then evaluate their suitability for fitting to your kinetic data. All of these software packages are accompanied by helpful manuals and extensive case studies that will not be replicated here. We provide just a few tips related specifically to the collection of stopped-flow data for DNA–protein association and dissociation.

a. The input for global fitting must be in the form of concentrations of reactants and/or products vs time. As your data has units of fluorescence or anisotropy, you must apply a conversion. Because the earliest time points in each data series are the noisiest (see Fig. 4), it is best to use the average value from the buffer control to determine the initial value of fluorescence or anisotropy. The ending value for the fluorescence or anisotropy can be accurately determined from the late time points in the averaged data. If you have performed a concentration series (varying either protein during association or competing DNA for dissociation) these end values should be the same for all concentrations tested, assuming you met the ideal conditions (concentrations of protein greater than fluorescent DNA or concentrations of competing ligand greater than 10-fold than complex of protein with fluorescent DNA.) Since you can accurately determine the concentration of fluorescent DNA (in a spectrophotometer), the difference between the end value and the initial value in fluorescence or anisotropy (delta F) can then be used to normalize all of your data to absolute concentrations. Even if you did not achieve the same delta F for all tested concentrations, you should have at least some data at sufficiently high concentrations to bind (or release) all of your fluorescent signal, and then use that delta F to normalize all of the samples. Remember, as before, that all solutions were diluted twofold in the stopped-flow instrument.

b. Because it is very easy to construct and test multiple models using global fitting, you should do exactly that. Come up with different kinetic schemes and see how well they fit, noting which variables are important for fitting and which appear to act as silent bystanders. Always start with the simplest model you can imagine and avoid over parameterization (using more variables than needed). As above, critical evaluation of the residuals is a quick and reliable method to evaluate fits, although you should remember that overfitting will generate flat residuals but meaningless rate constants. All global fitting software includes more sophisticated tools to aid in evaluating goodness-of-fit and variable-interdependencies.

9. Conclusions

It can be very exciting and informative to determine the rates and mechanisms for the association and dissociation of DNA–protein complexes. Applying these techniques to your favorite DNA-binding protein can reveal novel mechanistic insights, which can then be further probed and evaluated using mutants of the protein and/or altered DNA sequences. The insights derived from these types of experiments can often stand alone but are even more powerful when they provide deeper meaning to independent structural or cellular studies.

Acknowledgments

We would like to thank Dr. Annette Erbse, Director of the Biochemistry Core Facilities at the University of Colorado, Boulder, for help with training, instrument maintenance, and a critical reading of these methods. This work was supported by NIH-NCI R01 CA218255.

References

Bai, P. (2015). Biology of poly(ADP-ribose) polymerases: The factotums of cell maintenance. *Molecular Cell*, *58*(6), 947–958. https://doi.org/10.1016/j.molcel.2015.01.034.

Beck, C., Robert, I., Reina-San-Martin, B., Schreiber, V., & Dantzer, F. (2014). Poly (ADP-ribose) polymerases in double-strand break repair: Focus on PARP1, PARP2 and PARP3. *Experimental Cell Research*, *329*(1), 18–25. https://doi.org/10.1016/j.yexcr.2014.07.003.

Berg, O. G., Winter, R. B., & von Hippel, P. H. (1981). Diffusion-driven mechanisms of protein translocation on nucleic acids. 1. Models and theory. *Biochemistry*, *20*(24), 6929–6948. https://doi.org/10.1021/bi00527a028.

Biro, F. N., Zhai, J., Doucette, C. W., & Hingorani, M. M. (2010). Application of stopped-flow kinetics methods to investigate the mechanism of action of a DNA repair protein. *Journal of Visualized Experiments*, *37*, 1–9. https://doi.org/10.3791/1874.

Bock, F. J., & Chang, P. (2016). New directions in PARP biology. *FEBS Journal*, *283*, 4017–4031. https://doi.org/10.1111/febs.13737.

Crawford, K., Bonfiglio, J. J., Mikoč, A., Matic, I., & Ahel, I. (2018). Specificity of reversible ADP-ribosylation and regulation of cellular processes. *Critical Reviews in Biochemistry and Molecular Biology*, *53*(1), 64–82. https://doi.org/10.1080/10409238.2017.1394265.

de Murcia, J. M., Niedergang, C., Trucco, C., Ricoul, M., Dutrillaux, B., Mark, M., et al. (1997). Requirement of poly(ADP-ribose) polymerase in recovery from DNA damage in mice and in cells. *Proceedings of the National Academy of Sciences of the United States of America*, *94*(14), 7303–7307. https://doi.org/10.1073/pnas.94.14.7303.

Doucleff, M., & Clore, G. M. (2008). Global jumping and domain-specific intersegment transfer between DNA cognate sites of the multidomain transcription factor Oct-1. *Proceedings of the National Academy of Sciences*, *105*(37), 13871–13876. https://doi.org/10.1073/pnas.0805050105.

Dunphy, G., Flannery, S. M., Almine, J. F., Connolly, D. J., Paulus, C., Jønsson, K. L., et al. (2018). Non-canonical activation of the DNA sensing adaptor STING by ATM and IFI16 mediates NF-κB signaling after nuclear DNA damage. *Molecular Cell*, *71*(5), 745–760.e5. https://doi.org/10.1016/j.molcel.2018.07.034.

Ferreiro, D. U., & De Prat-Gay, G. (2003). A protein-DNA binding mechanism proceeds through multi-state or two-state parallel pathways. *Journal of Molecular Biology*, *331*(1), 89–99. https://doi.org/10.1016/S0022-2836(03)00720-4.

Fersht, A. (1998). *Structure and mechanism in protein science: A guide to enzyme catalysis and protein folding*. W.H. Freeman.

Franzese, E., Centonze, S., Diana, A., Carlino, F., Guerrera, L. P., Di Napoli, M., et al. (2019). PARP inhibitors in ovarian cancer. *Cancer Treatment Reviews*, *73*(October 2018), 1–9. https://doi.org/S0305737218302007.

Fried, M. G., & Crothers, D. M. (1984). Kinetics and mechanism in the reaction of gene regulatory proteins with DNA. *Journal of Molecular Biology*, *172*(3), 263–282. Retrieved from http://www.ncbi.nlm.nih.gov/pubmed/6319716.

Gharasoo, M., Thullner, M., & Elsner, M. (2017). Introduction of a new platform for parameter estimation of kinetically complex environmental systems. *Environmental Modelling and Software*, *98*, 12–20. https://doi.org/10.1016/j.envsoft.2017.09.005.

Gibson, B. A., & Kraus, W. L. (2012). New insights into the molecular and cellular functions of poly(ADP-ribose) and PARPs. *Nature Reviews Molecular Cell Biology*, *13*(7), 411–424. https://doi.org/10.1038/nrm3376.

Halford, S. E., & Marko, J. F. (2004). How do site-specific DNA-binding proteins find their targets? *Nucleic Acids Research*, *32*(10), 3040–3052. https://doi.org/10.1093/nar/gkh624.

Iwahara, J., Zweckstetter, M., & Clore, G. M. (2006). NMR structural and kinetic characterization of a homeodomain diffusing and hopping on nonspecific DNA. *Proceedings of the National Academy of Sciences of the United States of America*, *103*(41), 15062–15067. https://doi.org/10.1073/pnas.0605868103.

Jameson, D. M., & Ross, J. A. (2010). Fluorescence polarisation/anisotropy in diagnostics and imaging. *Chemical Reviews*, *110*(5), 2685–2708. https://doi.org/10.1021/cr900267p. Fluorescence.

Johnson, R. S., & Chester, R. E. (1998). Stopped-flow kinetic analysis of the interaction of *Escherichia coli* RNA polymerase with the bacteriophage T7 A1 promoter. *Journal of Molecular Biology*, *283*, 353–370.

Lieberman, B. A., & Nordeen, S. K. (1997). DNA intersegment transfer, how steroid receptors search for a target site. *Journal of Biological Chemistry*, *272*(2), 1061–1068. https://doi.org/10.1074/jbc.272.2.1061.

Liu, H., Zhang, H., Wu, X., Ma, D., Wu, J., Wang, L., et al. (2018). Nuclear cGAS suppresses DNA repair and promotes tumorigenesis. *Nature*, *563*, 131–136. https://doi.org/10.1038/s41586-018-0629-6.

Lukina, M. V., Koval, V. V., Lomzov, A. A., Zharkov, D. O., & Fedorova, O. S. (2017). Global DNA dynamics of 8-oxoguanine repair by human OGG1 revealed by stopped-flow kinetics and molecular dynamics simulation. *Molecular BioSystems*, *13*, 1954–1966. https://doi.org/10.1039/C7MB00343A.

Morales, J., Li, L., Fattah, F. J., Dong, Y., Bey, E. A., Patel, M., et al. (2014). Review of poly (ADP-ribose) polymerase (PARP) mechanisms of action and rationale for targeting in cancer and other diseases. *Critical Reviews in Eukaryotic Gene Expression*, *24*(1), 15–28. https://doi.org/10.1615/CritRevEukaryotGeneExpr.2013006875.

Rack, J. G. M., Perina, D., & Ahel, I. (2016). Macrodomains: Structure, function, evolution, and catalytic activities. *Annual Review of Biochemistry*, *85*(1), 431–454. https://doi.org/10.1146/annurev-biochem-060815-014935.

Rudolph, J., Mahadevan, J., Dyer, P. N., & Luger, K. (2018). Poly(ADP-ribose) polymerase 1 searches DNA via a "monkey Bar" mechanism. *eLife*, 7 e37818.

Ruusala, T., & Crothers, D. M. (1992). Sliding and intermolecular transfer of the lac repressor: Kinetic perturbation of a reaction intermediate by a distant DNA sequence. *Proceedings of the National Academy of Sciences of the United States of America*, *89*(11), 4903–4907. https://doi.org/10.1073/pnas.89.11.4903.

Thomas, C., & Tulin, A. V. (2013). Poly-ADP-ribose polymerase: Machinery for nuclear processes. *Molecular Aspects of Medicine*, *34*(6), 1–24. https://doi.org/10.1016/j.mam.2013.04.001.Poly-ADP-Ribose.

Zandarashvili, L., Esadze, A., Vuzman, D., Kemme, C. A., Levy, Y., & Iwahara, J. (2015). Balancing between affinity and speed in target DNA search by zinc-finger proteins via modulation of dynamic conformational ensemble. *Proceedings of the National Academy of Sciences*, *112*(37), E5142–E5149. https://doi.org/10.1073/pnas.1507726112.

CHAPTER TEN

Characterization of DNA bound cyclic GMP-AMP synthase using atomic force microscopy imaging

Alexander Lushnikov[a], Richard Hooy[b], Jungsan Sohn[b], Alexey Krasnoslobodtsev[a,c,*]

[a]Nanoimaging Core Facility at the University of Nebraska Medical Center, Omaha, NE, United States
[b]Department of Biophysics and Biophysical Chemistry, John Hopkins School of Medicine, Baltimore, MD, United States
[c]Department of Physics, University of Nebraska Omaha, Omaha, NE, United States
*Corresponding author: e-mail address: akrasnos@unomaha.edu

Contents

Abstract

The protocol described herein allows for acquiring topography images of DNA-protein complexes using Atomic Force Microscopy imaging. Since the very beginning of this method, AFM has been an indispensable tool for characterization of biomolecular complexes with exceptional capability of observing single complexes. This method can visualize structural characteristics of DNA-protein assemblies and evaluate differences between individual complexes. Although this protocol is generally applicable to a large number of various proteins complexed with DNA, we use cyclic G/AMP synthase (cGAS) enzyme as a case study for the protocol description.

Methods in Enzymology, Volume 625
ISSN 0076-6879
https://doi.org/10.1016/bs.mie.2019.07.031

1. Introduction

A large group of enzymes interacts directly with DNA substrates to perform their function. Examples of such enzymes are DNA polymerases (Garcia-Diaz & Bebenek, 2007), DNA repair enzymes (Evans & Nichols, 2008), restriction enzymes (Loenen, Dryden, Raleigh, & Wilson, 2014), DNA replicases (Johnson & O'Donnell, 2005), DNA ligase (Doherty & Suh, 2000) to mention a few. Cyclic GMP-AMP synthase (cGAS) enzyme initiates an immune response against cytosolic double-stranded DNA (dsDNA) (Hooy & Sohn, 2018; Yang, Wang, Ren, Chen, & Chen, 2017). The presence of dsDNA in cytosol is a sign of cellular damage or infection. cGAS finds and binds to dsDNA in the cytosol (Yang et al., 2017). Upon binding to DNA, cGAS changes its state and stimulates the formation of cyclic-GMP-AMP (cGAMP) which activates the stimulator of interferon genes (STING) initiating immune response (Yang et al., 2017).

While it is clear that cGAS operates in a coordinated fashion, initial stages of activation remain poorly defined. Specifically, it is not clear how cGAS's activity varies based on dsDNA length. The length of dsDNA molecules in the cytosol can vary dramatically. It was initially proposed that any length of dsDNA could activate cGAS to a similar degree, but later studies showed that the enzyme's activity is length dependent (Luecke et al., 2017). It still remains unclear how the dsDNA length affects the activity of cGAS. It appears that dsDNA length-dependent binding is coupled with dimerization state of cGAS (Andreeva et al., 2017; Hooy & Sohn, 2018; Luecke et al., 2017). However, the mechanisms by which cGAS dimerizes in a dsDNA length-dependent manner remain poorly explained.

To characterize cGAS-DNA complexes we have used Atomic Force Microscopy (AFM) imaging. A few methods can be used to structurally characterize DNA-enzyme interactions such as electron microscopy or X-ray crystallography. However, only AFM can provide structural information of individual complexes at a single molecule level without any labeling. Atomic Force Microscopy is a powerful tool for resolving aspects of molecular topology at specific locations. Structural peculiarities of DNA-enzyme complexes such as binding location, stoichiometry, and the efficiency of complex formation can be obtained from topographical AFM images. The morphological analysis performed on the images provides specific structural and functional aspects of the complexes, such as the multimerization state of proteins bound to DNA, or DNA conformational changes induced by the DNA-binding proteins. Herein, a method for analyzing a complex formed by a cGAS protein with DNA substrate using AFM imaging in air is presented. We describe technical aspects of sample preparation and imaging technique for a user with beginner AFM skills to get good quality images of DNA-enzyme complexes. Due to high sensitivity of AFM in z-dimension, very accurate height information can be obtained from topography images. Post-acquisition image processing can help visualize DNA-protein complexes in 3D, resolving, thus, fine structural features.

2. Equipment

1. MultiMode AFM Nanoscope IV system (Bruker Instruments, Santa Barbara, CA).
2. Sharp scissors for cutting mica sheets.
3. Tweezers for handling AFM probes.
4. Vacuum oven for storing APS-modified mica and samples deposited on APS-modified mica.
5. A vacuum pump connected to the vacuum oven.
6. Ultra high purity (99.999%) compressed argon cylinder.
7. Scotch tape for mica substrate preparation.
8. Double-sided adhesive tape for mounting samples on metal pucks.
9. 10 mm metal pucks for mounting mica substrates onto an AFM stage.
10. Disposable plastic cuvettes for mica modification with APS and sample storage.

3. Materials

1. All dsDNA substrates, oligonucleotides (including PCR primers) shorter than 100 bps were purchased from Integrated DNA Technologies, Inc. (Coralville, IA, USA) as described in (Hooy & Sohn, 2018).
2. dsDNA substrates longer than >100 bps were generated by PCR (Hooy & Sohn, 2018).
3. Muscovite mica sheets in "block" form were purchased from "Asheville Schoonmaker Mica", Newport News, VA
4. cGAS protein was prepared as described in (Hooy & Sohn, 2018).
5. AFM probes—RTESPA-300—were purchased from Bruker AFM Probes, Inc.,
6. All other reagents including buffer components were purchased from Sigma-Aldrich, Inc.
7. Buffers were prepared using ultrapure water (Milli-Q, 18 MΩ-cm).

4. Protocol

4.1 Protein preparation

1. Clone recombinant cGAS constructs into the pET28b vector (Novagen) with an N-terminal MBP-tag and a TEV protease cleavage site.
2. Express proteins using 200 mM IPTG at 16 °C for overnight in *E. coli* BL21 Rosetta 2.
3. Purify recombinant cGAS constructs using amylose affinity chromatography, cation-exchange, and size exclusion chromatography.
4. Freeze and store tag-free, purified cGAS proteins in −80 °C with a buffer containing 20 mM Tris HCl at pH 7.5, 300 mM NaCl, 10% glycerol, 5 mM DTT.
5. Prepare Tris reaction buffer as follows: 20 mM Tris-HCl with pH 7.5, 50 mM NaCl, 2 mM DTT, 1 mM EDTA.

4.2 Mica surface modification with 1-(3-aminopropyl)silatrane

There are several approaches for modification of mica to obtain overall positively charged surface such as mostly used Mg^{2+}, Ni^{2+} or other cations, and 3-aminopropyl-trietoxy silane (APTES) treatments (Bezanilla, Manne, Laney, Lyubchenko, & Hansma, 1995; Lyubchenko, Gall, & Shlyakhtenko, 2014; Piétrement et al., 2003; Vesenka et al., 1992). We use modification of mica substrates with 1-(3-aminopropyl)silatrane (APS) as the procedure is rather simple, reproducible and provides a functionalized substrate for sample deposition in a wide range of ionic conditions, temperatures, and pH values. APS allows to create positively charged surface for deposition of DNA and DNA-protein complexes via electrostatic attraction between negatively charged DNA backbone and positively charged APS-mica. This modification is very robust and reliable and has been extensively used in AFM studies of DNA and DNA-protein complexes. APS is an amorphous solid at normal conditions and is highly soluble in water. 1-(3-aminopropyl)silatrane is not commercially available and requires custom synthesis (Lyubchenko, Shlyakhtenko, & Gall, 2009; Shlyakhtenko, Gall, & Lyubchenko, 2013) (see Note).

1. Dissolve APS in DI water to prepare a stock solution of 50 mM.
2. Dilute stock solution immediately prior to surface modification 200–300 times. We routinely use solution diluted to the final concentration of about $C_{APS} = 167\,\mu M$. Use freshly diluted APS solution each time APS modification is needed.

3. Cut mica substrates into pieces of 1 cm by 3 cm size using sharp scissors.
4. Cleave mica substrates using scotch tape by peeling off layers on both sides. Usually this procedure is repeated several times until required quality of mica top layer is achieved. The quality of peeling is monitored by the appearance of mica replica on the scotch tape. Repeat cleaving procedure if any defects are visible and the surface is not glossy.
5. Place freshly cleaved mica substrates into a plastic cuvette filled with diluted APS solution and leave for 30 min to achieve optimal modification.
6. Rinse mica substrates with DI water a few times and dry using ultra high purity argon.
7. Store APS-modified mica in vacuum overnight before sample deposition. Functionalized mica can be stored for several days under vacuum before use.

4.3 Preparation of DNA-cGAS complexes

1. Mix appropriate amounts of cGAS protein and desired length DNA to prepare DNA-cGAS complexes. Ratios of DNA to protein are calculated using 18 base pairs as the size of binding site per 1 protein. Example used in this protocol for demonstration purposes is 1:5 ratio of protein to DNA (Fig. 2).
2. Incubate the mixtures for 10 min at room temperature in Tris reaction buffer. Proceed with deposition of complexes on APS-modified mica.

4.4 Deposition of DNA-cGAS protein complexes on APS-modified mica surface

1. Cut a small piece of APS-modified mica for sample deposition. Typical size necessary for AFM imaging is about 3 mm by 5 mm (Fig. 1).
2. Place mica piece on a hard surface.

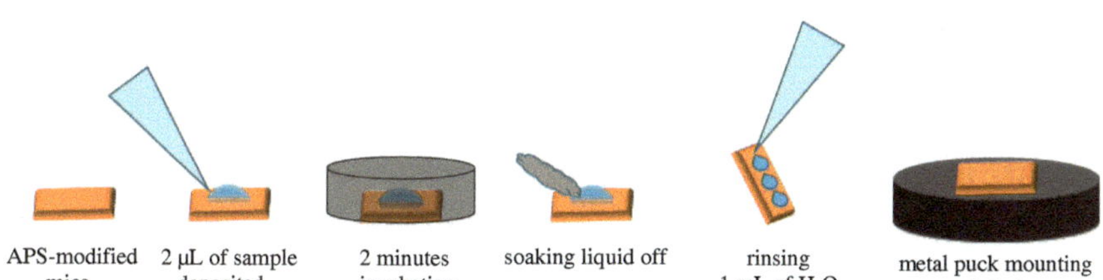

Fig. 1 Schematic representation of sample deposition procedure.

3. Deposit 2–5 μL of the DNA-protein complex solution in a drop-wise manner for a total of 2 min in a chamber with constant humidity.
4. Wash excess sample with several drops of DI water.
5. Gently dry the sample under a flow of ultra-high purity argon gas.
6. Store the deposited samples under vacuum in the vacuum oven overnight to remove excess liquid for imaging in air.

5. Atomic force microscopy imaging

5.1 Brief overview

This chapter is not intended to describe the details of AFM imaging and we will refer interested readers to the following reviews recently published on this topic (Lyubchenko et al., 2009, 2014). We will rather concentrate on the practical aspects of imaging that are likely to be of interest to non-specialists who wish to visualize interactions of enzymes with DNA using AFM topography imaging in air. Certain details of imaging in air, however, need to be mentioned. The imaging mode used in this protocol is TappingMode™—a Bruker-patented technique which allows for gentle imaging of biomolecular complexes. The acquisition of topography maps is achieved by tapping the sample's surface with an AFM probe while oscillating it at the resonant frequency. The interactions with underlying surface greatly affect the oscillation of the probe. By monitoring the probe's amplitude of oscillation a high-resolution 3D image of the sample's surface topography is produced. Sample preparation procedures described herein result in biomolecular complexes being dry and fixed on the substrate. Unlike larger complexes, the size of complex constituents, cGAS protein and DNA, are small (<10 nm) which contributes to complexes being stably fixed so that tapping does not distort and does not drag these complexes along the scanning lines. APS mica treatment is another significant requirement for stable fixation of biomolecular complexes without conformational distortion of DNA and proteins (Lyubchenko et al., 2009, 2014; Shlyakhtenko et al., 2013).

5.2 Atomic force microscopy imaging

1. Place the AFM probe in the probe holder. Probes, which have been found to work reliably for imaging of DNA-protein complexes in air, are RTESPA-300 from "Bruker AFM Probes". These probes have

a nominal spring constant of 40 N/m and a resonance frequency at ~300 kHz.

2. Mount the sample mica substrate onto a metal puck using double-sided adhesive tape.
3. Mount the metal puck onto the scanning stage of the AFM microscope. Scanning stage of Bruker's Nanoscope family of AFM is magnetic—designed to hold the metal pucks well during scanning.
4. Position and align the laser onto the AFM cantilever using laser adjustment knobs to maximize the sum.
5. Test if the laser is optimally reflected onto the quadrant detector by turning mirror's lever—adjust to maximize the sum.
6. Adjust the horizontal deflection to as close to 0 V settings as possible by using adjustment knobs.
7. Open imaging software and run the "Autotune" feature. This feature will find the resonance peak of the cantilever and adjust the settings to the resonance frequency of the probe. Make sure that the target amplitude is at ~0.5 V and the drive amplitude is no larger than ~10 mV.
8. Adjust the scan size to 100 × 100 nm and start approach of the probe to the surface by pressing the "Engage" button. Scanning will start automatically once the probe is engaged.
9. Change the scan size to 3 × 3 μm, the pixel size to 128, and the scan rate to 1.5 Hz. These settings will allow for fast surveying of the sample's area.
10. Reduce the set point until the surface is tracked well. Complexes will appear as bright spots.
11. Move to a different area if needed.
12. Proceed to the next step when optimal number of complexes is found within the size of the image.
13. Increase the number of pixels to 512 and decrease the scan rate to 1 Hz to acquire a topography image with high resolution.
14. Adjust the drive amplitude to a value that produces desired high quality appearance of complexes.
15. Repeat imaging for areas of interest within the 3 × 3 μm area by decreasing the scan size to 1 × 1 μm and 0.5 × 0.5 μm. Fig. 2, A shows a representative image obtained for cGAS-DNA complexes on a 0.5 × 0.5 μm area with 512 pixels.
16. If a large number of complexes is required for statistical analysis, repeat steps 8–15 several times.

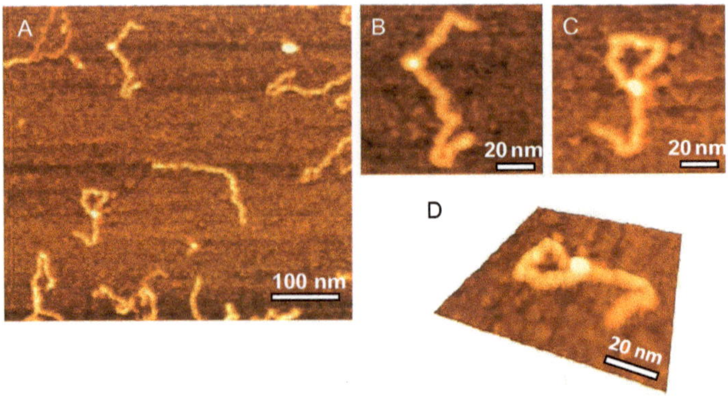

Fig. 2 AFM topography images. (A) 500 × 500 nm image of cGAS-DNA complexes. *Z*-scale is 1.5 nm, (B) and (C) 100 × 100 nm images of individual complexes, (D) 3D profile of the complex shown in C obtained with "Femtoscan Online" software package.

5.3 AFM image analysis

AFM topography images can be processed using a variety of different software suites including *WSxM*, *Gwyddion*, *ImageSXM* and *ImageJ*. We used the *FemtoScan Online* software package from Advanced Technologies Center, Moscow, Russia. This software has a comprehensive set of tools for analysis of sizes and stoichiometry of biomolecular complexes and it served well for analysis of cGAS-DNA complexes.

1. Open a file saved in Bruker imaging software. FemtoScan Online can read raw data file without any additional conversions.
2. Adjust image scan line to the same mean level for the whole image.
3. In the *Z*-scale settings, choose "brown" color palette and check "fixed scale" parameter.
4. Adjust the Z-scale so that mica substrate appears light brown and complexes are bright. Typically −0.75 nm to +0.75 nm scale works well with total Z-scale being 1.5 nm. Scale can be adjusted depending on the size of the complexes to achieve desired visual contrast between protein, DNA, and mica substrate.
5. For analysis of individual complexes select rectangular area around a complex and click "duplicate image." A separate window with the complex of interest will appear. Examples of 100 × 100 nm areas with individual complexes are shown in Fig. 2B and C.
6. Build a cross-section line using a "section" tool.
7. Measure the height of the complex as dY from the average background to the maximum of the line using two markers.

8. Measure the full width at half maximum (FWHM) as dX using two markers positioned at half maximum of the line.
9. Build a 3D figure of a complex by creating a 3D view in "Operations" option. Fig. 2D shows a representative 3D figure of one of the cGAS-DNA complexes.

6. Summary

Investigators who seek to begin using Atomic Force Microscopy imaging to visualize how proteins recognize specific DNA sequences and bind to DNA substrates may encounter a number of specific technical challenges. This chapter introduces specific details of AFM imaging and describes systematically the protocol and methods for obtaining good quality AFM topography images of biomolecular complexes using cGAS-DNA complexes as an example. Special attention should be paid to sample preparation procedures, the use of appropriate AFM probes, imaging mode, and imaging settings. These will contribute largely to reproducibility and quality of imaging. Finally, we note that the AFM approach described here is also broadly applicable to studying other nucleic-acid sensors such as Absent-In-Melanoma-2 (AIM2)-Like-Receptors and Retinoic-acid-Inducible-Gene-I (Rig-I)-Like-Receptors.

Note: Aminopropyl-silatrane has been synthesized by A. A. Gall (Cepheid, CA, USA) as previously described in (Lyubchenko et al., 2014) and kindly provided to us by Y. L. Lyubchenko (University of Nebraska Medical Center, NE, USA).

Acknowledgments

The work was partially supported by the following grants: NIH R01GM129342. AFM imaging of the DNA and DNA-protein structures was performed with the use of instruments of the Nanoimaging core facility at the University of Nebraska Medical Center. The Facility is partly supported with funds from the Nebraska Research Initiative (NRI).

References

Andreeva, L., Hiller, B., Kostrewa, D., Lassig, C., de Oliveira Mann, C. C., Jan Drexler, D., et al. (2017). cGAS senses long and HMGB/TFAM-bound U-turn DNA by forming protein-DNA ladders. *Nature, 549*(7672), 394–398.

Bezanilla, M., Manne, S., Laney, D. E., Lyubchenko, Y. L., & Hansma, H. G. (1995). Adsorption of DNA to mica, silylated mica, and minerals: Characterization by atomic force microscopy. *Langmuir, 11*(2), 655–659.

Doherty, A. J., & Suh, S. W. (2000). Structural and mechanistic conservation in DNA ligases. *Nucleic Acids Research*, *28*(21), 4051–4058.

Evans, T. C., & Nichols, N. M. (2008). DNA repair enzymes. *Current Protocols in Molecular Biology*, *84*(1), 3.9.1–3.9.12.

Garcia-Diaz, M., & Bebenek, K. (2007). Multiple functions of DNA polymerases. *Critical Reviews in Plant Sciences*, *26*(2), 105–122.

Hooy, R. M., & Sohn, J. (2018). The allosteric activation of cGAS underpins its dynamic signaling landscape. *eLife*, *7*, e39984.

Johnson, A., & O'Donnell, M. (2005). Cellular DNA replicases: Components and dynamics at the replication fork. *Annual Review of Biochemistry*, *74*(1), 283–315.

Loenen, W. A. M., Dryden, D. T. F., Raleigh, E. A., & Wilson, G. G. (2014). Type I restriction enzymes and their relatives. *Nucleic Acids Research*, *42*(1), 20–44.

Luecke, S., Holleufer, A., Christensen, M. H., Jønsson, K. L., Boni, G. A., Sørensen, L. K., et al. (2017). cGAS is activated by DNA in a length-dependent manner. *EMBO Reports*, *18*(10), 1707–1715.

Lyubchenko, Y. L., Gall, A. A., & Shlyakhtenko, L. S. (2014). Visualization of DNA and protein-DNA complexes with atomic force microscopy. *Methods in Molecular Biology*, *1117*, 367–384.

Lyubchenko, Y. L., Shlyakhtenko, L. S., & Gall, A. A. (2009). Atomic force microscopy imaging and probing of DNA, proteins, and protein DNA complexes: Silatrane surface chemistry. *Methods in Molecular Biology*, *543*, 337–351.

Piétrement, O., Pastré, D., Fusil, S., Jeusset, J., David, M.-O., Landousy, F., et al. (2003). Reversible binding of DNA on NiCl2-treated mica by varying the ionic strength. *Langmuir*, *19*(7), 2536–2539.

Shlyakhtenko, L. S., Gall, A. A., & Lyubchenko, Y. L. (2013). Mica functionalization for imaging of DNA and protein-DNA complexes with atomic force microscopy. *Methods in Molecular Biology*, *931*, 295–312.

Vesenka, J., Guthold, M., Tang, C. L., Keller, D., Delaine, E., & Bustamante, C. (1992). Substrate preparation for reliable imaging of DNA molecules with the scanning force microscope. *Ultramicroscopy*, *42–44*(Pt. B), 1243–1249.

Yang, H., Wang, H., Ren, J., Chen, Q., & Chen, Z. J. (2017). cGAS is essential for cellular senescence. *Proceedings of the National Academy of Sciences of the United States of America*, *114*(23), E4612–E4620.

CHAPTER ELEVEN

Preparation of filamentous proteins for electron microscopy visualization and reconstruction

Mariusz Matyszewski, Jungsan Sohn*
Department of Biophysics and Biophysical Chemistry, Johns Hopkins University School of Medicine, Baltimore, MD, United States
*Corresponding author: e-mail address: jsohn@jhmi.edu

Contents

Abstract

Cryo electron microscopy (cryo-EM) has become a mainstream tool for determining the structures of macromolecular complexes at the atomic resolution. It has many advantages over other techniques such as X-ray crystallography and nuclear magnetic resonance (NMR). However, it also entails several challenges, a major one being preparation of an ideal sample. Recent studies have identified that DNA sensors and inflammasomes often assemble into filamentous oligomers, which poses a unique set of challenges in preparing ideal samples for high-resolution reconstruction using cryo-EM. This chapter will discuss how to overcome several major issues in cryo-EM sample preparation including construct design, screening using negative stain (ns) EM, and tips on working with filamentous proteins.

Methods in Enzymology, Volume 625
ISSN 0076-6879
https://doi.org/10.1016/bs.mie.2019.06.007

1. Introduction

Thanks to the "resolution revolution," the use of cryo-EM in structural biology has skyrocketed (Kühlbrandt, 2014). Currently, there are almost 3000 high-resolution cryo-EM structures deposited in the PDB, 700 of which were deposited in 2018 alone (www.rcsb.org). While the majority of the structures were solved using the Single Particle Reconstruction (SPR) approach, there are many higher order protein complexes whose structures were determined via helical reconstructions. This is noteworthy as several DNA sensors and inflammasomes (at least their signaling ends) assemble into helical filaments (Lu et al., 2014; Lu et al., 2016; Li et al., 2018; Matyszewski et al., 2018; Morrone et al., 2015). The general principles of working with helical assemblies are similar to those of SPR projects (see elsewhere in this volume), but there are some key differences. By definition, helical samples have helical symmetry which can be used to improve final resolution, often requiring a smaller data set to achieve comparable results. Additionally, some SPR problems like preferential orientation do not apply to filaments as each filament contains all the views necessary for reconstruction (Egelman, 2007). Nevertheless, getting filamentous samples to the collection-ready stage can still be as challenging as in any SPR project. Here we share techniques and tips for preparing an ideal sample for cryo-EM reconstruction of filamentous proteins.

2. Construct design

The first step is to design a construct suitable for cryo-EM. In many instances of filamentous complexes found in inflammasomes and DNA sensors, only one domain of the protein is responsible for filamentation (Morrone et al., 2015; Shen et al., 2019; Tenthorey et al., 2017). Isolating such a domain can significantly simplify the challenge of getting a high-resolution structure of the minimal filament. On the other hand, working with the full-length protein can provide additional insight into the workings of the filament, but comes at a cost of a much more difficult system to work with (i.e., expression and purification). Moreover, the rest of the protein usually has regulatory roles or act as a scaffold for filament assembly, often resulting in disordered arrays surrounding the ordered filament core (Shen et al., 2019). Nevertheless, we found that even working with isolated polymerization domains poses many challenges.

2.1 Controlling oligomerization

One of the biggest problems working with filamentous protein is controlling their oligomeric state. While the final goal is to capture the protein as a filament, it is often advantageous to prepare it as a monomer and to later induce polymerization in a controlled manner. Monomeric constructs tend to be significantly more soluble than their filamented counterparts, allowing higher yields and better concentration control.

The most straightforward solution for the oligomerization problem is to use cleavable solubility tags. In many cases, large protein tags such as Maltose Binding Protein (MBP) can sterically interfere with filament assembly (Jin, Perry, Smith, Jiang, & Xiao, 2013), greatly reducing the intrinsic oligomerization activity. Combined with a protease cut site, such as Tabacco Etch Virus (TEV) protease, this not only prevents oligomerization, but also allows for temporal control over filamentation (Morrone et al., 2015). For the vast majority of inflammasome signaling components, the polymerization domain is located at the N-terminus (e.g., PYrin Domain, PYD or CAspase-Recruiting Domain, CARD). For best prospect of inhibiting polymerization, the tag should be placed at the N-terminus of either PYD or CARD with minimal linker (Fig. 1A; MBP-CARD/PYD construct).

2.2 Filament separation (bundling)

In many cases, despite controlling their polymerization using the methods described above, resulting filaments aggregate and "bundle up" (Fig. 1B). If simply lowering the concentration does not solve this issue, the next best option is to introduce another tag which would repulse other filaments, while still allowing polymerization. The identity of such repulsion tags can vary, but enhanced Green Fluorescence Protein (eGFP) has been used with great success (Li et al., 2018; Matyszewski et al., 2018). However, care should be taken not to affect the native form of the filament (Lu et al., 2015). To avoid this issue, the tag should ideally be placed as to mimic the rest of the native protein. If working with the full-length construct, the tag should be placed at the opposite terminus from the solubility tag (Fig. 1A; MBP-CARD/PYD-eGFP construct). The end result should be a double tagged protein, with filaments only forming after solubility tag is cleaved (Figs. 1B-b and c).

Note: the repulsion-tags should not be visible in the final reconstruction, as they should be flexibly attached to the filament core without affecting its

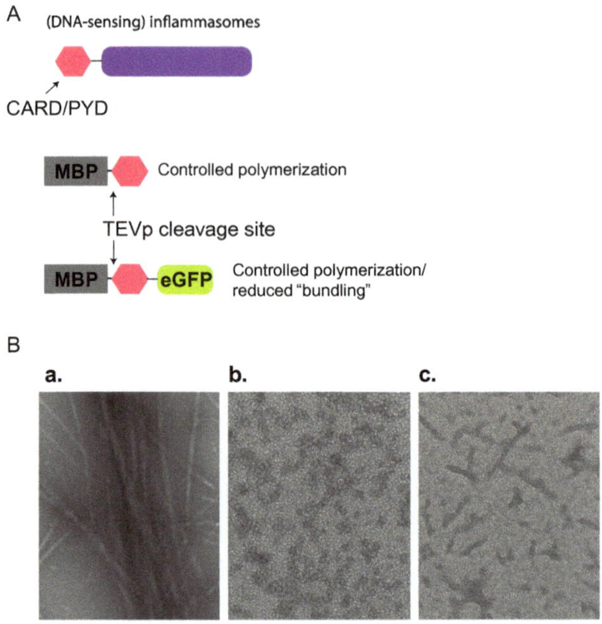

Fig. 1 Construct variants for filamentous protein. (A) Typical inflammasome receptors contain either PYD or CARD at the N-terminus, which is capable of auto-polymerization. Adding MPB with TEV protease cut-site to the N-terminus of CARD/PYD allows controlled polymerization. Adding eGFP at the C-terminus minimizes bundling. (B) nsEM micrographs of (a) auto-assembled $NLRC4^{CARD}$, (b) MBP-CARD of NRLC4 ($NLRC4^{CARD}$)-eGFP, and (c) $NLRC4^{CARD}$-eGFP after removing MBP via TEVp.

native architecture. Otherwise, the presence of the tag in the final reconstruction could suggest that the native symmetry has been tempered with, especially for truncated constructs (Lu et al., 2015). An even less intrusive way of preventing bundling is temporal control. Since bundling requires pre-formed filaments, decreasing the incubation time before freezing or staining samples might reduce aggregation. However, the efficacy of this approach will vary by samples and is usually best when combined with other approaches.

2.3 Buffer components

While most buffer components do not affect the quality of the sample in cryo-EM, there are a few things to avoid, or be cautious about, in the final sample buffer.

1. Salt: High salt concentrations have been shown to reduce contrast, and thus Signal-to-Noise Ratio (SNR) of cryo-EM micrographs. Typically, concentrations below 300 mM should be used (Drulyte et al., 2018).

2. Glycerol: Reduces the SNR and should be kept at or below 5% concentration (Drulyte et al., 2018) for cryo-EM data collection.
3. Detergent: While detergents above critical micelle concentration will add to the background, use of detergent can also be beneficial in changing particle distribution, especially when it comes to air-water interface interactions. Since it can affect how the sample freezes, only introduce it if the particle distribution problem occurs. It will often require screening many detergents before the best one is found.
4. Phosphate: Interferes with nsEM. Uranyl will react with phosphate to form crystals, making it impossible to screen with phosphate buffers present. It can be safely reintroduced if only doing cryo-EM.

3. Negative stain electron microscopy

Imaging the purified protein sample by Negative Stain Electron Microscopy (nsEM) is a critical step even if the final goal is a cryo-EM reconstruction. While nsEM is unlikely to provide anything more than a symmetry estimate, it can provide plenty of qualitative information on how the sample filaments behave. The most crucial piece of information is if the protein actually polymerizes into a filament vs. disordered aggregates (this cannot be distinguished by size-exclusion chromatography). This is especially critical when testing new constructs and buffer conditions, as nsEM is a much higher throughput method than cryo-EM. However, one has to keep in mind that conditions might not transfer to cryo-EM; freezing is much different than staining a sample.

Once filamentation has been confirmed, one could potentially try to reconstruct the nsEM filaments. Usually it is at least worth to check for symmetry by collecting a power spectrum. However, in some cases the helical rise might be too small to obtain a power spectrum by nsEM.

In most cases, fully optimizing a sample in nsEM is not necessary, as conditions (primarily protein concentrations and some buffer components) will vary between nsEM and cryo-EM, unless one wants to perform nsEM reconstruction first. It is worth stopping once a suitable range of conditions has been found where protein filaments form without excessive bundling.

3.1 Staining protocol

3.1.1 Equipment required

- Carbon coated copper grids (glow discharged)
- EM tweezers
- Parafilm

- Blotting paper
- Uranyl based stains (Either formate or acetate recommended) at ~0.75% (keep on ice and away from light)
- Aspirator (or fume hood if slowly drying the samples instead)

3.1.2 Protocol (*Fig.* 2)

1. Prepare (at least) 5–10 μL samples, giving them enough time to polymerize if using cleavable tags (e.g., ~30 min upon TEV addition 1:10 M ratio against target proteins).
2. Lay out a piece of parafilm on the benchtop. Approximately 3 × 5 in should have a plenty of room for 6–8 samples. Use a razor blade on the corners of the film to adhere it to the benchtop.

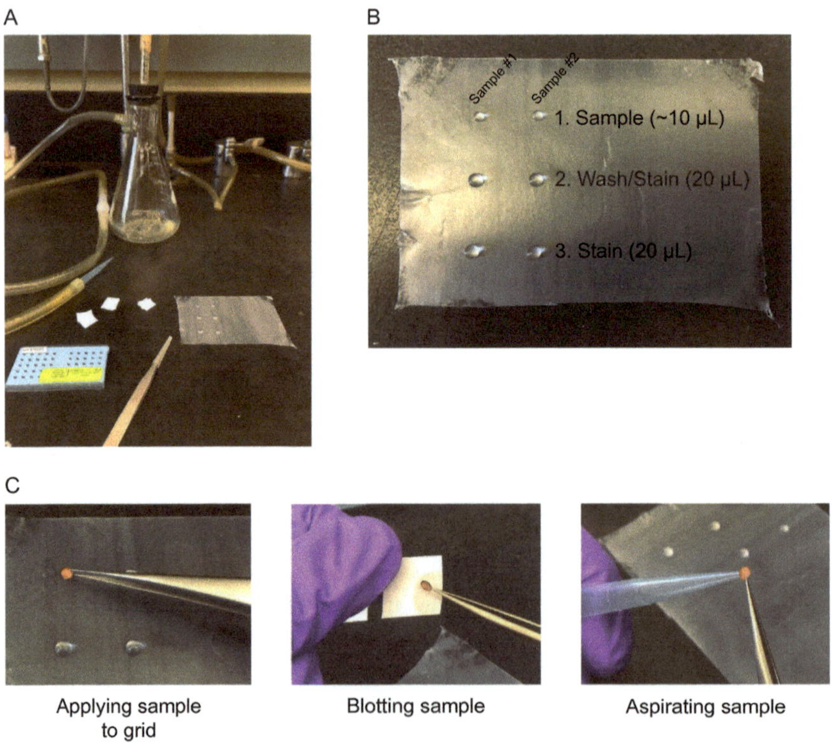

Fig. 2 Negative Staining Equipment and Setup. (A) Equipment setup. (B) Suggested layout of sample and stain droplets. (C) Sample handling examples. When blotting, always try to blot from the side of the grid first. If slow drying is preferred, blot most of the dye before moving the sample to the fumehood.

3. Place sample droplets on the parafilm. If excessive bubbles are forming, try using 1 μL less than the total volume.
4. Place grids carbon side down into the sample. Let them sit for approximately 2 min each.
5. While the samples are incubating, place two 20 μL drops of stain per sample.
6. After 2 min, blot the excess sample from the grid and transfer to the first drop. Let this grid float for about 30 s, or until all the other samples have been transferred. The grid might sink if the buffer contained detergent, but this is normal.
7. Move the grid into the second drop for another 30 s.
8. If using the aspirator, remove the stain from the sample and dry it out before transferring to a storage box.

   If using the drying method, blot away the excess stain and let the grid dry out in a fume hood. This might take 30 min to 1 h.

3.1.3 Variations

- Instead of using two drops of stain, the first one can be replaced by buffer as a wash step
- Aspiration produces lighter stain. If darker stain is preferred, use the fume hood to dry out the sample instead. This will also create a bigger stain gradient, and often leaves more protein on the grid.

4. Cryo-EM sample preparation

Finding the right conditions for freezing cryo-EM grids is a significant part of the project. While the basic idea is the same as nsEM, freezing in ice entails many more variables and challenges. The key challenges are ice thickness and sample distribution.

Since cryo-EM sample preparation has been covered many times, and in much more detail in other chapters in this volume. We will only focus on differences relevant to filamentous samples.

4.1 Grid type

Many labs choose C-Flat or Quantifoil grids as they are a very popular choice in SPR. Their advantages are still useful for filaments but are not critical. While lacey grids might reduce data collection speed due to their random nature, final reconstructions are rarely affected by grid type (e.g., 3.4 Å lacey Matyszewski et al., 2018 vs. 3.6 Å Quantifoil Li et al., 2018).

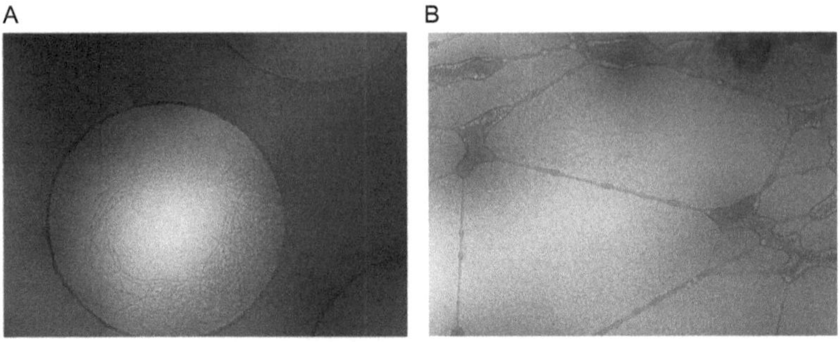

Fig. 3 C-Flat vs. Lacey Grids. (A) C-Flat grid with filaments preferentially adhering to carbon over ice-filled holes. (B) Lacey carbon grid has much less carbon to adhere to, thus forcing filaments into ice. Some filament bundling is present in both samples.

Additionally, C-Flat grids tend to have much more carbon which can lead to filaments adhering to the carbon rather than ice-filled holes (Fig. 3).

4.2 Sample concentration and distribution

Sample distribution between nsEM and cryo-EM will differ significantly. Often, higher concentration of sample is required to create an even distribution of particles in cryo-EM. This is mainly due to a larger amount of sample being blotted away, as well as a lack of carbon to adhere to. Filaments are big particles, and their presence can be checked for even before the sample ice thickness has been optimized simply by imaging at higher defocus levels (usually about $-5\,\mu m$). In many cases, changing concentration is fairly straightforward, but be aware that it does not necessarily have a linear correlation with sample distribution. Additionally, distribution of filaments over the grid might not be uniform. Fortunately, in cryo-EM the whole grid does not have to be perfect; a portion of a grid can provide more than enough data in most cases.

If increasing protein concentration causes filament bundling, it is possible to modify grid properties instead. One common approach is to change glow-discharging conditions, as it will affect how the sample adheres to grid (Passmore & Russo, 2016). A more drastic approach is to introduce a Thin Carbon Layer (TCL), which will cause the grid to behave more like a standard nsEM sample. However, it comes at a cost of lower signal-to-noise ratio. Either option will allow for working with lower concentration of protein by causing more filaments to adhere to the grid. Additionally, if repulsion tags were not already being used, it can be beneficial to introduce them now, as well as trying to freeze earlier.

4.3 Other parameters and tips

One thing to keep in mind is that buffer composition can still be changed. If the ice is not forming well, lowering glycerol or detergent could help. If filaments are bundling, changing the buffer can help to separate them again. When changing the buffer, it is usually worthwhile to go back to nsEM and test how different components affect filament formation. Be sure to keep the overall components within working limits for cryo-EM.

As for humidity and temperature settings of the blotting machine, in most cases setting humidity to 100% and temperature to 4 °C is most advantageous. Both will prevent sample evaporation after blotting. Additionally, drain time setting should ideally be kept to a minimum, if not eliminated, to reduce the exposure to the air-water interface. Since the blotting chamber will usually be at high humidity, blotting paper condition should be monitored. New paper will absorb differently than paper that has been in the machine for a few minutes. This can be a source of ice thickness variability, so make sure to keep a record of the blotting paper "age" when making grids.

References

Drulyte, I., et al. (2018). Approaches to altering particle distributions in cryo-electron microscopy sample preparation. *Acta Crystallographica. Section D, Structural Biology*, *74*(Pt 6), 560–571.

Egelman, E. H. (2007). The iterative helical real space reconstruction method: Surmounting the problems posed by real polymers. *Journal of Structural Biology*, *157*(1), 83–94.

Jin, T., Perry, A., Smith, P., Jiang, J., & Xiao, T. S. (2013). Structure of the absent in melanoma 2 (AIM2) pyrin domain provides insights into the mechanisms of AIM2 autoinhibition and inflammasome assembly. *The Journal of Biological Chemistry*, *288*(19), 13225–13235.

Kühlbrandt, W. (2014). The resolution revolution. *Science*, *343*(6178), 1443–1444.

Li, Y., et al. (2018). Cryo-EM structures of ASC and NLRC4 CARD filaments reveal a unified mechanism of nucleation and activation of caspase-1. *Proceedings of the National Academy of Sciences*, *115*(43), 10845–10852.

Lu, A., et al. (2014). Unified polymerization mechanism for the assembly of ASC-dependent inflammasomes. *Cell*, *156*(6), 1193–1206.

Lu, A., et al. (2015). Plasticity in PYD assembly revealed by cryo-EM structure of the PYD filament of AIM2. *Cell Discovery*, *1*, 15013.

Lu, A., et al. (2016). Molecular basis of caspase-1 polymerization and its inhibition by a new capping mechanism. *Nature Structural & Molecular Biology*, *23*(5), 416–425.

Matyszewski, M., Zheng, W., Lueck, J., Antiochos, B., Egelman, E. H., & Sohn, J. (2018). Cryo-EM structure of the NLRC4CARD filament provides insights into how symmetric and asymmetric supramolecular structures drive inflammasome assembly. *The Journal of Biological Chemistry*, *293*(52), 20240–20248.

Morrone, S. R., Matyszewski, M., Yu, X., Delannoy, M., Egelman, E. H., & Sohn, J. (2015). Assembly-driven activation of the AIM2 foreign-dsDNA sensor provides a polymerization template for downstream ASC. *Nature Communications*, *6*, 7827.

Passmore, L. A., & Russo, C. J. (2016). Specimen preparation for high-resolution Cryo-EM. *Methods in Enzymology*, *579*, 51–86.

Shen, C., et al. (2019). Molecular mechanism for NLRP6 inflammasome assembly and activation. *Proceedings of the National Academy of Sciences*, *116*(6), 2052–2057.

Tenthorey, J. L., et al. (2017). The structural basis of flagellin detection by NAIP5: A strategy to limit pathogen immune evasion. *Science*, *358*(6365), 888–893.

CHAPTER TWELVE

Cryo-EM studies of NAIP–NLRC4 inflammasomes

Nicole Haloupek[a,*], Patricia Grob[b], Jeannette Tenthorey[a,†], Russell E. Vance[a,b], Eva Nogales[a,b,c,*]

[a]Department of Molecular and Cell Biology, UC Berkeley, Berkeley, CA, United States
[b]Howard Hughes Medical Institute, UC Berkeley, Berkeley, CA, United States
[c]Molecular Biophysics and Integrated Bioimaging Division, Lawrence Berkeley National Lab, Berkeley, CA, United States
*Corresponding authors: e-mail address: nhaloupek@gmail.com; enogales@lbl.gov

Contents

† Present address: Basic Sciences Division, Fred Hutchinson Cancer Research Center, Seattle, WA, United States.

Methods in Enzymology, Volume 625
ISSN 0076-6879
https://doi.org/10.1016/bs.mie.2019.04.030

Abstract

The NAIP–NLRC4 family of inflammasomes are components of the innate immune system that sound a molecular alarm in the presence of intracellular pathogens. In this chapter, we provide an in-depth guide to using cryo-electron microscopy (cryo-EM) to investigate these inflammasomes, focusing especially on the techniques we used in our recent structural analysis of the NAIP5–NLRC4 inflammasome. We explain how to circumvent specific obstacles we encountered at each step, from sample preparation through data processing. The methods described here will be useful for further studies of the NAIP5–NLRC4 inflammasome and related supracomplexes involved in innate immune surveillance; they may also be useful for unrelated complexes that present similar issues, such as preferential orientations and compositional heterogeneity.

1. Introduction

NAIP–NLRC4 inflammasomes are protein complexes composed of a single NAIP (NLR family, apoptosis inhibitory protein) bound to a variable number of NLRC4 (NLR family, CARD [Caspase Activation and Recruitment Domain]-containing 4) subunits. Their role in the innate immune system is to detect the presence of intracellular pathogens and initiate a signaling cascade leading to pyroptosis, a form of programmed cell death that causes inflammation, and inflammatory cytokine release that initiates an immune response to restrict the proliferation of the pathogen (Bergsbaken, Fink, & Cookson, 2009; Jones, Vance, & Dangl, 2016).

The open-ring-shaped structures assumed by these inflammasomes are nucleated by a bacterial ligand bound to a NAIP, with the ligand preference being dictated by the specific NAIP isoform that is engaged. NAIP5, the sensor on which this chapter will primarily focus, detects the bacterial protein flagellin, whereas its paralogs NAIP1 and NAIP2 detect the needle and inner rod proteins, respectively, of bacterial type III secretion systems (Kofoed & Vance, 2011; Rayamajhi, Zak, Chavarria-Smith, Vance, & Miao, 2013; Yang, Zhao, Shi, & Shao, 2013; Zhao et al., 2011). Other than detecting different ligands, these inflammasomes appear to work in similar ways.

Structural studies have uncovered much information about the mechanism of inflammasome formation and activation, from revealing how NAIPs bind their ligands to hinting how, once assembled, inflammasomes exert their effects. Ligand binding by a NAIP and the consequent inflammasome formation are only the first steps in a larger pathway. Structural and biochemical work has shown that the CARD of NLRC4 binds procaspase-1, juxtaposing the proenzymes within the oligomeric complex to facilitate their dimerization and maturation into active caspase-1 (Lu et al., 2016; Shi, 2004). The active caspase-1 then cleaves the inflammatory cytokines pro-interleukin-1β and pro-interleukin-18 into their mature forms (Broz & Dixit, 2016; Martinon, Burns, & Tschopp, 2002). Caspase-1 also cleaves gasdermin D, the N-terminal portion of which forms pores in the plasma membrane, triggering the release of mature cytokines and pyroptosis (Aglietti et al., 2016; Chen et al., 2016; Ding et al., 2016; Liu et al., 2016; Ruan, Xia, Liu, Lieberman, & Wu, 2018; Sborgi et al., 2016).

Recently, we published the complete structure of the flagellin-bound NAIP5–NLRC4 inflammasome as well as a partial inflammasome structure consisting of NAIP5 bound to flagellin and two NLRC4 subunits at a higher overall resolution of 5.2 Å (Tenthorey et al., 2017). Shortly thereafter, Yang et al. reported the structure of NAIP5 bound to the NAIP5-activating portion of flagellin and one NLRC4 subunit at 4.3 Å resolution (Yang et al., 2017). Because the two published structures contained slightly different components and were obtained in different ways, as we will explain at later points in this chapter, they should be considered complementary to one another. There is likely room for improvements in resolution for these structures, especially in regions where the resolution of both structures suffered due to flexibility, such as the NAIP5 BIR (baculovirus inhibitor-of-apoptosis repeat) domains, where some conflict in the modeling of the structures exists between the two studies.

In this chapter, we will give a detailed description of our methods, focusing on less conventional technical aspects that were designed by our group specifically for work on the NAIP5–NLRC4 inflammasome. We also provide suggestions that may help future researchers improve on what we and others have done so far in the study of inflammasomes and other challenging samples being studied by cryo-EM.

2. Preparing NAIP–NLRC4 inflammasomes for EM

Cryo-electron microscopy (cryo-EM) has generally been the method of choice for structural studies of inflammasomes. The structure of inactive NLRC4 lacking its CARD has been solved using X-ray crystallography (Hu et al., 2013), but full inflammasome complexes do not seem amenable to crystallographic studies. First, it has not been possible, in our experience, to produce enough of these complexes for systematic crystallization trials.

Second, inflammasomes are constitutionally heterogeneous (containing variable numbers of NLRC4 subunits) and conformationally heterogeneous, characteristics that do not typically bode well for crystallography.

The cryo-electron microscopist has many tools for dealing with such heterogeneity; however, it can still pose a significant challenge. In our cryo-EM studies of the NAIP5–NLRC4 inflammasome, we also encountered other difficulties, including sample sensitivity to the air–water interface, preferred orientations of the inflammasomes on EM grids, and low numbers of particles per micrograph. In this section, we describe these issues in further detail. In Sections 3–5 of this chapter, we present solutions to these problems, many of which may also be applicable to other, unrelated complexes, and provide in-depth explanations of our group's protocols so that they may be replicated and, we hope, improved upon.

2.1 Heterogeneity-related obstacles

Garnering structural information about inflammasomes has not proven trivial. The constitutional and conformational heterogeneity that in part limit the use of crystallography in studying these inflammasomes are also problematic—although not intractably so—in cryo-EM studies. Cryo-EM relies on averaging two-dimensional projections of individual molecules (particles) to achieve high-resolution structures. We have observed that NAIP5–NLRC4 inflammasomes tend to exhibit CARD-mediated ring stacking in solution, and that this stacking is haphazard, leading to a variety of conformations that complicate averaging and severely restrict resolution. Not only does the stacking limit resolution, but we also believe the stacked rings are unlikely to be a biologically relevant form of the complex, since stacking occludes the CARDs of NLRC4, which are necessary for interaction with (and, thus, activation of) procaspase-1.

2.2 Sensitivity to the air–water interface and preferred orientations

NAIP–NLRC4 inflammasomes also appear highly sensitive to the air–water interface, posing a particular problem for cryo-EM studies. In cryo-EM, thinner ice generally yields structures of higher resolution because buffer components scatter electrons, blurring particle images in areas where ice is thick. However, very thin ice is not suitable for studying NAIP–NLRC4 inflammasomes because they are large, disk-shaped complexes, so ice that is somewhat thicker than what is usually desirable for cryo-EM is needed to obtain projections of the side views of the complexes. Therefore, a careful balance must be struck to obtain ice that is as thin as possible to support high-resolution reconstructions while also being just thick enough to support the side views. NAIP5–NLRC4 inflammasome grids with ice that is too thin are easy to recognize under the electron microscope: side views will be confined to the edges of holes (where the ice is typically thicker) or will simply not be seen at all.

2.3 Low particle number

Additionally, simply getting inflammasomes to adhere to EM grids proved problematic in our studies. Incubating sample droplets on the grids for longer periods of time than are often used increases the particle concentration, but in our experiments, water evaporation from the sample droplet limited our incubation time, even when incubation was done in a humidity chamber. We have found that over longer incubation periods, the inflammasomes degraded into indistinct "exploded" particles, likely due to exposure to the air–water interface, especially as the water evaporates.

3. Biochemical preparation

The first step toward obtaining any high-resolution structure using cryo-EM is to biochemically optimize the sample. Despite the fact that improved methods for particle picking and sorting are continuously being developed, sample impurities can lead to reduced resolution, inaccuracies, or, at worst, failure to obtain a structure at all. Concentration is also a concern; if the concentration is too low and few particles adhere to the EM grid, it is difficult to collect enough particle images to enable high-resolution reconstructions.

To achieve high yield and concentration for our inflammasome preparations, we experimented with several heterologous expression systems, including insect and mammalian cells. We found that co-expression of inflammasome components in HEK-293T cells achieved sufficient yields; this system also had the advantage of recapitulating the known biological properties of NAIP–NLRC4 inflammasomes (Kofoed & Vance, 2011).

To achieve sufficient purity, we isolated inflammasomes from these cells using a two-step purification protocol. We first immuno-affinity purified NAIP5 using an N-terminal FLAG tag and commercially available resin containing a highly specific anti-FLAG antibody. In principle, this purification should yield a mixture of monomeric NAIP5, NAIP5 bound to its ligand flagellin, and a mixture of partially or fully assembled NAIP5–NLRC4 inflammasomes. To separate the fully assembled complexes, we subjected our FLAG-purified NAIP5 to size-exclusion chromatography, which separates proteins based on their hydrodynamic radius (roughly, size). This allowed us to isolate large (>1 MDa) protein complexes that, as assessed by Western blot, contained NAIP5, flagellin, and a stoichiometric excess of NLRC4.

For the reasons previously discussed, with respect to both heterogeneity (mainly concerning ring stacking) and concentration, the biochemical preparation of NAIP–NLRC4 inflammasomes requires care. Further, these inflammasomes contain flexible elements, which hamper efforts to achieve high resolution if not managed. In this section, we discuss methods for mitigating these problems, some of which may be applicable to other complexes being studied using cryo-EM, and provide our protocol for the biochemical preparation of NAIP5–NLRC4 inflammasomes.

3.1 Preventing constitutional heterogeneity

As previously discussed, the tendency of NAIP–NLRC4 inflammasome rings to stack is a major problem for EM studies, and each group studying these complexes using cryo-EM has taken a different approach to addressing it.

In one structural study, Hu et al. used a truncated NLRC4 lacking the stacking-mediating CARD ($NLRC4^{\Delta CARD}$), thus completely eliminating the problem (Hu et al., 2015). Using radial averaging, the group was able to determine the structure of active $NLRC4^{\Delta CARD}$ at high resolution, revealing that the global conformation of $NLRC4^{\Delta CARD}$ in assembled inflammasomes (as observed in vitreous ice in cryo-EM) differs from the previously published structure of crystallized monomeric $NLRC4^{\Delta CARD}$. Because of the radial averaging, they were not able to solve the structure of the NAIP or its ligand as these proteins were each present in only one copy per ring and the average was thus dominated by NLRC4.

Deleting the CARD is only one option for preventing ring stacking, and it does have drawbacks. The most obvious of these is that CARD deletion makes it impossible to visualize the CARDs. However, despite not deleting the CARDs, our group and others have not been able to attain inflammasome structures with high resolution in the CARD region, possibly because of flexibility between the CARDs and the rest of the structure or inherent flexibility within the CARDs themselves as they exist in the inflammasome structure. Other drawbacks of deleting the CARDs are that it is conceivable that the same structure would not form in the presence of CARDs and that it would not be possible to perform structural studies of the inflammasomes with their partner, caspase-1, the binding of which is mediated by the CARDs.

Instead of using NLRC4$^{\Delta CARD}$, our group used NLRC4 with an N-terminal green fluorescent protein (GFP) linked to the CARDs to sterically block CARD–CARD interactions (Tenthorey et al., 2017). This reduced but did not fully prevent ring stacking. With the addition of two targeted mutations in the CARD (F79A and D83A) aimed at preventing stacking based on what was known about the CARD–CARD interface, we observed that stacking was nearly completely eliminated (see Fig. 1). We note that the introduction of these CARD mutations prevents recruitment of caspase-1 by the CARDs, so future attempts to visualize the entire flagellin–NAIP–NLRC4–caspase-1 holocomplex will need to rely on non-mutated NLRC4.

A further source of heterogeneity in NAIP–NLRC4 inflammasome preparations is the number of NLRC4 subunits each ring contains. While we ultimately dealt with this computationally (see Section 5.2), another strategy, employed by Yang et al., is to prevent full rings from forming by mutating the NLRC4 oligomerization "donor" surface such that while NLRC4 can still be recruited to NAIP, NLRC4 cannot recruit additional NLRC4 protomers (Yang et al., 2017). These are only two possible options, and it is conceivable that future studies will include better ways of dealing with the constitutional heterogeneity observed with these inflammasomes.

Note that we did attempt to reduce the problem of heterogeneity of NAIP5–NLRC4 inflammasomes due to different number of NLRC4 molecules by crosslinking the sample with glutaraldehyde prior to grid application. This produced no noticeable improvement in our results, leading us to use computational methods to deal with the heterogeneity instead (see Section 5). However, it may be worth attempting to stabilize the complex with other crosslinkers in future studies.

3.2 Preventing conformational heterogeneity

In addition to compositional heterogeneity, we have observed that some parts of the NAIP5–NLRC4 inflammasome exhibit marked conformational heterogeneity. The NLRC4 CARD, in particular, has never been a well-resolved domain in any of our inflammasome structures, instead appearing as a blurry density at the center of the ring. We attempted adding caspase-1 to inflammasomes prepared with wild-type GFP-NLRC4 in the hope that it

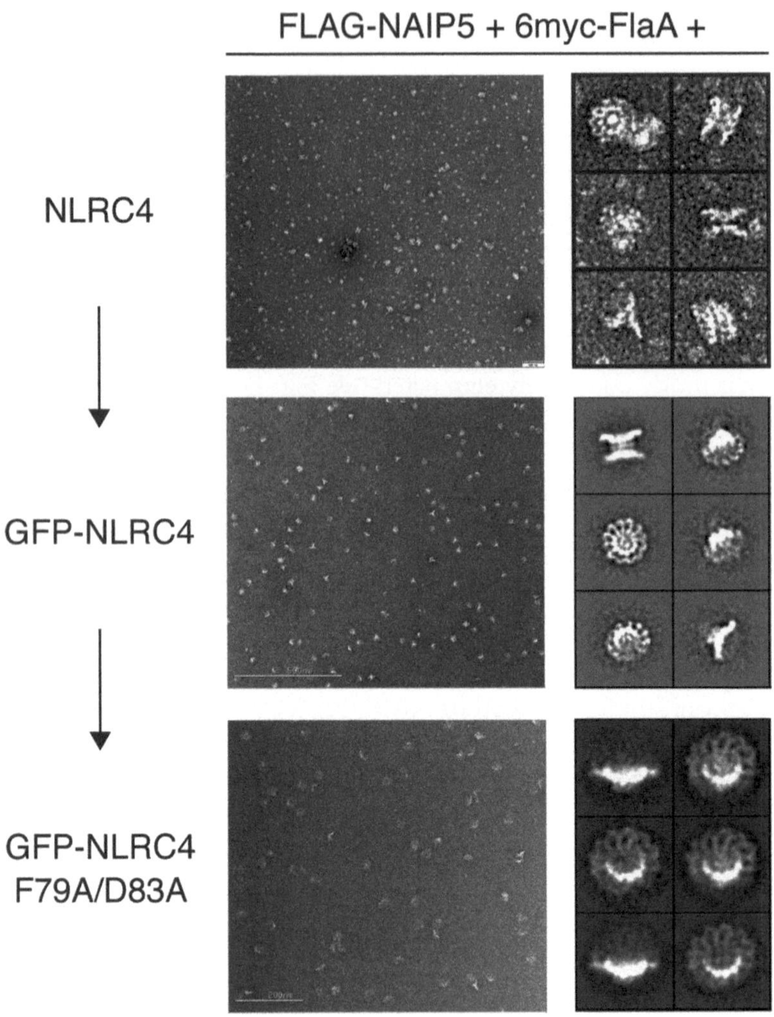

Fig. 1 Biochemical optimization. Micrographs of negatively stained NAIP5–NLRC4 inflammasomes (left) and 2D class averages (right) reveal the presence of stacked rings. Only when two CARD mutations and an N-terminal GFP tag are added to NLRC4 (bottom) is ring stacking reduced to a manageable level. See Section 3.1 for more information. *Part of this figure is reused with publisher's permission from Tenthorey, J. L., Haloupek, N., López-Blanco, J.R., Grob, P., Adamson, E., Hartenian, E., et al. (2017). The structural basis of flagellin detection by NAIP5: A strategy to limit pathogen immune evasion.* Science, *358, 888–893.*

might stabilize the CARDs, but this did not improve resolution in the CARD region in our preliminary tests. Still, we believe the caspase-1-bound inflammasome may be an interesting structural target on its own and is worth pursuing in further studies.

Another apparently flexible region is the large portion of flagellin that does not bind NAIP5. One solution to this problem is to use truncated flagellin containing only the NAIP5–binding domain, which was the strategy employed by Yang et al. (2017). We opted to use full-length flagellin, which allowed us to visualize slightly more of the subunit at a cost to overall resolution (Tenthorey et al., 2017). Areas in the N-terminal region of NAIP5, including the three BIR domains and the linkers connecting them, also appear relatively flexible. This is particularly troublesome since the BIRs are very similar to one another, making unambiguous assignment of each one to a portion of the density a challenge.

3.3 Protocol for biochemical preparation of NAIP5–NLRC4 inflammasomes

3.3.1 Equipment

- CO_2 incubator set to 37 °C and humidified with sterile water
- Biosafety cabinet
- Light microscope, GFP laser optional
- Centrifuge, preferably one that allows chilling
- Microcentrifuge that allows chilling (or one used in cold room)
- Tube rotator (in cold room)
- High performance liquid chromatography (HPLC) system (in cold room or deli case refrigerator)
- Superose6 10/300 GL size exclusion column

3.3.2 Materials

- HEK293T cells or other easily transfected cell line
- DMEM (Gibco)
- Fetal bovine serum (Gibco)
- 200 mM L-glutamine (Gibco)
- 10,000 U/mL Penicillin/Streptomycin (Gibco)
- Phosphate-buffered saline, pH 7.4 (Gibco)
- Trypsin-EDTA (Gibco)
- Tissue culture-treated plates
- Sterile serological pipettes
- Sterile Pasteur pipettes
- Cell scraper
- Lipofectamine 2000 (Invitrogen)
- Optimem (Gibco)
- GFP-NLRC4-F79A/D83A, FLAG-NAIP5, and 6myc-FlaA plasmids, purified from *E. coli* using a midiprep or maxiprep kit

- 15- and 50-mL conical tubes
- Eppendorf tubes
- HEPES, pH 7.6
- NaCl
- KCl
- $MgCl_2$
- Glycerol
- Triton X-100
- NP-40
- DTT
- Trehalose
- 100 mM glycine, pH 3.5
- NaN_3 (optional)
- Protease inhibitor tablets (Roche)
- FLAG M2 agarose resin (Sigma)
- FLAG peptide (Sigma)
- 10 mL chromatography column, with top cap and stopcock
- Ring stand with column clamp
- 500 mL 0.2 μm filter
- HPLC loading needle
- 1 mL HPLC sample loop
- HPLC manual injection syringe

3.3.3 Protocol

1. Grow HEK293T cells on sterile, tissue culture-treated plates in DMEM supplemented with 10% fetal bovine serum (FBS), 2 mM L-glutamine, and 100 U/mL penicillin/streptomycin. These cells double in number roughly every 24 h, and should be passaged when they reach confluency (i.e., when they cover most of the plate's surface area). To passage cells, aspirate the media, gently rinse with PBS, aspirate PBS, incubate in trypsin at 37 °C for several minutes, and resuspend in supplemented DMEM. Cells can then be pelleted, resuspended in fresh media, and diluted up to 10-fold for further growth in fresh plates. Cells are grown in a humidified CO_2 incubator at 37 °C and are handled exclusively in a biosafety cabinet.
2. Prepare HEK293T for expressing inflammasome components by harvesting confluent plates and re-plating in 15-cm plates at a 1:4 dilution 1 day prior to transfection. Typically, 6–12 plates are required for sufficient yield. Cells should be roughly 50% confluent the following day.

3. Transiently express inflammasome components (GFP-NLRC4-F79A/D83A, FLAG-NAIP5, and 6myc-FlaA, all in the mscv2.2 vector backbone) via transfection with Lipofectamine 2000. The following recipe is for a single 15-cm plate; scale up accordingly. Mix 16 μg of NLRC4 plasmid, 16 μg of NAIP5 plasmid, and 8 μg of FlaA plasmid with 2 mL of Optimem. In a separate tube, mix 160 μL of Lipofectamine 2000 with 1840 μL of Optimem and incubate at room temperature for 5 min. Add the diluted Lipofectamine to the plasmid mix, incubate another 20 min at room temperature, and gently add the DNA–Lipofectamine mix to the 15-cm plate of HEK293T cells. Swirl to mix and incubate for 48 h at 37 °C.
4. Transfection efficiency can (optionally) be assessed via GFP fluorescence, as all plasmids contain either a GFP fusion or an IRES-GFP marker in the mscv2.2 plasmid backbone. GFP fluorescence should be visible 24 h after transfection and increases in intensity thereafter.
5. 48 h after transfection, harvest transfected cells by aspirating the media and incubating 5 min in 15 mL of cold PBS. Use a cell scraper to gently dislodge cells from the plate. Transfer cells to a 50-mL conical tube and pellet at 1000 × *g* for 10 min at 4 °C.
6. Prepare 25 mL of cold lysis buffer (50 mM HEPES, pH 7.6; 150 mM NaCl; 10 mM KCl; 5 mM $MgCl_2$; 5% glycerol; 1% Triton X-100; 1 × protease inhibitor cocktail).
7. Prepare 500 mL of cold size exclusion buffer (50 mM HEPES, pH 7.6; 150 mM NaCl; 10 mM KCl; 5 mM $MgCl_2$; 5% glycerol; 0.02% NP-40). Filter this buffer with a 0.2-μm filter and keep it under vacuum for 10 min with occasional tapping to de-gas the buffer.
8. Prepare 1.6 mL of cold elution buffer (size exclusion buffer supplemented with 0.15 mg/mL FLAG peptide).
9. Aspirate PBS from pelleted cells and resuspend in 7.5 mL of cold lysis buffer. To lyse, incubate with gentle agitation or rotation for 30 min at 4 °C.
10. Pellet cell debris at 16,000 × *g* for 30 min at 4 °C in micro-centrifuge.
11. Working at 4 °C, rinse chromatography column with PBS and add 200 μL of FLAG M2 agarose resin. Allow the storage buffer to drain and rinse with PBS. Add PBS as soon as buffer stops flowing to avoid drying the resin. Wash resin three times with 200 μL of 100 mM glycine, pH 3.5, letting the resin drain completely between each wash. Wash the resin with 1 mL of PBS to remove glycine. Finally, equilibrate

with 5 mL of cold lysis buffer. Cap the bottom of the chromatography column with closed stopcock.

12. Add supernatant from lysed cells to equilibrated resin. Be careful to avoid taking the cell pellet.
13. Cap the top of the chromatography column and incubate rotating at 4 °C for 2 h.
14. Working at 4 °C, place chromatography column into clamp on ring stand, remove top cap, and open stopcock to allow the lysate to drain from the resin.
15. Wash the resin with 20 mL of cold size exclusion buffer, allowing resin to drain at ~1 mL/min.
16. Once the wash has completely drained, close the stopcock and add 100 μL of cold elution buffer. Incubate 3 min at 4 °C before opening stopcock to collect elution fraction in a cold Eppendorf tube.
17. Repeat the elution process for a total of 8 fractions. Pool fractions 3–7 and add 1 mM DTT. FLAG resin can (optionally) be regenerated for another use by repeating the glycine washes and storing in PBS containing 0.02% sodium azide.
18. Pellet pooled eluate in micro-centrifuge at 16,000 × *g* for 15 min at 4 °C. Remove cleared eluate to fresh Eppendorf tube, being careful to avoid pelleted aggregates.
19. Working at 4 °C on the HPLC, equilibrate Superose6 size exclusion column with 1.5 column volumes of Milli-Q water followed by 1.5 column volumes of size exclusion buffer. Note that only filtered, degassed buffers should be used. The Superose6 column volume is 24 mL, and the max pressure is 1.5 MPa. Typically, this column allows only low flow rates (<0.5 mL/min).
20. Using an HPLC manual injection syringe, equilibrate the 1 mL sample loop and injection port with size exclusion buffer.
21. Load cleared eluate into the injection syringe, being careful to avoid air bubbles, and then inject eluate into sample loop.
22. Using the HPLC pumps, inject sample from sample loop onto column and then elute with 1.25 column volumes of size exclusion buffer, collecting 0.5 mL fractions. Follow the elution of inflammasome complexes via UV absorbance at 280 nm. Note that large protein complexes enter resin poorly and are eluted faster than individual protein components. Assembled inflammasomes should elute beginning at ~0.35 column volumes, whereas unassembled components will elute much later.

23. Add 20% trehalose to the peak elution fraction and use for cryo-EM analysis. Store the sample on ice and prepare grids as soon as possible as described in the protocol in Section 4.2. If the sample must be stored, immediately flash-freeze small aliquots in liquid nitrogen and keep at −80 °C.

4. Preparation of cryo-EM grids

As with most samples being prepared for cryo-EM analysis, an initial assessment of the biochemical preparation's potential can be conducted using negatively stained EM techniques. Less concentrated samples can generally be used for negatively stained EM grid preparation than can be used for cryo-EM grid preparation, and compared to screening cryo-EM grids, checking negatively stained EM grids is fast and simple. Although these negatively stained samples will not give rise to high-resolution structures, preparing them is worthwhile. A sample that has noticeable impurities or does not appear concentrated enough in negatively stained EM will almost certainly not look better in cryo-EM. Once a sample looks free of such problems in negatively stained EM, screening for optimal cryo-EM grid preparation conditions can begin.

With some samples, including the NAIP5–NLRC4 inflammasomes as we prepared them, one does not have the luxury of finding the right cryo-EM parameters through brute-force screening of dozens of conditions, particularly since these inflammasomes are time-consuming to prepare. In our experience, inflammasomes are delicate (i.e., they cannot be repeatedly flash-frozen and thawed, and even one round of freezing and thawing is not optimal), and it is essential that consideration is devoted to selecting grid preparation parameters that are likely to yield grids suitable for data collection.

Some groups using cryo-EM to study NAIP–NLRC4 inflammasomes have used conventional grid preparation methods. However, their preparations have been different from ours in several ways, using methods such as blocking full inflammasome formation (mutating the NLRC4 "donor" surface to prevent large rings from assembling) and deleting the NLRC4 CARD. We did not use these methods (see Section 3.1, for more information) and encountered difficulties preparing our inflammasome grids conventionally. In this section, we share how we overcame two major problems—that of missing side views and that of low particle density—using easily fabricated tools. The technique we describe may be useful for preparing other, unrelated complexes as well. We also provide our full protocol for preparing NAIP–NLRC4 cryo-EM grids.

4.1 Minimizing preferred orientations and increasing particle density

One of the first difficulties we encountered when preparing NAIP5–NLRC4 inflammasome grids is a common one: low particle density. When it is not possible to further concentrate a sample prior to grid application or when increasing concentration does not sufficiently increase the particle density seen on the grid, as in our case, the electron microscopist has many options. Often, applying a supporting layer of continuous carbon to the cryo-EM grid and glow discharging the grid just prior to sample application increases particle density. We found this to be necessary with our NAIP5–NLRC4 inflammasome samples, but it was not until we introduced our new technique that we were able to increase particle density sufficiently to make 3D reconstruction practical.

We originally conceived of this technique to solve a second problem familiar to cryo-electron microscopists: preferred orientations. We saw many circular top views of the NAIP5–NLRC4 inflammasome in our micrographs, but relatively few of the crescent-shaped side views, particularly in micrographs from cryo-EM grids rather than micrographs from negatively stained EM grids. Inflammasomes are roughly disk-shaped complexes and much longer in one dimension than the other, so we hypothesized that contact with the air–water interface during the longer sample application time required to reach high particle density on the grid for cryo-EM was causing inflammasomes that were originally in the side-view orientation to rotate toward a top-view orientation—or to denature, since we did observe denatured particles in micrographs from many cryo-EM grids.

Incubating our inflammasomes on grids in a humidity chamber, as is often done when samples must be incubated on cryo-EM grids for long periods of time (on the order of minutes), was not sufficient to prevent this problem. Thus, we developed a simple apparatus we call the grid floater. The experimental setup contains both the grid floater and a Teflon well plate, a machined piece of Teflon with smooth holes drilled into it to serve as wells for liquid samples (see Fig. 1). Teflon was our material of choice because it is easy to work with, inexpensive, and not prone to reacting with the biological material or accumulating residues. In our protocol, a small volume (~20 μL) of prepared inflammasome samples is aliquoted into a

Teflon well, forming a droplet with a meniscus high above the Teflon. The grid is then rested (or "floated"), carbon-side down, on the droplet for incubation. With experience, some particularly dexterous individuals may be able to capture the grid with tweezers when the time for plunge-freezing comes, but we preferred to use the grid floater, which could be any device that allows a pair of tweezers pre-loaded with a grid to be aligned precisely atop the sample droplet. Using the grid floater ensures that the grid will be gripped perfectly by the rim every time. In practice, we used a ramp-shaped object made of many layers of crumpled aluminum foil as our grid floater (see Fig. 2). This inexpensive, improvised device proved easily adjustable by bending to accommodate different droplet heights (which vary due to volume dispensed, surface tension, etc.) as well as durable. Clearly, though, many other objects could be used as support for the tweezers.

After incubating the sample on the grid, we rinsed the surface of the grid for 10 s by gently touching it, sample side down, to a 20-μL droplet of EM washing buffer (see Section 4.2, for recipe) contained in another well. The main purpose of this step is to reduce the concentration of glycerol, which is present at 5% in the size exclusion buffer (see Section 3.3.3, for recipe), replacing it with a small amount of trehalose. Both glycerol and trehalose are cryoprotectants, but glycerol reduces contrast in the images more than trehalose does, so it is desirable to remove at least some of the glycerol. Since rinsing leaves some residual glycerol, we tried rinsing with up to three droplets of buffer of up to 30 μL each, but this did not appear to improve contrast compared to using just one smaller droplet.

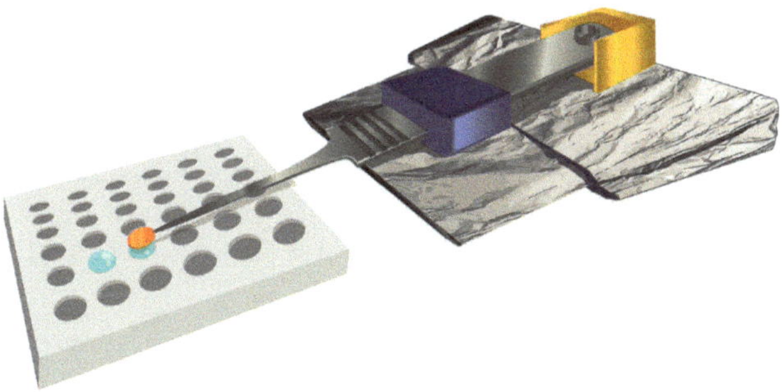

Fig. 2 Depiction of the grid floater and Teflon well plate in use. The edge of the EM grid is gripped by Vitrobot tweezers and rests, carbon-side down, on a sample droplet. After incubation, the downward-facing surface of the grid is gently rinsed in the second droplet, which contains only buffer, and then the tweezers are loaded in the Vitrobot. See Section 4.2 for the full protocol.

Using the grid floater just described, the particles adhering to the grid are most at risk of coming into contact with the air–water interface during the moment it takes to blot and plunge-freeze the grid. Of course, it is important not to blot excessively at this stage, recalling that the vitreous ice must be thick enough to still contain particles in the side-view orientation. Those who have experience blotting and plunge-freezing by hand will find that it is easy to do so using this setup, although we did use a Vitrobot in our studies.

We found that using this grid floater, the preferred orientation problem was much reduced, although still not totally eliminated, and since we were able to incubate our samples on the grids longer without worrying about evaporation of the droplet, we were able to increase the number of particles on the grid as well. (Note that no further improvement in particle density was seen after 10–15 min of incubation.) This setup could thus be useful for any samples requiring prolonged incubation times on cryo-EM grids.

It should be mentioned that there are alternatives to using the grid floater that may also increase particle number and/or reduce the preferred orientation problem, although they may not be as simple or inexpensive. Namely, using grid coatings such as streptavidin monolayers (with compatible biotinylated samples) and graphene or graphene derivatives, to name just two types, may be worth exploring (Han et al., 2016; Hu et al., 2015; Palovcak et al., 2018; Russo & Passmore, 2014, 2016; Wang, Ounjai, & Sigworth, 2008). For the most part, these alternative methods should not be incompatible with grid floating, and we encourage those working with difficult samples to try them in combination.

4.2 Protocol for NAIP5–NLRC4 inflammasome cryo-EM grid preparation

4.2.1 Equipment

- 0.2 μm filter
- Grid floater (see Section 4.1)
- Teflon well plate stored in Milli-Q water (see Section 4.1)
- Micropipettes
- C-flat holey carbon grids (Protochips) with thin layer of amorphous carbon deposited as described by Passmore and Russo (2016)
- Solarus Plasma Cleaner (Gatan)
- Vitrobot Mark II (FEI) or other plunge-freezing apparatus for cryo-EM, including tweezers
- Cryo-EM grid boxes

- Aluminum foil
- Ultrasonic bath sonicator
- Glass beaker large enough to store Teflon well plate

4.2.2 Materials

- Milli-Q water
- HEPES, pH 7.6
- NaCl
- KCl
- $MgCl_2$
- Trehalose
- NP-40
- Nitrogen gas
- Dry liquid nitrogen
- 100% ethanol
- ES 7 × cleaning solution

4.2.3 Protocol

1. Prepare EM washing buffer (50mM HEPES, pH 7.6, 150mM NaCl, 10mM KCl, 5mM $MgCl_2$, 1% trehalose, 0.02% NP-40). Do not shake or vortex mix after adding NP-40. Slowly pass through a 0.2-μm filter. EM washing buffer without trehalose and NP-40 may be prepared in advance and stored, but the trehalose and NP-40 must be added just prior to use.
2. Remove Teflon well plate from Milli-Q water and dry with nitrogen gas. Using a micropipette, rinse all wells to be used with EM washing buffer.
3. Test the height and alignment of the grid floater by adding a droplet of EM washing buffer to one of the wells in the Teflon well plate and positioning the tip of the Vitrobot tweezers over the droplet. Ensure that a grid held by the tweezers would rest atop the droplet.
4. Plasma clean grids 8s in air using a Solarus Plasma Cleaner (Gatan) operated at 10W.
5. Set up Vitrobot according to manufacturer's instructions at 22°C and 100% relative humidity.
6. Add 20μL droplet of inflammasomes (prepared as described in Section 3.3.3) and 20μL droplet of EM washing buffer to separate wells in the Teflon well plate.

7. Grip one of the plasma-cleaned grids by the edge using Vitrobot tweezers and place it, carbon-side down, on the droplet of inflammasomes. Incubate for 10 min.
8. Use the tweezers to remove the grid from the inflammasome droplet, then immediately rinse the grid by gently touching the sample side to the droplet of EM buffer for 10 s. Do not submerge the grid or otherwise allow buffer to come in contact with the back of the grid.
9. Immediately load the tweezers on the Vitrobot and plunge-freeze the grid in liquid ethane with no delay between blotting and freezing. Store the grids in plastic cryo-EM grid boxes in dry liquid nitrogen until use.
10. Clean and store Teflon well plate. To clean, soak overnight in 1% *v*/v ES 7 × cleaning solution diluted in Milli-Q water (or other dilute, phosphate-free detergent), then rinse the plate 20 times with Milli-Q water. Ensure that cleaning solution and water enter all the wells. Submerge well plate in 100% ethanol in a glass beaker, ensure ethanol enters all wells, cover the beaker with aluminum foil, and sonicate in an ultrasonic bath sonicator for 30 min. Rinse the plate 20 more times with Milli-Q water and store fully submerged in Milli-Q in a beaker covered with aluminum foil. Ensure that water fills all wells before storage.

5. Cryo-EM data collection and analysis

Even after undertaking diligent work in preparing and screening samples, care is required during data collection and processing. The cryo-EM data collection and processing pipeline we used for the NAIP5–NLRC4 inflammasome (Fig. 3) is similar in many respects to that used for most asymmetric biological samples. First, micrographs that are good—a judgment based on criteria such as having an abundance of intact particles in many views and lacking ice contamination—are selected for use. Particles are then picked from these micrographs by hand or using software, and various preprocessing steps, such as corrections for contrast transfer function and drift, are conducted. Two-dimensional (2D) classification is often used at this stage to remove particles that do not align with others and for initial analysis. Next, particles are typically further sorted into three-dimensional (3D) classes, and the class(es) deemed biologically relevant and likely to yield high-resolution reconstructions are refined and further polished. Finally, depending on resolution and other factors, various types of structural modeling may also be done.

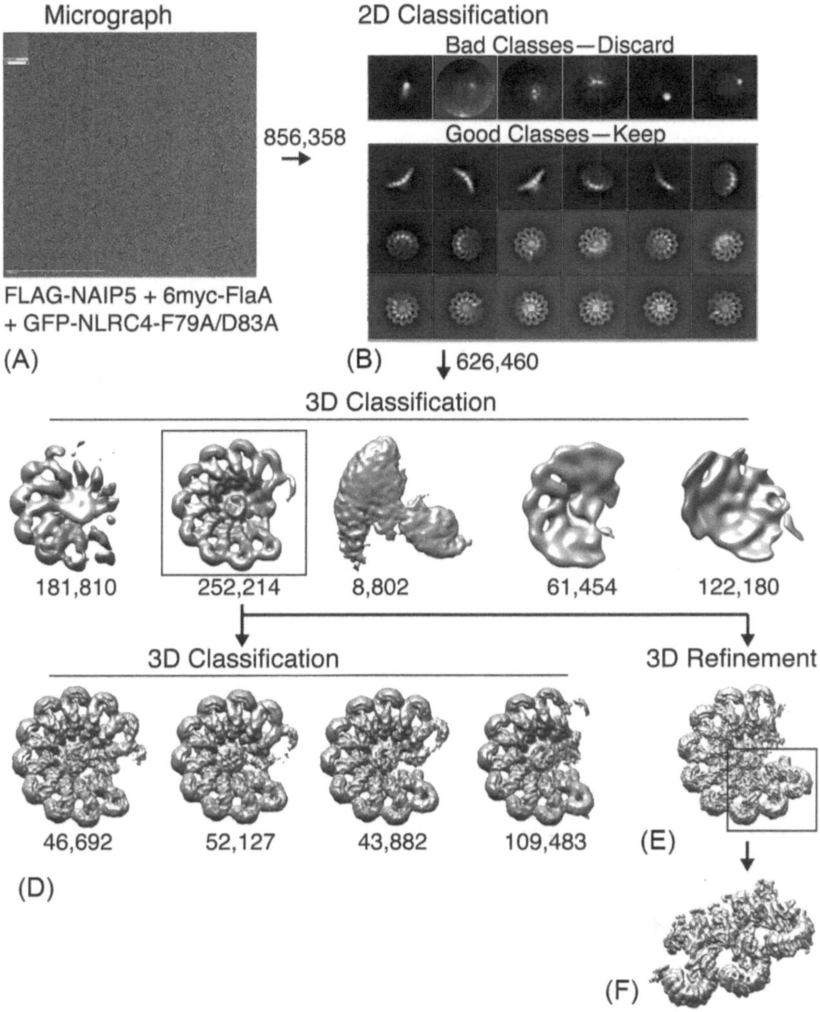

Fig. 3 Data processing pipeline leading to the reconstruction of the NAIP5–NLRC4 inflammasome in Tenthorey et al. Numbers in this figure refer to numbers of particles. (A) A representative micrograph. Note the presence of both crescent-shaped side views and circular top views. (B) Representative 2D class averages. Particles in bad classes—those that were blurry and/or did not resemble particles—were discarded at this stage. (C) The results of RELION's 3D classification into five classes. This step yielded only one class (boxed) with potential for producing a high-resolution structure. (D) The results of further 3D classification of the best 3D class from (C). This classification step mainly sorted particles on the basis of number of NLRC4 subunits. Further refinement of any of these classes did not yield better results than refinement of the pooled classes shown in (E). (F) The result of focused refinement of NAIP5, flagellin, and the first two NLRC4 subunits using the same particles used in (E). For full details on the data processing, including the rationale behind each step, see Section 5. *This figure is reused with publisher's permission from Tenthorey, J. L., Haloupek, N., López-Blanco, J.R., Grob, P., Adamson, E., Hartenian, E., et al. (2017). The structural basis of flagellin detection by NAIP5: A strategy to limit pathogen immune evasion.* Science, *358, 888–893.*

Some aspects of this process are typically consistent from sample to sample, but there are many additional considerations for each case. In this section, we focus on the specific strategies we used for the NAIP5–NLRC4 inflammasome that we believe may be useful to others studying it in the future. Some of these methods may also be useful for dealing with other samples sharing similar issues, such as preferred views and severe compositional and conformational heterogeneity. We do not aim to provide a full guide to the standard cryo-EM methods we employed, which have been extensively described in other texts referenced throughout this chapter.

5.1 Data collection

Given the relatively high resolutions of currently available structures of NAIP–NLRC4 inflammasomes and their components, those seeking new structures should undoubtedly plan to use a microscope equipped with a direct electron detector. The use of these detectors has revolutionized structural biology over the past several years, making it possible to use cryo-EM to solve structures of biological macromolecules at unprecedentedly high resolution (Passmore & Russo, 2016).

Electron microscopes and the software used to operate them have also become much more advanced in recent years, and many steps that until recently vexed researchers can now be done automatically. Even the newest electron microscopes with the highest level of automation, however, should not be left to run unsupervised for prolonged periods if a data collection session is to be successful, especially when working with difficult samples. During data collection, it is essential to consistently assess whether micrographs being collected are suitable for use and to adjust the microscope or change to new areas of the grid accordingly.

For the NAIP5–NLRC4 inflammasome and other samples that exhibit preferred orientations, a special consideration is whether the underrepresented views (in this case, side views) are present. If not, in the case of these inflammasomes, it is likely necessary to move to an area with thicker ice to see all views. To limit the resolution-diminishing impact of greater ice thickness, choose the grid areas with the thinnest ice that contains side views (see Fig. 3A for a micrograph from an optimal grid area). Note that to best visualize particles in thicker ice, the defocus may need to be higher magnitude than is typically used for cryo-EM; we used a range of $-1.8\,\mu m$ through $-4.0\,\mu m$. Using a phase plate and/or energy filter (if available) may alleviate this problem, but we did not have a chance to experiment with either for our work.

To truly maximize the potential of a microscope session, when possible, one should initiate some on-the-fly, preliminary data processing as each micrograph is collected. Early-stage processing (e.g., particle picking and contrast transfer function estimation) can now be conducted rapidly with GPU-based programs such as Gautomatch and Gctf (Zhang, 2016; Zhang & Zhang, n.d.). A new GPU-based program, Warp, offers another option for quick preprocessing (Tegunov & Cramer, 2018). Even programs capable of 2D and 3D classification and refinement, including CryoSPARC and RELION, can now be run on GPUs and provide feedback on the quality of data as it is being collected (Kimanius, Forsberg, Scheres, & Lindahl, 2016; Punjani, Rubinstein, Fleet, & Brubaker, 2017).

5.2 Micrograph selection and particle picking

Although options exist for automatic micrograph selection, it is our experience that it is well worth the time to check each micrograph manually. The usual considerations, such as avoiding micrographs with excessive ice contamination or drift that cannot be adequately corrected, apply to the NAIP–NLRC4 inflammasomes. However, there are some additional characteristics to note about micrographs of these complexes.

During data collection, the microscopist should have attempted to find grid areas that have ice just thick enough to support side views (see Section 5.1). Some images from these areas may have ice that is too thick to produce high-resolution reconstructions; these images will appear dimmer and the particles will not seem as crisp. (Be sure to adjust display contrast and brightness to avoid throwing away potentially useful micrographs.) Other images from areas directly adjacent to ideal ones may suffer the opposite problem: the ice may be slightly too thin, resulting in images with few side views—or, at worst, images containing many damaged particles that appear fuzzy or distorted. These micrographs should also be discarded.

Particle picking is another process with many options for full automation. With the NAIP5–NLRC4 inflammasome, automatic picking may be sufficient for real-time assessment of data quality during a session. However, for this sample, we found that the best option when seeking high resolution is to pick particles manually or to use manual picking to generate 2D class averages equally representing all views of the complex and to use these class averages as templates for automatic picking. This is because the side views resemble the top views so little that non-template-based automatic picking methods can often be made to select either side or top views, but not both. We tried RELION and FindEM for template-based picking, with comparable results (Roseman, 2004; Scheres, 2012).

At minimum, if automatic picking is used, particle picks from several micrographs of various defocus values from different grid areas should be examined to ensure few to no good particles are missed while undesirable particles (such those that appear "exploded" or indistinct) and non-particles are not selected. Inspecting micrographs and picked particle can even be combined: using RELION, one can easily open micrographs, add or remove particle picks, and discard any micrographs that are not of acceptable quality using the graphical user interface.

5.3 Sorting particles in 2D and 3D

Methods for sorting particles after extracting them from micrographs are as varied as the samples to which they are applied. In most cases, 2D classification is done after particle extraction and preprocessing. In principle, 3D classification can be done without classifying in two dimensions first, but in our case including the preliminary 2D step led to better results.

2D classification can yield interesting results on its own. For example, we observed early on that NAIP5–NLRC4 inflammasomes vary in the number of subunits they contain. 2D results can also be used as an indicator of the potential of a dataset. For the NAIP5–NLRC4 inflammasome, for instance, the number of particles going into each 2D class is of interest—if there is a dearth of side views at this stage, it should not come as a surprise if the 3D reconstruction is of low quality or exhibits strong resolution anisotropy. Also, if 2D class averaging yields classes in which internal structural details such as alpha helices can be seen, it is more likely that the data contains enough high-resolution information to generate a top-quality reconstruction. (As with micrograph inspection, it may be necessary to adjust display contrast and brightness when checking 2D classes for these details.) Often, particles sorted into 2D classes that appear hazy or do not resemble the sample of interest are discarded at this stage; we did this with our NAIP5–NLRC4 inflammasome data (see Fig. 3B).

The standard next step, 3D class averaging, is a stage of processing that allows the user considerable flexibility and can make the difference between solving a high-resolution structure and achieving nothing. (Note that in this section we generally refer to 3D classification practices in RELION and CryoSPARC, the two programs of which we made the most use). Hypothetically, one could proceed directly to 3D refinement using all particles, but conducting 3D classification first offers an opportunity not only to rid the dataset of more of low-quality (junk) particles, but also to sort out heterogeneity and sometimes even to glean some insights into the biology. For example, in the case of our NAIP5–NLRC4 inflammasome data, we immediately observed that, no matter what processing parameters we used, each "non-junk" class consisted of an open ring of identical-looking subunits nucleated by a single larger subunit. The smaller subunits were later confirmed to be NLRC4 monomers, whereas the larger subunit was NAIP5 bound to flagellin.

It is always prudent to try 3D classification using a variety of parameters. Varying the number of classes, for instance, sometimes yields dramatically different results. With too few classes, junk may not be sorted out, and if the sample is heterogeneous, dissimilar particles will be forced into classes with each other. With too many classes, particles that do not actually represent different states may be sorted apart from one another, resulting in a lower particle number available for refinement and sometimes, thus, a final reconstruction of lower resolution. It is therefore critical to attempt the process several times, with different parameters, and ideally with more than one program to see which yields the best results after the most promising classes (i.e., those that achieve the highest resolution in the areas of interest) are refined.

Due to heterogeneity, the 3D classification process is especially important when working with NAIP5–NLRC4 inflammasomes. It may be worthwhile to select the best classes from the first round of three-dimensional classification and subject those classes to further rounds of classification. This strategy did not yield any reconstructions with higher resolution in our case, possibly because of dwindling particle numbers as the classification process went on—with a greater starting number of particles, the results may have been better. However, the process did help confirm our suspicions that these inflammasomes exist in a variety of states, varying especially in the number of NLRC4 subunits each ring contains (see Fig. 3D).

We also noticed that in no case did we find 3D classes in which the inflammasome rings fully closed, with NLRC4's donor surface contacting NAIP5—rather, the inflammasomes appeared to twist into short helices, a finding supported but not conclusively shown by some further structural modeling our group conducted. This finding was not central to our results; nevertheless, it is in alignment with some other observations and may have functional implications (Diebolder, Halff, Koster, Huizinga, & Koning, 2015; Li et al., 2018; Matyszewski et al., 2018). Thus, extensive three-dimensional classification can provide biological insights, even when the result is not a high-resolution structure—although, of course, that is usually one of the goals of any structural study.

5.4 Refinement and modeling

Once a 3D class of interest has been selected, 3D refinement of particles from the class is needed to produce a structure with the highest resolution possible. As with 3D classification, it is advisable to run the refinement several times using different parameters, as the results may vary substantially in quality. Overall resolution may vary from refinement to refinement as parameters are manipulated, and local resolution should also be checked—we paid closest attention to refinements that had the best local resolution in areas we believed to be most biologically relevant, such as the interface between NAIP5 and flagellin.

It is advisable to check for resolution anisotropy (better resolution in one direction than another) at this stage and, if it is observed, to determine whether it is present to such a degree that it would hinder any further analysis. If resolution anisotropy is beyond a manageable level, first ensure that side views are not being accidentally discarded during sorting (see Section 5.3). On this issue, a useful diagnostic is to run 2D class averaging on the particles that went into the 3D refinement in question and check whether the proportion of side views is on par with that of the entire dataset. If the proportion of side views is much lower in the set of particles going into the refinement, redoing the 3D classification with different parameters may help. If the proportion of side views was very low from the beginning, as seen in the original 2D class averages, more data may need to be collected and pooled with the existing dataset to ameliorate the resolution anisotropy. To minimize the additional labor and risk of error when collecting additional data to pool with the first dataset, it is easiest to use the same data collection parameters (e.g., pixel size) each time.

Optimally, the data to be added would come from grids with a greater proportion of side views. But even if all conceivable means to increase the proportion of side views have already been explored (see Section 4.1, for our solutions), increasing the absolute number of side views simply by collecting more data may still improve overall resolution and reduce or resolve anisotropy as the improvements in resolution in one direction from greater and greater numbers of top views come to a halt. We found this to be the case with the number of particles used throughout our analysis (under one million), although this brute-force method may not be seen as the optimal way to make use of microscope time if there is concurrently the problem of low number of particles per micrograph (see Section 4.1, for our suggested solution to that issue). After collecting this additional data, there was no noticeable resolution anisotropy in our final reconstructions, despite the fact that we were not able to totally eliminate the problem of preferred orientations.

It is also a good idea to try the refinement in more than one program. We obtained results of similar resolution using RELION and CryoSPARC, but RELION provided more useful results because, at the time of our work, RELION but not CryoSPARC offered masked local refinement. Thus, using RELION, one could focus the refinement on the region of greatest interest. In the case of the NAIP5–NLRC4 inflammasome structure, we made extensive use of masking. Since the particles varied in the number of NLRC4 subunits they contained, we found (as would be expected) areas of much lower resolution farther from the nucleation point of the ring due to variable occupancy.

For this reason, we applied a mask around the region containing only NAIP5, flagellin, and the first two NLRC4 subunits at the start of local searches during refinement in RELION. The rationale for using a mask around this particular region was that we could capture the structures of all three unique components of the NAIP5–NLRC4 inflammasome as well as all the possible interfaces (NAIP5–flagellin, NAIP5–NLRC4, and NLRC4–NLRC4) while ignoring the most heterogeneous areas. This allowed us to pool particles from classes that had different numbers of NLRC4 subunits, increasing our final resolution. We also attempted locally masked refinement with different numbers of subunits, without the flexible portion of flagellin, and without the CARDs, as well as masking during 3D classification, but did not achieve better results. To generate masks, we segmented the areas of interest from the full EM density of the inflammasome in Chimera and saved them separately, then used them as inputs for RELION's automatic mask generation tool with varying parameters (Pettersen et al., 2004; Scheres, 2012). We ran refinements and classifications with masks of varying tightness before discovering the one that worked best in our case. Now that local masking during refinement is an option in CryoSPARC, it would be interesting to apply the program to our data again.

It is worth noting that even when masked local refinement is not done, choice of the overall mask used during refinement can make a great difference in the resolution of the resulting structure. A too-tight mask that cuts into the density can lead to falsely inflated calculated resolution, whereas one that is too loose may result in resolution that is lower than could be achieved with an optimal mask. Thus, it is generally worth experimenting with different masks during refinement, regardless of the sample and regardless of whether locally masked refinement is being done.

Once refinement and any postprocessing are complete, the choice of structural modeling methods depends on a multitude of factors, perhaps most important of which is resolution. Resolution dictates whether a model can be built into the EM density de novo, for example, and has a major impact on how confident one can be in a model. Since the resolution of our NAIP5–NLRC4 inflammasome structure varied widely from region to region, we filtered the structure by local resolution, using different modeling methods in different areas. We refer any readers interested in our full modeling methods to our recent paper, where they are described in extensive detail (Tenthorey et al., 2017). However, it is our hope that future researchers studying these inflammasomes will improve on the resolution we achieved, necessitating different modeling techniques, a full explanation of which is well beyond the scope of this chapter.

6. Concluding remarks

Despite the existence of some published structures, our understanding of inflammasome structures and their functional implications remains far from complete. We hope that the methods described in this chapter will help others step beyond the frontiers of our current knowledge by increasing the resolution of existing structures and by investigating other states, such as the caspase-1-bound inflammasome. We also hope this chapter will find its way into the hands of researchers struggling with other difficult samples and serve as a starting point to guide their studies.

Acknowledgments

We are grateful to Tom Houweling and Abhiram Chintangal for computational support, Basil Greber and Robert Louder for EM discussions, Sarah Sterling for her work on the Teflon well plate, and Andreas Martin for advice. We thank Ashley Truxal for illustrating Fig. 2. This work was funded by NIH grants AI075039 and AI063302 (R.E.V.). N.H. and J.L.T. were supported by the NSF Graduate Research Fellowship Program. R.V. and E.N. are Howard Hughes Medical Institute Investigators.

References

Aglietti, R. A., Estevez, A., Gupta, A., Ramirez, M. G., Liu, P. S., Kayagaki, N., et al. (2016). GsdmD p30 elicited by caspase-11 during pyroptosis forms pores in membranes. *Proceedings of the National Academy of Sciences of the United States of America*, *113*, 7858–7863. https://doi.org/10.1073/pnas.1607769113.

Bergsbaken, T., Fink, S. L., & Cookson, B. T. (2009). Pyroptosis: Host cell death and inflammation. *Nature Reviews. Microbiology*, *7*, 99.

Broz, P., & Dixit, V. M. (2016). Inflammasomes: Mechanism of assembly, regulation and signalling. *Nature Reviews. Immunology*, *16*, 407–420.

Chen, X., He, W.-T., Hu, L., Li, J., Fang, Y., Wang, X., et al. (2016). Pyroptosis is driven by non-selective gasdermin-D pore and its morphology is different from MLKL channel-mediated necroptosis. *Cell Research*, *26*, 1007–1020. https://doi.org/10.1038/cr.2016.100.

Diebolder, C. A., Halff, E. F., Koster, A. J., Huizinga, E. G., & Koning, R. I. (2015). Cryoelectron tomography of the NAIP5/NLRC4 inflammasome: Implications for NLR activation. *Structure*, *23*, 2349–2357. https://doi.org/10.1016/j.str.2015.10.001.

Ding, J., Wang, K., Liu, W., She, Y., Sun, Q., Shi, J., et al. (2016). Pore-forming activity and structural autoinhibition of the gasdermin family. *Nature*, *535*, 111.

Han, B.-G., Watson, Z., Kang, H., Pulk, A., Downing, K. H., Cate, J., et al. (2016). Long shelf-life streptavidin support-films suitable for electron microscopy of biological macromolecules. *Journal of Structural Biology*, *195*, 238–244. https://doi.org/10.1016/j.jsb.2016.06.009.

Hu, Z., Yan, C., Liu, P., Huang, Z., Ma, R., Zhang, C., et al. (2013). Crystal structure of NLRC4 reveals its autoinhibition mechanism. *Science*, *341*, 172–175.

Hu, Z., Zhou, Q., Zhang, C., Fan, S., Cheng, W., Zhao, Y., et al. (2015). Structural and biochemical basis for induced self-propagation of NLRC4. *Science*, *350*, 399–404.

Jones, J. D. G., Vance, R. E., & Dangl, J. L. (2016). Intracellular innate immune surveillance devices in plants and animals. *Science*, *354*, aaf6395-1–aaf6395-8.

Kimanius, D., Forsberg, B. O., Scheres, S. H. W., & Lindahl, E. (2016). Accelerated cryo-EM structure determination with parallelisation using GPUs in RELION-2. *eLife*, *5*, e18722. https://doi.org/10.7554/eLife.18722.

Kofoed, E. M., & Vance, R. E. (2011). Innate immune recognition of bacterial ligands by NAIPs dictates inflammasome specificity. *Nature*, *477*, 592–595. https://doi.org/10.1038/nature10394.

Li, Y., Fu, T.-M., Lu, A., Witt, K., Ruan, J., Shen, C., et al. (2018). Cryo-EM structures of ASC and NLRC4 CARD filaments reveal a unified mechanism of nucleation and activation of caspase-1. *Proceedings of the National Academy of Sciences of the United States of America*, *115*, 10845–10852. https://doi.org/10.1073/pnas.1810524115.

Liu, X., Zhang, Z., Ruan, J., Pan, Y., Magupalli, V. G., Wu, H., et al. (2016). Inflammasome-activated gasdermin D causes pyroptosis by forming membrane pores. *Nature*, *535*, 153.

Lu, A., Li, Y., Schmidt, F. I., Yin, Q., Chen, S., Fu, T.-M., et al. (2016). Molecular basis of caspase-1 polymerization and its inhibition by a new capping mechanism. *Nature Structural & Molecular Biology*, *23*, 416–425. https://doi.org/10.1038/nsmb.3199.

Martinon, F., Burns, K., & Tschopp, J. (2002). The Inflammasome: A molecular platform triggering activation of inflammatory caspases and processing of proIL-β. *Molecular Cell*, *10*, 417–426. https://doi.org/10.1016/S1097-2765(02)00599-3.

Matyszewski, M., Zheng, W., Lueck, J., Antiochos, B., Egelman, E. H., & Sohn, J. (2018). Cryo-EM structure of the $NLRC4^{CARD}$ filament provides insights into how symmetric and asymmetric supramolecular structures drive inflammasome assembly. *Journal of Biological Chemistry*, *293*, 20240–20248. https://doi.org/10.1074/jbc.RA118.006050.

Palovcak, E., Wang, F., Zheng, S. Q., Yu, Z., Li, S., Betegon, M., et al. (2018). A simple and robust procedure for preparing graphene-oxide cryo-EM grids. *Journal of Structural Biology*, *204*, 80–84. https://doi.org/10.1016/j.jsb.2018.07.007.

Passmore, L. A., & Russo, C. J. (2016). Specimen preparation for high-resolution cryo-EM. *Methods in Enzymology*, *579*, 51–86. https://doi.org/10.1016/bs.mie.2016.04.011.

Pettersen, E. F., Goddard, T. D., Huang, C. C., Couch, G. S., Greenblatt, D. M., Meng, E. C., et al. (2004). UCSF chimera—A visualization system for exploratory research and analysis. *Journal of Computational Chemistry*, *25*, 1605–1612. https://doi.org/10.1002/jcc.20084.

Punjani, A., Rubinstein, J. L., Fleet, D. J., & Brubaker, M. A. (2017). cryoSPARC: Algorithms for rapid unsupervised cryo-EM structure determination. *Nature Methods*, *14*, 290.

Rayamajhi, M., Zak, D. E., Chavarria-Smith, J., Vance, R. E., & Miao, E. A. (2013). Mouse NAIP1 detects the type III secretion system needle protein. *Journal of Immunology (Baltimore, Md. : 1950)*, *191*, 3986–3989. https://doi.org/10.4049/jimmunol.1301549.

Roseman, A. M. (2004). FindEM—A fast, efficient program for automatic selection of particles from electron micrographs. *Journal of Structural Biology*, *145*, 91–99. https://doi.org/10.1016/j.jsb.2003.11.007.

Ruan, J., Xia, S., Liu, X., Lieberman, J., & Wu, H. (2018). Cryo-EM structure of the gasdermin A3 membrane pore. *Nature*, *557*, 62–67. https://doi.org/10.1038/s41586-018-0058-6.

Russo, C. J., & Passmore, L. A. (2014). Controlling protein adsorption on graphene for cryo-EM using low-energy hydrogen plasmas. *Nature Methods*, *11*, 649.

Russo, C. J., & Passmore, L. A. (2016). Progress towards an optimal specimen support for electron cryomicroscopy. *Current Opinion in Structural Biology*, *37*, 81–89. https://doi.org/10.1016/j.sbi.2015.12.007.

Sborgi, L., Rühl, S., Mulvihill, E., Pipercevic, J., Heilig, R., Stahlberg, H., et al. (2016). GSDMD membrane pore formation constitutes the mechanism of pyroptotic cell death. *The EMBO Journal*, *35*, 1766–1778. https://doi.org/10.15252/embj.201694696.

Scheres, S. H. W. (2012). RELION: Implementation of a Bayesian approach to cryo-EM structure determination. *Journal of Structural Biology*, *180*, 519–530. https://doi.org/10.1016/j.jsb.2012.09.006.

Shi, Y. (2004). Caspase activation: Revisiting the induced proximity model. *Cell*, *117*, 855–858. https://doi.org/10.1016/j.cell.2004.06.007.

Tegunov, D., & Cramer, P. (2018). Real-time cryo-EM data pre-processing with warp. *bioRxiv*. 338558. https://doi.org/10.1101/338558.

Tenthorey, J. L., Haloupek, N., López-Blanco, J. R., Grob, P., Adamson, E., Hartenian, E., et al. (2017). The structural basis of flagellin detection by NAIP5: A strategy to limit pathogen immune evasion. *Science*, *358*, 888–893.

Wang, L., Ounjai, P., & Sigworth, F. J. (2008). Streptavidin crystals as nanostructured supports and image-calibration references for cryo-EM data collection. *Journal of Structural Biology*, *164*, 190–198. https://doi.org/10.1016/j.jsb.2008.07.008.

Yang, X., Yang, F., Wang, W., Lin, G., Hu, Z., Han, Z., et al. (2017). Structural basis for specific flagellin recognition by the NLR protein NAIP5. *Cell Research*, *28*, 35.

Yang, J., Zhao, Y., Shi, J., & Shao, F. (2013). Human NAIP and mouse NAIP1 recognize bacterial type III secretion needle protein for inflammasome activation. *Proceedings of the National Academy of Sciences of the United States of America*, *110*, 14408–14413. https://doi.org/10.1073/pnas.1306376110.

Zhang, K. (2016). Gctf: Real-time CTF determination and correction. *Journal of Structural Biology*, *193*, 1–12. https://doi.org/10.1016/j.jsb.2015.11.003.

J. (Kai) Zhang, Jack (Kai) Zhang. (n.d.). http://www.mrc-lmb.cam.ac.uk/kzhang/, Accessed 8 May 2019.

Zhao, Y., Yang, J., Shi, J., Gong, Y.-N., Lu, Q., Xu, H., et al. (2011). The NLRC4 inflammasome receptors for bacterial flagellin and type III secretion apparatus. *Nature*, *477*, 596–600.

CHAPTER THIRTEEN

An integrative protocol for the structure determination of the mouse ASC-PYD filament

Adam Mazur[a], Petr Broz[b], Sebastian Hiller[a,*]
[a]Biozentrum, University of Basel, Basel, Switzerland
[b]Department of Biochemistry, University of Lausanne, Lausanne, Switzerland
*Corresponding author: e-mail address: sebastian.hiller@unibas.ch

Contents

Abstract

Several inflammasomes that activate as part of the eukaryotic innate immune response contain long helical filaments of the adaptor protein ASC as a central structural element. Here, we describe a technical protocol that has enabled the structure determination of the filament of the ASC pyrin domain. The protocol integrates data from cryo-electron microscopy and solid-state NMR spectroscopy into a single simulated annealing protocol to determine structural coordinates that fit all input data optimally. The structure shows that the ASC pyrin domain filament is formed by helical stacking of individual pyrin domains forms and that the CARD domains are flexibly attached to the filament outside. An artificial perturbation of the input data shows that the integrated structure determination protocol can allow high quality structures even at resolutions of the electron density map as low 8 Å. The protocol is extendable to other structural input data from biochemical or biophysical experiments.

Methods in Enzymology, Volume 625
ISSN 0076-6879
https://doi.org/10.1016/bs.mie.2019.04.033

1. Introduction

Inflammasomes are large multi-protein complexes that play a key structural role in the eukaryotic innate immune response (Broz & Dixit, 2016; Hoss, Rodriguez-Alcazar, & Latz, 2017). Inflammasome assembly is initiated by specific receptor proteins that sense distinct signals ranging from pathogen surface molecules to toxic substances in the cellular cytosol (Lamkanfi & Dixit, 2012). Upon activation of a typical receptor, a single macromolecular inflammasome per cell is formed, which contains large filaments of the adaptor protein ASC (apoptosis-associated speck-like protein containing a caspase recruitment domain) as the central architectural component (Lu et al., 2014; Masumoto et al., 1999). These inflammasome filaments are strongly branched, forming a μm-sized "speck" in the cellular cytosol. The main function achieved by this polymerization of ASC is the amplification of the original signal (Dick, Sborgi, Ruhl, Hiller, & Broz, 2016). Inflammasome activation results in the maturation of large amounts of caspase-1, which in turn leads to the activation of the pro-inflammatory cytokine interleukin-1β and of gasdermin-D, eventually causing pyroptotic cell death (Mulvihill et al., 2018; Sborgi et al., 2016; Schett, Dayer, & Manger, 2016; Sims & Smith, 2010). The protein ASC, which forms the structural core of the inflammasome, is a bipartite protein comprised of two independent ~100-residue domains connected by an unstructured linker: a pyrin domain (PYD) and a caspase recruitment domain (CARD). An atomic level characterization of ASC structure and dynamics and its assembly to a filament is essential for understanding inflammation processes.

Structure determination of long, insoluble protein filaments is possible by either of two experimental methods, solid-state NMR spectroscopy (He et al., 2016; Vasa et al., 2015; Wasmer et al., 2008) and cryo-electron microscopy (cryo-EM) (Chou & Pollard, 2019; Fitzpatrick et al., 2017; Guerrero-Ferreira et al., 2018; Yu, Qu, & Modis, 2018). While each of these methods is capable of determining filament structures, the techniques are also mutually complementary: cryo-EM delivers an electron density map of the well-ordered, coherently repetitive regions of a filament, in typical cases at atomic or near-atomic resolution. Dynamic and disordered fragments of the polypeptide chain are, however, not resolved. Solid-state NMR spectroscopy provides local structural information for both the well-ordered and dynamic structural motifs. When the two distinct sources of experimental data are combined toward a single hybrid structure determination, the local character of NMR-derived experimental restraints complements the global filament structure information available from cryo-EM. A recent review summarizes combined use of cryo-EM and NMR methods (Cuniasse, Tavares, Orlova, & Zinn-Justin, 2017). Successful applications have been reported for the Shigella type-III secretion needle (Demers et al., 2014), the amyloid-β(1–42) fibril structure (Gremer et al., 2017), the tubular HIV-1 capsid assembly (Perilla et al., 2017), the dodecameric aminopeptidase TET2 (Gauto et al., 2018) and, by the present authors and colleagues, the mouse ASC-PYD filament (Ravotti et al., 2016; Sborgi et al., 2015). The methods to combine data differ considerably among these studies, mainly due to the use of various isotope labeling schemes, which hence leads to different collection schemes of solid-state NMR restraints. Here, we describe methodological details underlying the structure determination protocol of the mouse ASC-PYD filament, with a focus on the computational aspects of the work.

2. Experimental data

2.1 Sample preparation and characterization

The mouse ASC-PYD in its monomeric state is soluble to mM concentrations in aqueous solutions at conditions of pH 4.0 (de Alba, 2009). At physiological conditions around pH 7.0, in turn, the protein shows a high propensity for filament formation. Consequently, upon a pH jump from 4.0 to 7.0, the oligomerization of ASC-PYD leads to the formation of macroscopic filaments with lengths in the range of 500–2000 nm, as observed by negative-stained transmission electron microscopy (TEM) (Sborgi et al., 2015). The filaments are macroscopically homogeneous, in contrast to filaments of the entire ASC protein, which are heavily branched, resembling in their morphology the native inflammasome speck. The microscopic homogeneity of the ASC-PYD filaments was assessed by solid-state NMR spectroscopy on a uniformly ^{13}C-labeled sample. 2D [^{13}C,^{13}C]-dipolar assisted rotational resonance (DARR) and 2D NCA correlation spectra of the ASC-PYD and the full-length ASC filaments showed narrow peaks, with average ^{13}C line widths of around 0.5 ppm (Sborgi et al., 2015). In these spectra, only a single set of peaks was observed, which is a direct evidence for microscopic structural homogeneity and the absence of polymorphism. In addition, the observation of a single set of resonances implies symmetry equivalence of all ASC-PYD monomers in the multimeric filament form. Notably, the 2D

[^{13}C,^{13}C]-DARR and 2D NCA spectra of ASC-PYD and full-length ASC filaments superimposed virtually perfectly, thus indicating that the rigid core of the ASC filaments is formed solely by the ASC-PYD domain (Sborgi et al., 2015). These filament preparations are thus suitable for structural analysis.

2.2 Solid-state NMR spectroscopy

[*U*-^{15}N,^{13}C]-labeled samples of ASC-PYD filaments were packed into 3.2mm ZrO_2 rotors (Bruker Biospin) by ultracentrifugation using a custom-made filling device. The spectra were recorded on a Bruker Avance II+ spectrometer operating at 850MHz ^{1}H frequency. The sample temperature was maintained at 288K during the experiments. The backbone assignment was based on analysis of a standard set of experiments: NCOCA, NCACO, NCACB and CANCO supplemented by N(CO)CACB and CAN(CO)CA to complete the assignment in the spectrally overlapped regions (Ravotti et al., 2016). All spectra were recorded at 17kHz magic angle spinning (MAS) frequency. The side chain assignment was based on the analysis of 3D CCC and N(CA)CBCX and 2D DARR spectra. Long-range spin contacts were obtained from 2D CHHC (Lange, Luca, & Baldus, 2002) and PAR (De Paëpe, Lewandowski, Loquet, Böckmann, & Griffin, 2008) experiments recorded at 15kHz MAS frequency. The MAS frequency was decreased for these experiments to avoid overlap with the sideband in the aromatic region of the spectra. The spectra were processed using TopSpin (Bruker Biospin) software using shifted squared cosine window function and were analyzed with the CCPN package (Vranken et al., 2005).

On the basis of these spectra, the backbone nuclei of residues 3–85 were sequence-specifically assigned. Chemical shifts depend on the local protein structure and therefore, the secondary chemical shifts of α- and β-carbons strongly correlate with protein secondary structure (Wishart & Sykes, 1994). The analysis of secondary chemical shifts revealed the presence of six distinct α-helices in the filament form of ASC-PYD (Fig. 1A). In contrast, the resonances of the N-terminal residues 1–3 and the C-terminal residues 86–91 were detected only weakly or not at all in the dipolar-transfer-based NMR experiments and they were also absent in an INEPT-type [^{15}N, ^{1}H]-correlation spectrum, strongly indicating that these signals are line-broadened because of intermediate time scale dynamics. Subsequently, unambiguous side chain assignments were obtained for 80% of carbon nuclei. Thereby, the unassigned spins were either not observed, most probably due to local

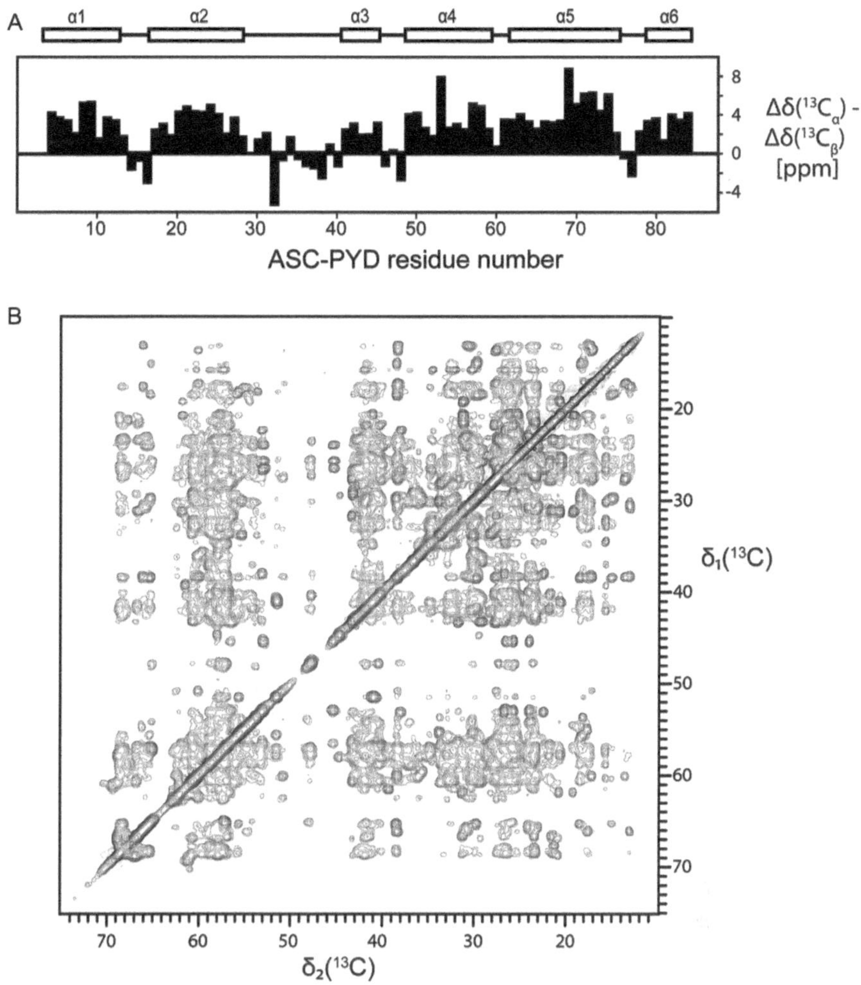

Fig. 1 Solid-state NMR spectroscopy of ASC-PYD in its filament form. (A) Secondary chemical shifts $\Delta\delta = \delta_{\text{observed}} - \delta_{\text{random coil}}$ of C_α minus C_β atoms. These values are strongly correlated with the protein secondary structure. Consecutive stretches of $\Delta\Delta\delta > +3$ ppm indicate α-helices. The analysis reveals the presence of six α-helices, as indicated on top. (B) Aliphatic region of a 2D [^{13}C, ^{13}C]-PAR spectrum, recorded on a sample of [*U*-^{13}C]-labeled ASC-PYD on an 850 MHz spectrometer at 15 kHz MAS frequency. With line widths of around 0.5 ppm, substantial spectral overlap is observed.

dynamics, or an unambiguous assignment was not possible due to spectral overlap. Although the long-range CHHC and PAR spectra (Fig. 1B) encode interspin distance information and thus structural information, the direct conversion to distance restraints is limited by the ambiguity of the cross-peak assignment.

2.3 Cryo-electron microscopy

Protein filaments were imaged using an FEI Titan Krios electron microscope operated at an acceleration voltage of 300 keV. Image acquisition was done using a 4k × 4k K2 Summit direct electron detector with a back thinned CMOS chip, operated in electron counting mode. Images were recorded in movie mode format and were automatically drift corrected and averaged using 2dx_automator (Scherer et al., 2014) employing MOTIONCORR (Li et al., 2013). The complete data registration resulted in 150 images. The microscope was operating in low-dose mode at a nominal magnification of 22,500 × with effective pixel size of 1.34 Å. The cumulated dose was below ~20 $e^-/Å^2$ distributed over 30 frames recorded over 5 s.

Data acquisition was followed by image processing. In the first step, the defocus was determined using CTFFIND3 (Mindell & Grigorieff, 2003) and the contrast transfer function (CTF) of the microscope was corrected by multiplication with the calculated CTF. The filament sections were segmented in total 21,138 overlapping square boxes of 320 × 320 pixel size, using e2helixboxer from EMAN2 package (Tang et al., 2007). The 3D reconstruction of electron density was done by iterative helical real space reconstruction (IHRSR) algorithm (Egelman, 2000) as implemented in SPIDER program (Frank et al., 1996; Frank, Shimkin, & Dowse, 1981). The helical parameters C3 point group symmetry with 53° right-handed rotation and 14.2 Å rise per subunit were imposed during the IHRSR calculations and a B-factor of $-150\,Å^2$ was used. As the outcome, a high-resolution electron density was obtained (Sborgi et al., 2015), with individual side chains being visible (Fig. 2). The Fourier shell correlation (FSC) between the refined model (see below) and electron density map was calculated and used to determine the resolution of the map to 3.7 Å (Fig. 3).

3. Structure determination protocol

3.1 Structure of the ASC-PYD monomer

In the first step of the structure calculation of the ASC-PYD filament, a *de novo* model of the monomeric ASC-PYD subunit was manually built into the electron density. To this end, the portion of the electron density map corresponding to one monomer was extracted as a separate density and the ASC-PYD polypeptide chain was built into this density map using the program Coot (Emsley, Lohkamp, Scott, & Cowtan, 2010). This model

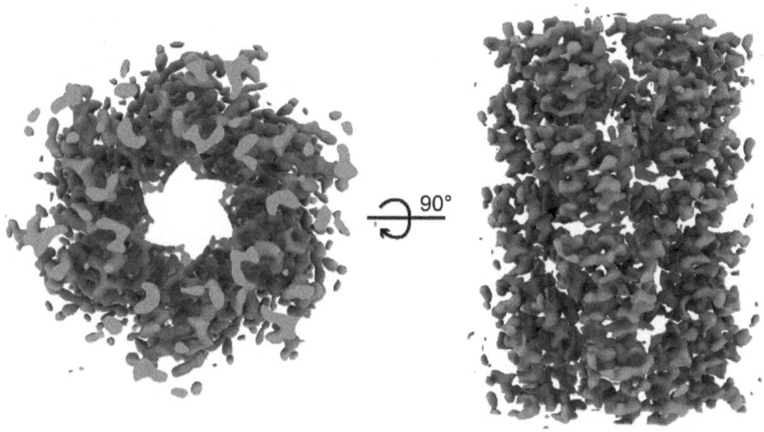

Fig. 2 Electron density of the ASC-PYD filament at 3.7 Å resolution. 150 images were recorded on an FEI Titan Krios cryo-electron microscope with K2 summit direct electron detector with a back thinned CMOS camera. Subsequent image processing consisted of contrast transfer function correction, drift correction, averaging and segmentation. The 3D electron density map was obtained by iterative helical real space reconstruction.

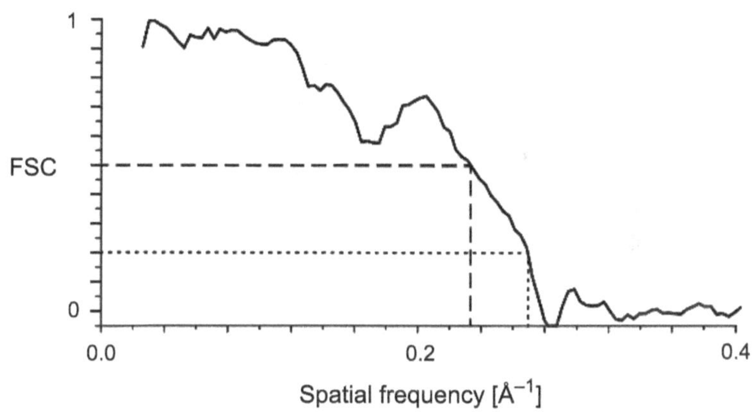

Fig. 3 Fourier Shell Correlation (FSC) between the electron density map and the refined structural model. FSCs of 0.2 (dotted line) and 0.5 (dashed line) correspond to resolutions of 3.7 and 4.3 Å, respectively.

building was supported by knowledge of the positions of α-helices from the solid-state NMR experiments (Fig. 1), and the correct orientation of the polypeptide chain was confirmed by examining local structural features of residues F23, R38, Y59, Y60 and Y64, which had well-resolved electron density of their bulky side chains. The single subunit structure was subsequently transferred into one position of the full-length ASC-PYD filament electron density map and then positioned by translation and rotation operations according to the filament symmetry parameters into 15 symmetrically equivalent positions (Model 1; Fig. 4).

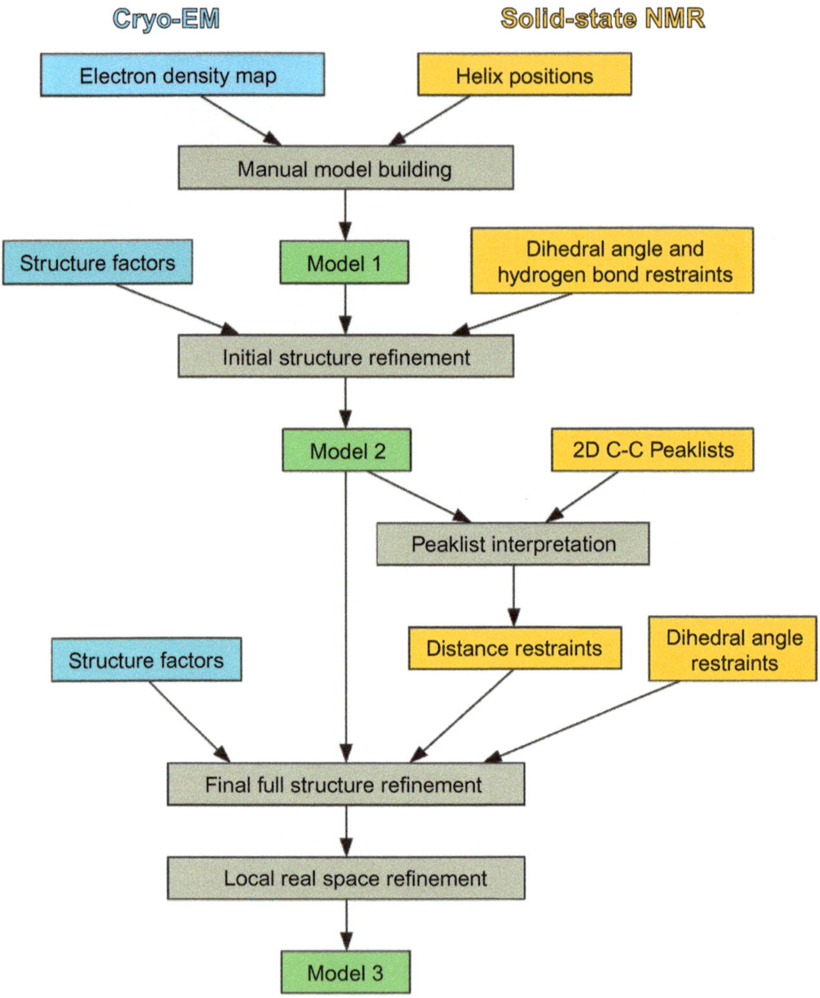

Fig. 4 Flowchart of the protocol for *de novo* ASC-PYD filament structure determination. Yellow and blue boxes represent experimental data derived from solid-state NMR and cryo-EM measurements, respectively. Gray-colored boxes describe model building and refinement steps. Intermediate and final structural models are marked in the green boxes.

3.2 Refinement of the ASC-PYD filament structure

Subsequently, model 1 was refined by a simulated annealing protocol using the program XPLOR-NIH (Schwieters, Kuszewski, Tjandra, & Clore, 2003) with experimental restraints derived from solid-state NMR and cryo-EM experiments. Generally in these calculations, the symmetry equivalence of all monomers in the filament model was enforced by applying noncrystallographic symmetry (NCS) potential. The electron density map was positioned in an artificial crystal lattice (space group P1, unit cell $a = b = c = 107.2$ Å, $\alpha = \beta = \gamma = 90°$) and Fourier transformation was applied to convert the real space density map to reciprocal space representation. This transformation, which fully preserves the data information content, was necessary to be able to combine the cryo-EM and NMR data in a single simultaneous structure refinement protocol with the software XPLOR-NIH. The conversion was performed using the *em_map_to_hkl.inp* module from CNS program (Brunger, 2007; Brunger et al., 1998). Both the amplitudes and phases of the generated structure factors were used as experimental diffraction data. Structural restraints for the backbone dihedral angles φ and ψ were derived from the solid-state chemical shifts by the software TALOS+ (Shen & Bax, 2015; Shen, Delaglio, Cornilescu, & Bax, 2009). Only restraints for residues classified as "good" quality by TALOS+ were kept, all others were disregarded. For the resulting six α-helices, hydrogen bond restraints with a connection of $i,i+4$ residues were defined for residues 4–10, 17–24, 41–42, 49–55, 62–71 and 79–90.

With these input data, the structure determination was accomplished by simulated annealing method consisting of (i) initial energy minimization, (ii) high temperature molecular dynamics (2000 steps), (iii) 1000 steps of cooling during which the temperature parameter was gradually decreased from 3000 to 100 K, and finally (iv), energy minimization of the system. In this protocol, it is essential, that the high-temperature step, in which the molecular system efficiently samples its conformational space, is sufficiently long to reach the proximity of the global energy minimum. The subsequent cooling scheme should be slow enough to ensure the establishment of thermal equilibrium at every intermediate step, allowing the system to relax into the global energy minimum. This protocol was repeated 100 times from the same starting configuration (Model 1, Fig. 4), but with different state of random number generator used for initial velocity assignment, and the 20 lowest energy structures from these 100 calculations were selected. At this step, an ensemble of models consistent with cryo-EM map, hydrogen bond restraints and backbone φ, ψ dihedral angle restraints had thus been generated (Model 2; Fig. 4). So far, the protocol did not exploit the structural information of the CHHC and PAR peak lists, since their manual interpretation was impeded by substantial chemical shift overlap.

3.3 Automatic interpretation of CHHC and PAR solid-state data

In the next step, we aimed at interpreting the through-space correlation experiments, the 2D [^{13}C, ^{13}C]-CHHC and [^{13}C, ^{13}C]-PAR spectrum (Fig. 1). While these spectra contained a total of 2483 cross peaks, their manual interpretation lead to only a total of 13 spectrally unambiguous cross-peak interpretations. This low number is due to the high degree of chemical shift overlap in a ^{13}C-dimension and the resulting high combinatorial ambiguity in cross-peak interpretation. Notably, in addition to the chemical shift ambiguity, any interpretation is ambiguous with respect to their relative molecule position, i.e., their emergence from either intra- or inter-subunit contacts. High degree of spectral crowding in uniformly ^{13}C-labeled samples is a well-known phenomenon, which can be diminished by adopting sparse labeling schemes (Castellani et al., 2002; Hong, 1999) or by mixed ^{13}C- and ^{15}N-labeled-samples for fibrils with dominant β-sheet structure (Van Melckebeke et al., 2010). The first approach requires laborious and expensive sample preparations. Using mixed ^{13}C and ^{15}N labeling is presumably not effective for α–helical filaments, because nitrogen atoms are mainly part of the protein backbone and thus buried deeply inside the monomeric subunit and are not strongly involved in inter-subunit contacts. However, the availability of a high-quality structural model (Model 2) provided us with a third alternative, namely, the automated, structure-based interpretation of these spectra.

3.3.1 Automatic peak list interpretation

2D [^{13}C,^{13}C]-CHHC and [^{13}C,^{13}C]-PAR spectra were automatically peak-picked and the peak lists were interpreted using the CANDID algorithm (Herrmann, Güntert, & Wüthrich, 2002) as implemented in the CYANA software package (Güntert, 2004). The CANDID algorithm was originally developed to enable efficient NMR protein structure determinations by automatic interpretation of the NOESY spectra. CANDID introduces two powerful concepts: network-anchoring and constraint-combination. Network-anchoring relies on the fact that any network of correct NOE cross-peak assignments form a self-consistent set. Constraint-combination rules reduce the impact of experimental uncertainties and chemical shift assignment imperfections on the resulting structure. The CHHC and

PAR peak list data is conceptionally equivalent to solution NMR NOESY experiments, and we thus could use the CANDID software to automatically assign the corresponding cross peaks. In the first step, the structural model 2 was reduced to a trimer, which is the smallest multimeric assembly retaining all possible inter-subunit contacts. Then, the last cycle (cycle 7) of automatic structure calculation in CYANA was directly executed, omitting the initial six steps of the classical protocol. The calculation was performed with a ^{13}C chemical shift tolerance of 0.2 ppm, automatic integral calibration and using the trimeric model 2 as a structural template. Uniform 7.0 Å upper distance limit were applied to all observed cross peaks, since solid-state NMR cross-peak integrals are complicated functions of inter-spin distance, local environment and experimental settings (De Paëpe et al., 2008). CANDID algorithm derives the peak interpretation using exclusively distance information from the structural model. The intra- or intermolecular origin of the distances is irrelevant and both categories are treated equally. The automated peak interpretation was successful and 90% of the observed cross-peaks were assigned and subsequently converted to a total of 674 unique distance restraints with 65 restraints being unambiguously classified as intermolecular.

3.3.2 Validation of the peak interpretation procedure

When considering the dense carpet of 2D correlation cross-peaks in the 2D [^{13}C, ^{13}C] solid-state NMR spectra (Fig. 1), and the strong chemical shift ambiguity, it might be suspected that the automatic peak interpretation algorithm will support any given input structure with possible peak interpretations. To address this issue, we wanted to quantify the uniqueness of the information content of the peak lists specific for the ASC-PYD filament. We thus repeated the CANDID calculations using the peak lists from a 2D [^{13}C, ^{13}C] experiment that was recorded on a sample of microcrystalline ubiquitin instead of the correct peak lists, but with all other parameters identical, i.e., the ASC-PYD structural model 2 and resonance assignment. In this control calculation, the CANDID peak interpretation routine should interpret less peaks than for the original calculation, because many possible peak interpretations will not fit to the chemical shift assignments and the structural template model. Indeed, while 86% of the cross peaks of the ASC filament spectra are interpreted with chemical shift tolerances of 0.2 ppm (Fig. 5), only 35% of the ubiquitin peak list were interpreted. Even with very broad chemical shift tolerances of 0.4 ppm, the ubiquitin peak list remained significantly non-interpreted with just 70% assigned cross peaks. These findings confirm the high level of compatibility among the structure model and the ASC-PYD peak lists.

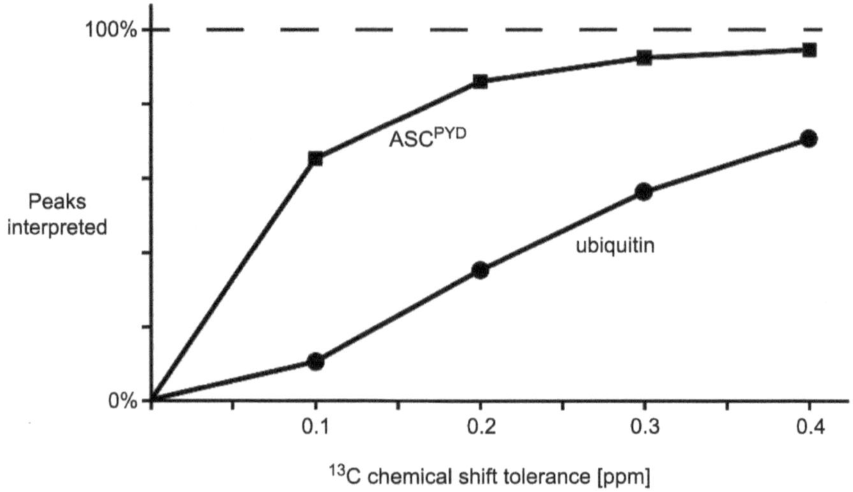

Fig. 5 Coverage of automated cross peak interpretation with the algorithm CANDID. 2D [^{13}C, ^{13}C]-PAR spectrum of ASC-PYD (squares) and of ubiquitin (circles) were interpreted using the initially refined ASC-PYD structure model (Model 2). The level of peak list interpretation is significantly higher for ASC-PYD data than ubiquitin, confirming that the ASC-PYD peak list indeed contains non-trivial structural information.

3.4 Second-stage full structure refinement of the ASC-PYD filament

Based on the CANDID-interpreted peak lists and the resulting distance constraints, a second round of the structure refinement was performed (Fig. 4). This step was using as input (i) the fundamental constraints on subunit symmetry, (ii) the structure factors from the cryo-EM map transformation, (iii) the TALOS+ dihedral angle restraints and (iv) the distance restraints from CYANA/CANDID automatic assignment. Hydrogen bond distance restraints were not used during this structure calculation. The simulated annealing protocol was carried out with 10,000 steps of high temperature dynamics at 3000 K, subsequent cooling in 5000 steps and final energy minimization. The agreement between the structural models and the experimentally derived structure factors was evaluated by calculation of R-factors. To assess potential overfitting of the data, the structure factors were randomly assigned to two distinct categories: "work" and "free," with 95% and 5% partitioning, respectively. The free set was not used during the refinement to assess extent of possible overfitting. The values of the R_{work} and R_{free} were 0.32 and 0.33, respectively, confirm that the data were not overfitted. As in the previous stage, 100 structures were initially

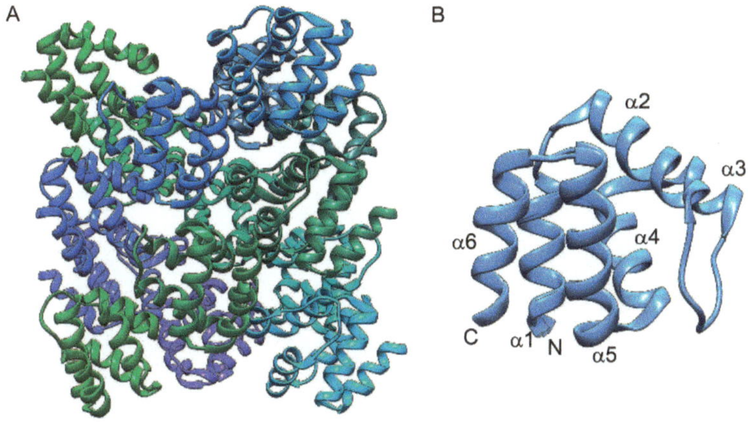

Fig. 6 Final structure of ASC-PYD filament (Model 3, PDB 2N1F) and model for the full ASC filament. (A) Ribbon representation of the filament. (B) Single subunit structure with annotated secondary structure elements.

generated starting from the previously refined configuration (Model 2, Fig. 4) and now the 10 lowest energy structures were selected. The structure was inspected graphically and positions of S16, K21, K22, L32, R41, T69 and R74 side-chains were further refined locally against cryo-EM electron density map. The final bundle of 10 conformers represents the final structure of the ASC-PYD filament (Model 3, Fig. 4). The structure (Fig. 6) shows a helical arrangement of the ASC-PYD subunits, maintained by electrostatic and hydrophobic interactions among three interfaces (Sborgi et al., 2015).

4. Effect of the cryo-EM electron density resolution on the structure quality

The quality of structures generally depends on the quality of the input data. Since for pure cryo-EM structure determinations an atomic model can often not be reliably built for resolutions below around 4 Å, we were intrigued to find out how the combined approach of integrating solid-state NMR data improves this situation. As an approach toward identifying the resolution limit, we investigated the influence of the cryo-EM resolution on the final structure. To this end, a set of artificial electron density maps with reduced resolutions were prepared by exclusion of the corresponding high-resolution structure factors in reciprocal space representation up to a specified resolution limit. Correct and unambiguous placement of the structural

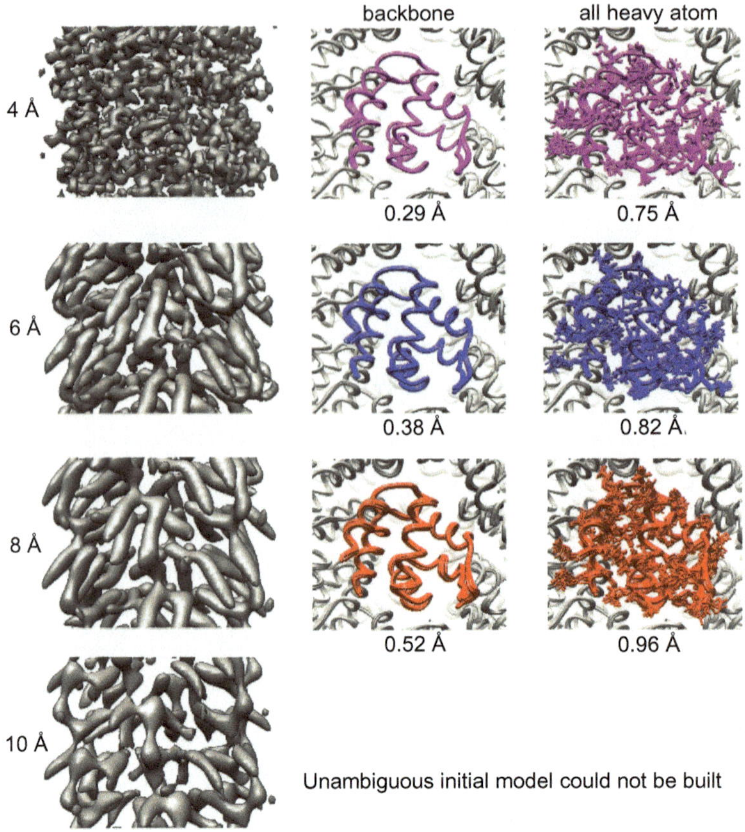

Fig. 7 Impact of the electron density resolution on the ASC-PYD structure. Left column: Electron density maps artificially truncated to lower resolution in 2 Å resolution intervals. Center and right column: Backbone and all heavy atoms structures of the final ASC-PYD filament after fully automated refinement (Model 3). The initial model could be built for electrons density maps up to 8 Å resolution.

model onto the electron density maps was possible for maps with resolution of 4, 6 and 8 Å, but not for the map with 10 Å. In this map, the position of the secondary structure elements is unclear (Fig. 7). Then, in order to quantify the effect of the cryo-EM maps resolution on the quality of the final structure, we rerun the refinement steps, starting from the Model 2 structures, with the input EM maps corresponding to resolutions of 4, 6 and 8 Å. The structures were superimposed on the backbone atoms and root-mean-square deviations (RMSDs) of their atomic positions were calculated for backbone and all non-hydrogen atoms. Average pairwise RMSD is routinely used as a quantitative measure of structural bundle precision.

The cryo-EM map resolution has an impact on the quality; the lower the resolution, the less well defined is the resulting structural bundle both in terms of backbone and all-heavy atom RMSD values (Fig. 7). Importantly, the fold of the ASC-PYD filament remains virtually identical in the tested range of 4–8 Å resolution. This means that by integrating solid-state NMR data, reliable results can be obtained also in regions where local or global EM resolution is low, under the assumption that the solid-state NMR data is not severely affected by the same processes that lead to a low quality electron density map. Solid-state NMR data may thus be particularly valuable in cases where electron densities are lowly resolved.

5. Summary

The protocol presented here combines experimental data from two distinct sources into a single hybrid structure calculation. By combining the local character of NMR-derived experimental restraints with the global information from cryo-EM, the protocol exploits the mutual complementarity of cryo-EM and solid-state NMR spectroscopy. Such protocols may become intrinsically valuable for filaments where large parts of the assembly are dynamic or disordered fragments of the polypeptide chain. In the present application, the structure of the ASC filament was determined to atomic resolution, highlighting the helical stacking architecture of the ASC-PYD domain in the filament core and the presence of the flexibly unfolded CARD domains on the filament outside.

Acknowledgment

We thank Lorenzo Sborgi, Francesco Ravotti, Venkata Dandey, Edward Egelman, Henning Stahlberg and Beat Meier for experimental help and discussions.

References

Broz, P., & Dixit, V. M. (2016). Inflammasomes: Mechanism of assembly, regulation and signalling. *Nature Reviews. Immunology*, *16*, 407–420.

Brunger, A. T. (2007). Version 1.2 of the crystallography and NMR system. *Nature Protocols*, *2*, 2728–2733.

Brunger, A. T., Adams, P. D., Clore, G. M., DeLano, W. L., Gros, P., Grosse-Kunstleve, R. W., et al. (1998). Crystallography & NMR system: A new software suite for macromolecular structure determination. *Acta Crystallographica. Section D: Biological Crystallography*, *54*, 905–921.

Castellani, F., van Rossum, B., Diehl, A., Schubert, M., Rehbein, K., & Oschkinat, H. (2002). Structure of a protein determined by solid-state magic-angle-spinning NMR spectroscopy. *Nature*, *420*, 98–102.

Chou, S. Z., & Pollard, T. D. (2019). Mechanism of actin polymerization revealed by cryo-EM structures of actin filaments with three different bound nucleotides. *Proceedings of the National Academy of Sciences of the United States of America*, *116*, 4265–4274.

Cuniasse, P., Tavares, P., Orlova, E. V., & Zinn-Justin, S. (2017). Structures of biomolecular complexes by combination of NMR and cryoEM methods. *Current Opinion in Structural Biology*, *43*, 104–113.

de Alba, E. (2009). Structure and interdomain dynamics of apoptosis-associated speck-like protein containing a CARD (ASC). *Journal of Biological Chemistry*, *284*, 32932–32941.

De Paëpe, G., Lewandowski, J. R., Loquet, A., Böckmann, A., & Griffin, R. G. (2008). Proton assisted recoupling and protein structure determination. *Journal of Chemical Physics*, *129*, 245101.

Demers, J. P., Habenstein, B., Loquet, A., Kumar Vasa, S., Giller, K., Becker, S., et al. (2014). High-resolution structure of the Shigella type-III secretion needle by solid-state NMR and cryo-electron microscopy. *Nature Communications*, *5*, 4976.

Dick, M. S., Sborgi, L., Ruhl, S., Hiller, S., & Broz, P. (2016). ASC filament formation serves as a signal amplification mechanism for inflammasomes. *Nature Communications*, *7*, 11929.

Egelman, E. H. (2000). A robust algorithm for the reconstruction of helical filaments using single-particle methods. *Ultramicroscopy*, *85*, 225–234.

Emsley, P., Lohkamp, B., Scott, W. G., & Cowtan, K. (2010). Features and development of coot. *Acta Crystallographica. Section D: Biological Crystallography*, *66*, 486–501.

Fitzpatrick, A. W. P., Falcon, B., He, S., Murzin, A. G., Murshudov, G., Garringer, H. J., et al. (2017). Cryo-EM structures of tau filaments from Alzheimer's disease. *Nature*, *547*, 185–190.

Frank, J., Radermacher, M., Penczek, P., Zhu, J., Li, Y., Ladjadj, M., et al. (1996). SPIDER and WEB: Processing and visualization of images in 3D electron microscopy and related fields. *Journal of Structural Biology*, *116*, 190–199.

Frank, J., Shimkin, B., & Dowse, H. (1981). Spider—A modular software system for electron image processing. *Ultramicroscopy*, *6*, 343–357.

Gauto, D., Estrozi, L., Schwieters, C., Effantin, G., Macek, P., Sounier, R., et al. (2018). Integrated NMR and cryo-EM atomic-resolution structure determination of a half-megadalton enzyme complex. bioaRxiv, https://doi.org/10.1101/498287.

Gremer, L., Scholzel, D., Schenk, C., Reinartz, E., Labahn, J., Ravelli, R. B. G., et al. (2017). Fibril structure of amyloid-beta(1-42) by cryo-electron microscopy. *Science*, *358*, 116–119.

Guerrero-Ferreira, R., Taylor, N. M. I., Mona, D., Ringler, P., Lauer, M. E., Riek, R., et al. (2018). Cryo-EM structure of alpha-synuclein fibrils. *eLife*, *7*, e36402.

Güntert, P. (2004). Automated NMR structure calculation with CYANA. *Methods in Molecular Biology*, *278*, 353–378.

He, L., Bardiaux, B., Ahmed, M., Spehr, J., Konig, R., Lunsdorf, H., et al. (2016). Structure determination of helical filaments by solid-state NMR spectroscopy. *Proceedings of the National Academy of Sciences of the United States of America*, *113*, E272–E281.

Herrmann, T., Güntert, P., & Wüthrich, K. (2002). Protein NMR structure determination with automated NOE assignment using the new software CANDID and the torsion angle dynamics algorithm DYANA. *Journal of Molecular Biology*, *319*, 209–227.

Hong, M. (1999). Determination of multiple phi-torsion angles in proteins by selective and extensive (13)C labeling and two-dimensional solid-state NMR. *Journal of Magnetic Resonance*, *139*, 389–401.

Hoss, F., Rodriguez-Alcazar, J. F., & Latz, E. (2017). Assembly and regulation of ASC specks. *Cellular and Molecular Life Sciences*, *74*, 1211–1229.

Lamkanfi, M., & Dixit, V. M. (2012). Inflammasomes and their roles in health and disease. *Annual Review of Cell and Developmental Biology*, *28*, 137–161.

Lange, A., Luca, S., & Baldus, M. (2002). Structural constraints from proton-mediated rare-spin correlation spectroscopy in rotating solids. *Journal of the American Chemical Society*, *124*, 9704–9705.

Li, X., Mooney, P., Zheng, S., Booth, C. R., Braunfeld, M. B., Gubbens, S., et al. (2013). Electron counting and beam-induced motion correction enable near-atomic-resolution single-particle cryo-EM. *Nature Methods*, *10*, 584–590.

Lu, A., Magupalli, V. G., Ruan, J., Yin, Q., Atianand, M. K., Vos, M. R., et al. (2014). Unified polymerization mechanism for the assembly of ASC-dependent inflammasomes. *Cell*, *156*, 1193–1206.

Masumoto, J., Taniguchi, S., Ayukawa, K., Sarvotham, H., Kishino, T., Niikawa, N., et al. (1999). ASC, a novel 22-kDa protein, aggregates during apoptosis of human pro-myelocytic leukemia HL-60 cells. *Journal of Biological Chemistry*, *274*, 33835–33838.

Mindell, J. A., & Grigorieff, N. (2003). Accurate determination of local defocus and specimen tilt in electron microscopy. *Journal of Structural Biology*, *142*, 334–347.

Mulvihill, E., Sborgi, L., Mari, S. A., Pfreundschuh, M., Hiller, S., & Müller, D. J. (2018). Mechanism of membrane pore formation by human gasdermin-D. *EMBO Journal*, *37*, e98321.

Perilla, J. R., Zhao, G., Lu, M., Ning, J., Hou, G., Byeon, I. L., et al. (2017). CryoEM structure refinement by integrating NMR chemical shifts with molecular dynamics simulations. *Journal of Physical Chemistry B*, *121*, 3853–3863.

Ravotti, F., Sborgi, L., Cadalbert, R., Huber, M., Mazur, A., Broz, P., et al. (2016). Sequence-specific solid-state NMR assignments of the mouse ASC PYRIN domain in its filament form. *Biomolecular NMR Assignments*, *10*, 107–115.

Sborgi, L., Ravotti, F., Dandey, V. P., Dick, M. S., Mazur, A., Reckel, S., et al. (2015). Structure and assembly of the mouse ASC inflammasome by combined NMR spectroscopy and cryo-electron microscopy. *Proceedings of the National Academy of Sciences of the United States of America*, *112*, 13237–13242.

Sborgi, L., Ruhl, S., Mulvihill, E., Pipercevic, J., Heilig, R., Stahlberg, H., et al. (2016). GSDMD membrane pore formation constitutes the mechanism of pyroptotic cell death. *EMBO Journal*, *35*, 1766–1778.

Scherer, S., Kowal, J., Chami, M., Dandey, V., Arheit, M., Ringler, P., et al. (2014). 2dx_automator: Implementation of a semiautomatic high-throughput high-resolution cryo-electron crystallography pipeline. *Journal of Structural Biology*, *186*, 302–307.

Schett, G., Dayer, J. M., & Manger, B. (2016). Interleukin-1 function and role in rheumatic disease. *Nature Reviews. Rheumatology*, *12*, 14–24.

Schwieters, C. D., Kuszewski, J. J., Tjandra, N., & Clore, G. M. (2003). The Xplor-NIH NMR molecular structure determination package. *Journal of Magnetic Resonance*, *160*, 65–73.

Shen, Y., & Bax, A. (2015). Protein structural information derived from NMR chemical shift with the neural network program TALOS-N. *Methods in Molecular Biology*, *1260*, 17–32.

Shen, Y., Delaglio, F., Cornilescu, G., & Bax, A. (2009). TALOS+: A hybrid method for predicting protein backbone torsion angles from NMR chemical shifts. *Journal of Biomolecular NMR*, *44*, 213–223.

Sims, J. E., & Smith, D. E. (2010). The IL-1 family: Regulators of immunity. *Nature Reviews. Immunology*, *10*, 89–102.

Tang, G., Peng, L., Baldwin, P. R., Mann, D. S., Jiang, W., Rees, I., et al. (2007). EMAN2: An extensible image processing suite for electron microscopy. *Journal of Structural Biology*, *157*, 38–46.

Van Melckebeke, H., Wasmer, C., Lange, A., Ab, E., Loquet, A., Bockmann, A., et al. (2010). Atomic-resolution three-dimensional structure of HET-s(218-289) amyloid fibrils by solid-state NMR spectroscopy. *Journal of the American Chemical Society*, *132*, 13765–13775.

Vasa, S., Lin, L., Shi, C., Habenstein, B., Riedel, D., Kuhn, J., et al. (2015). Beta-helical architecture of cytoskeletal bactofilin filaments revealed by solid-state NMR. *Proceedings of the National Academy of Sciences of the United States of America*, *112*, E127–E136.

Vranken, W. F., Boucher, W., Stevens, T. J., Fogh, R. H., Pajon, A., Llinas, M., et al. (2005). The CCPN data model for NMR spectroscopy: Development of a software pipeline. *Proteins*, *59*, 687–696.

Wasmer, C., Lange, A., Van Melckebeke, H., Siemer, A. B., Riek, R., & Meier, B. H. (2008). Amyloid fibrils of the HET-s(218-289) prion form a beta solenoid with a triangular hydrophobic core. *Science*, *319*, 1523–1526.

Wishart, D. S., & Sykes, B. D. (1994). The 13C chemical-shift index: A simple method for the identification of protein secondary structure using 13C chemical-shift data. *Journal of Biomolecular NMR*, *4*, 171–180.

Yu, Q., Qu, K., & Modis, Y. (2018). Cryo-EM structures of MDA5-dsRNA filaments at different stages of ATP hydrolysis. *Molecular Cell*, *72*, 999–1012.

CHAPTER FOURTEEN

Protein interactions of the inflammasome adapter ASC by solution NMR

Pedro Diaz-Parga[a,b], Eva de Alba[a,*]

[a]Department of Bioengineering, School of Engineering, University of California, Merced, CA, United States
[b]Quantitative Systems Biology Graduate Program, University of California, Merced, CA, United States
*Corresponding author: e-mail address: edealbabastarrechea@ucmerced.edu

Contents

Methods in Enzymology, Volume 625
ISSN 0076-6879
https://doi.org/10.1016/bs.mie.2019.07.008

Abstract

ASC (apoptosis-associated speck-like protein containing a CARD) is a modular protein that functions as an adapter of the inflammasome, a multi-protein complex that triggers the inflammatory response in the presence of infection or cell damage. ASC bridges the inflammasome components (PYD-containing sensors and procaspase 1) via homotypic interactions mediated by its two death domains, PYD and CARD. The self-assembly and oligomerization of multiple copies of these three proteins result in the activation of procaspase 1, in turn rendering different cytokines functional. An in-depth understanding of ASC binding capabilities is crucial to decipher the molecular mechanisms of its role in inflammasome formation. In this chapter, we discuss the use of solution NMR to identify specific interacting surfaces of the inflammasome adapter ASC, and describe detailed protocols to perform NMR titrations with Death Domains to obtain apparent dissociation constants of the resulting complexes. The incorporation of NMR restraints in molecular docking to obtain models of these protein assemblies is presented.

1. Introduction

In the presence of danger signals resulting from infection or tissue damage, sensor proteins self-assemble and oligomerize with the protein ASC (Apoptosis-associated speck like protein containing a CARD) (De Alba, 2009; Masumoto et al., 1999; Srinivasula et al., 2002; Stehlik et al., 2003) that tethers the former to procaspase 1 (Broz & Dixit, 2016; Faustin et al., 2007; Guo, Callaway, & Ting, 2015; He, Hara, & Núñez, 2016; Malik & Kanneganti, 2017; Martinon, Burns, & Tschopp, 2002; Schroder & Tschopp, 2010). This large multiprotein complex, known as the inflammasome (Martinon et al., 2002), results in the activation of procaspase 1. The active caspase cleaves pro-inflammatory cytokines (Dinarello, 2011; Thornberry et al., 1992; Tominaga et al., 2000; Van Damme et al., 1985) into their functional form, which upon secretion to the extracellular milieu trigger the inflammatory response (Ding et al., 2016; Fink & Cookson, 2005, 2006; He et al., 2015; Liu et al., 2016; Rathkey et al., 2017; Shi et al., 2015). Different types of inflammasome sensor proteins become active in response to different stimuli (Fritz, Ferrero, Philpott, & Girardin, 2006; Gross et al., 2009; Ichinohe, Lee, Ogura, Flavell, & Iwasaki, 2009; Inohara & Nuñez, 2003; Joly et al., 2009; Kanneganti et al., 2006; Meylan, Tschopp, & Karin, 2006; Platnich et al., 2018; Proell, Riedl, Fritz, Rojas, & Schwarzenbacher, 2008; Xu et al., 2014). In particular, danger-signal sensing proteins from the AIM family sense viral and bacterial double-stranded DNA (Bürckstümmer et al., 2009;

Fernandes-Alnemri, Yu, Datta, Wu, & Alnemri, 2009; Matyszewski, Morrone, & Sohn, 2018; Rathinam et al., 2010; Sohn & Hur, 2016; Stratmann, Morrone, Van Oijen, & Sohn, 2015). The sensor, the adapter (ASC) and procaspase 1 are modular proteins that contain Death Domains (Martinon & Tschopp, 2005; Park, Lo, et al., 2007; Park, Logette, et al., 2007; Stehlik & Reed, 2004). Protein-protein interactions between these domains are critical driving forces in inflammasome formation (Bertin & DiStefano, 2000; Chinnaiyan, O'Rourke, Tewari, & Dixit, 1995; Chou, Matsuo, Duan, & Wagner, 1998; Dorfleutner, Chu, & Stehlik, 2015; Duan & Dixit, 1997; Fairbrother et al., 2001; Hofmann, 1997; Jin & Xiao, 2015; Kersse, Verspurten, Berghe, & Vandenabeele, 2011; Kwon et al., 2012; Masumoto, Taniguchi, Nakayama, Ayukawa, & Sagara, 2001; Park, Lo, et al., 2007; Park, Logette, et al., 2007; Weber & Vincenz, 2001). Detailed definition of the factors involved in these protein binding events is an essential step toward the understanding of the molecular mechanisms that lead to inflammasome assembly, and thus to the onset of inflammation.

Protein-protein interactions can be studied using a variety of biochemical and biophysical methods that provide different levels of information, ranging from the mere identification of the interaction to the detailed structural analysis of the interacting surfaces involved in binding (Phizicky & Fields, 1995; Rao, Srinivas, Sujini, & Kumar, 2014). To name some of the most commonly used techniques, co-immunoprecipitation provides basic information on the identification of the interaction. Electrophoresis and size exclusion chromatography can be used to determine the size of the resulting complex and native electrospray ionization is currently applied to investigate the stoichiometry of native protein assemblies and to probe with accuracy the mass of protein-ligand interactions (Heck, 2008; Leney & Heck, 2017; Smits & Vermeulen, 2016). Fluorescence anisotropy experiments allow measurements of the volume of the complex and binding constant under optimum conditions, when the overall fluorescence intensity of the fluorophore does not change with binding (Raines, 2015). Surface plasmon resonance is used to quantify the *on* and *off* rates of the binding process (Douzi, 2017). The latter parameters can also be obtained by isothermal titration calorimetry, which is considered to be the most quantitative technique, providing information on the thermodynamic properties of the interaction (Doyle, 1997; Leavitt & Freire, 2001). When detailed structural information on protein interactions and assemblies is necessary, NMR, X-ray crystallography, and cryo-EM are the techniques of choice.

In addition to structural data, solution NMR can also be used to study the dynamics of complex formation (Barbato, Ikura, Kay, Pastor, & Bax, 1992; De Alba, Baber, & Tjandra, 1999; De Alba & Tjandra, 2000; Ishima & Torchia, 2000; Mittermaier & Kay, 2006; Tjandra, Feller, Pastor, & Bax, 1995; Wagner, 1993) and to measure the affinity of the interaction (Arai, Ferreon, & Wright, 2012; Fielding, 2003; Williamson, 2013).

NMR is a very versatile technique that provides a large variety of information on protein complexes. Atomic-resolution three-dimensional (3D) structures of protein complexes typically rely on the measurement of interatomic distances that can be obtained from different type of NMR parameters. The most commonly used experiment is based on the quantification of the Nuclear Overhauser Effect. Less frequently, the complex can be attached to paramagnetic centers and distances are obtained from pseudocontact shifts and from the measurement of magnetic relaxation enhancement caused by the paramagnetic metal (Clore & Iwahara, 2009). In addition, residual dipolar couplings of protein complexes partially oriented with respect to the external magnetic field (De Alba & Tjandra, 2002, 2003; Tjandra & Bax, 1997) can provide the relative orientation of the molecular components of the complex (Jensen, Ortega-Roldan, Salmon, Van Nuland, & Blackledge, 2011). In all cases, this structural information is obtained after careful design of specific NMR experiments that probe the various parameters for which different sample preparations are necessary. However, the experimental 3D structure determination of the complex might not be strictly required or could be recalcitrant to obtain due to the particular characteristics of the complex in solution. In the latter scenario, NMR techniques that probe chemical shift perturbations upon complex formation can be adapted to different solution conditions, are relatively faster than structure determination, and provide detailed information on the pertinent interacting surfaces. By mapping the chemical shift information onto the 3D structure of a protein, the interacting surfaces are identified, and can also be used as experimental restraints in molecular docking calculations to obtain meaningful models of the protein complex (Fielding, 2003; Williamson, 2013).

NMR chemical shift perturbation data is particularly useful in the study of protein-protein interactions of Death Domains and other heavily oligomerizing or self-associating proteins, which might fail to produce high quality crystals for X-ray crystallographic studies. This chapter describes the main steps followed in the NMR study of ASC self-association and its interaction with other Death Domains, including protein overexpression by recombinant technologies, purification protocols, the specific experimental conditions of solution NMR experiments to probe the interactions, as well as the interpretation of the results and their application to molecular docking to generate models of protein ensembles (De Alba, 2009; Nambayan, Sandin, Quint, Satyadi, & De Alba, 2019; Oroz, Barrera-Vilarmau, Alfonso, Rivas, & De Alba, 2016).

2. Overexpression and purification of ASC, ASC^{PYD} and ASC^{CARD} in minimal media

Death Domains have a strong tendency to self-associate and interact with other Death Domains typically through homotypic interactions. Thus, overexpression of these protein-protein interacting domains often times results in the formation of inclusion bodies. Different purification tags, including tags that help decrease self-association, might not be desirable for certain experiments specifically aimed at oligomerization studies. In the latter case, denaturing conditions are commonly used to solubilize the overexpressed protein from inclusion bodies. In addition, certain affinity tags, such as a Histidine tag, allow the protein to be purified under denaturing conditions. In this section we describe the expression and purification of ASC and other Death Domains in minimal media for isotopic enrichment in spin-1/2 magnetically active nuclei.

2.1 Equipment

1. Autoclave
2. Dialysis cassettes with a cutoff of 3 kDa
3. Filters of 0.2 μm and 0.45 μm
4. HisTrap Histidine-tagged protein purification columns (5 mL)
5. HPLC with UV detection
6. Homogenizer
7. Laboratory balance
8. Large scale incubator with temperature control
9. Lyophilizer
10. pH meter
11. Preparative centrifuge and rotor for large volume flasks (500 mL)
12. Reverse phase column (C4)
13. Ultracentrifuge
14. UV–Vis spectrophotometer

2.2 Chemicals

1. Acetonitrile (HPLC grade)
2. Ammonium Chloride, ^{15}N labeled ($^{15}NH_4Cl$)
3. Ampicillin
4. Calcium chloride
5. Glucose, ^{13}C uniformly labeled
6. Guanidinium hydrochloride
7. Hydrochloric acid
8. Imidazole
9. IPTG (isopropyl β-D-thiogalacto-pyranoside)
10. Magnesium sulfate
11. Potassium phosphate monobasic
12. Sodium chloride
13. Sodium hydroxide
14. Sodium phosphate dibasic
15. Trifluoroacetic acid
16. Tris (2-carboxyethyl) phosphine, TCEP
17. Tris-HCl pH 8.0
18. Vitamin solution 100X

2.3 Protein expression in minimal media and preparation for purification

1. Prepare a 10X M9 media salts (0.2L) by dissolving in H_2O, KH_2PO_4 (6 g), Na_2HPO_4 (13.6 g) and NaCl (1 g). Adjust the pH to 7.1 using a diluted solution of NaOH. Filter the resulting solution through a 0.2 μm filter to sterilize.
2. Prepare 1 L of M9 media by mixing:
   867 mL H_2O (autoclaved)
   100 mL M9 10X salts (as described in 1 above)
   10 mL vitamins solution 100X
   2 mL of 1 M $MgSO_4$ (0.2 μm filtered)
   1 mL Ampicillin (from a stock of 100 mg/mL, 0.2 μm filtered)
   0.1 mL of 1 M $CaCl_2$ (0.2 μm filtered)
   3 g of uniformly labeled ^{13}C Glucose dissolved in 10 mL of distilled water (0.2 μm filtered)
   1 g $^{15}NH_4Cl$ in 10 mL of distilled water (0.2 μm filtered)
3. Remove 250 mL of the 1 L M9 media and pour into a 500 mL flask. Inoculate with bacteria using a previously prepared glycerol stock. Incubate

bacteria overnight at 37 °C. At an OD_{600} of ~1.5, pour the 250 mL of grown bacteria into the remaining 750 L of M9 media.

4. Incubate 1 L of cell culture at 37 °C for 30 min and induce protein expression at OD_{600} of 0.6–0.7 for 4 h by adding 1 mM IPTG.
5. Centrifuge bacteria at 8 K rpm for 30 min at 4 °C. Discard supernatant and resuspend cell pellets on ice (4 °C) using resuspension buffer (20 mM Tris-HCl pH 8.0, 500 mM NaCl, 5 mM imidazole and 6 M guanidinium hydrochloride).
6. Transfer resuspended bacteria to homogenizer tubes containing silica beads and homogenize five times for 30 s per run. Transfer supernatant to ultracentrifuge tubes and ultracentrifuge at 35 K for 30 min at 4 °C. Remove supernatant and filter using 0.45 μm filters.

2.4 Purification protocol

1. Prepare the necessary buffers for affinity and reverse phase chromatography:
   - Resuspension buffer (20 mM Tris-HCl pH 8, 500 mM NaCl, 5 mM imidazole and 6 M guanidinium hydrochloride, 0.2 μm filtered)
   - Elution buffer (20 mM Tris-HCl pH 8, 500 mM NaCl, 500 mM imidazole and 6 M guanidinium hydrochloride, 0.2 μm filtered)
   - Equilibration buffer (95% H_2O 5% Acetonitrile, 0.1% Trifluoroacetic acid)
   - Elution buffer (95% Acetonitrile, 5% H_2O, 0.1% Trifluoroacetic acid)
2. Purify by affinity chromatography using a His-Bind resin and a 45 min elution buffer gradient (0–100%).
3. Decrease the pH of the eluted protein solution to 3.8 prior to dialysis.
4. Prepare the dialysis buffer (1 L H_2O, 5 mM TCEP at pH 3.8). Place the acidified protein solution inside dialysis cassettes, which will be submerged in the dialysis buffer. Add stir bar and place on top of magnetic stirrer and stir at 4 °C. Change dialysis buffer three times every 2–3 h for proper buffer equilibration.
5. Purify the resulting protein solution using reverse phase chromatography in a C4 column utilizing the prepared equilibration buffer (95% H_2O, 5% acetonitrile, 0.1% trifluoroacetic acid) and elution buffer (95% Acetonitrile, 5% H_2O, 0.1% Trifluoroacetic acid).
6. Lyophilize the eluted protein solution to remove the organic solvents.
7. Store at 4 °C.

3. Aspects to consider in the design of solution NMR studies of ASC, ASC^{CARD} and ASC^{PYD}

3.1 Factors related to protein solubility (pH, protein and salt concentration, temperature)

ASC has a high propensity to self-associate because it bears two protein-protein interacting domains; the Death Domains PYD and CARD at its N- and C-terminus, respectively. Its ability to oligomerize is an inherent property that governs the role of ASC in inflammasome assembly (De Alba, 2009; Martinon et al., 2002). However, protein oligomerization poses significant challenges for solution NMR studies. Previous analyses indicate that ASC is predominantly insoluble at neutral pH, showing a significant increase in solubility at acidic pH (pH ~4) (De Alba, 2009). The predominant species in solution at neutral pH are oligomeric species that typically form filamentous macrostructures (Fig. 1). In contrast,

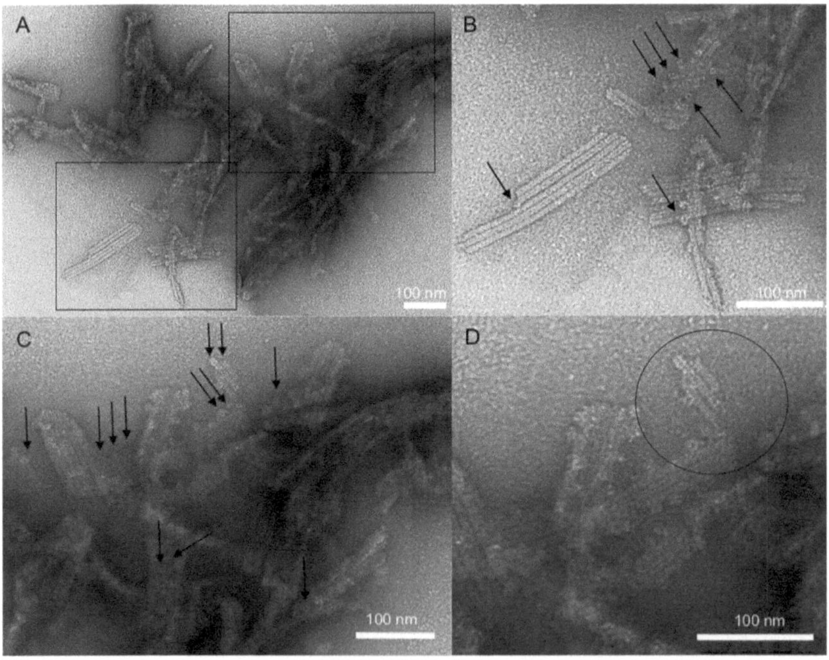

Fig. 1 Filaments of human full-length ASC. Transmission electron micrographs of ASC filaments formed at pH 8.2 (A). Rectangles show zoomed areas of (A) composing panels, (B) bottom rectangle, and (C) upper rectangle. Arrows point to ring-shaped structures (with a diameter ~7 nm) laterally attached to the filaments. An additional zoom of the upper-central region of (C) forms (D). The circle shows stacked rings forming the filament. Bar length is indicated.

monomeric ASC is prevalent under acidic conditions, making it suitable for solution NMR studies (De Alba, 2009).

Solution NMR experiments typically require protein concentrations in the millimolar range, although studies at submillimolar concentrations are also possible if the magnet is equipped with a cryoprobe. Even at low pH, ASC concentration for NMR studies needs to be carefully monitored. ASC oligomerization at low pH is minimal at 200 μM leading to a signal intensity decay to 98% after 3 h, and reaching a plateau of ~75% of the original signal after 3 days (De Alba, 2009). However, oligomerization increases substantially with time at 700 μM concentration even at low pH (Fig. 2). Under these conditions, the signal to noise ratio decreases to ~80% after 3 h and to ~30% within 5 days (De Alba, 2009). ASC oligomers are most likely too large to give rise to NMR signals because of the slower tumbling associated to high molecular weight species, which explains the absence of additional peaks upon self-association.

There could exist certain concerns in relation to the physiological relevance of studies done at acidic pH. It is important to mention in this regard that the structure of ASC (De Alba, 2009) has been shown to match the cryo-EM electron density maps of the individual domains (Li et al., 2018; Lu et al., 2014), which suggests that acidic conditions influence the

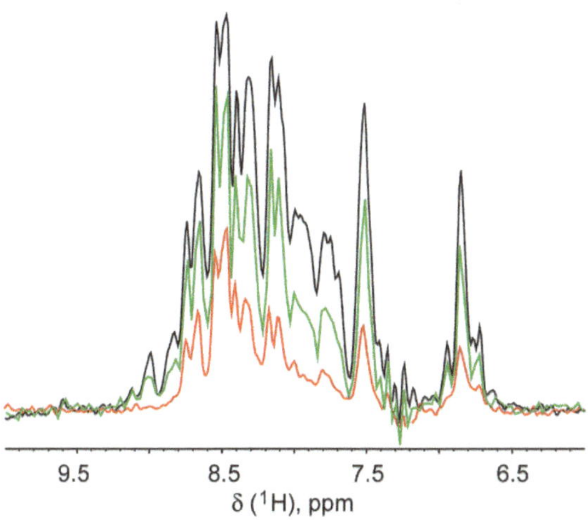

Fig. 2 ASC self-associates in solution. ^{1}H amide chemical shift region of the one-dimensional projection of a [^{1}H–^{15}N]-HSQC spectrum of 700 μM ^{15}N-, ^{13}C-labeled ASC at pH 3.8. The black, green and red lines correspond to spectra acquired ~1 h, ~2 h and 5 days after sample preparation, respectively.

oligomerization equilibrium but not the structure. In addition, self-association studies of ASC^{PYD} show that the binding interfaces at pH values 3.8 and 5.8 are equivalent (Oroz et al., 2016), pointing out that binding occurs at low pH and that it does not change significantly at higher pH (Fig. 3).

In addition, the presence of 100 mM NaCl has been shown to enhance ASC^{PYD} self-association (Oroz et al., 2016) (Fig. 4) and to favor the

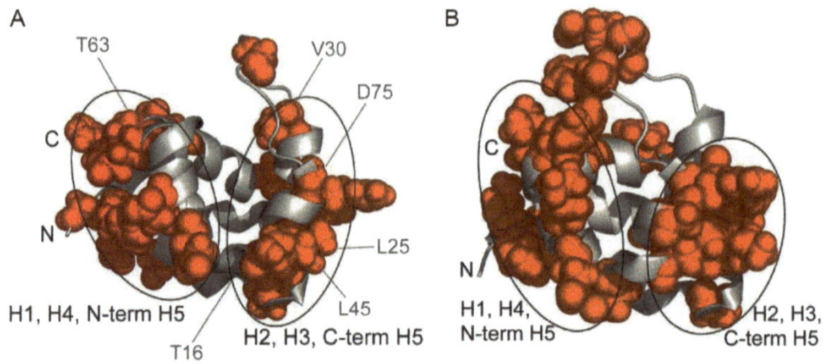

Fig. 3 ASC^{PYD} mode of interaction at different pH values. ASC^{PYD} residues with the largest chemical shift differences upon self-association (shown in red) mapped onto ASC^{PYD} 3D structure (De Alba, 2009) at pH 3.8 (A) and pH 5.8 (B).

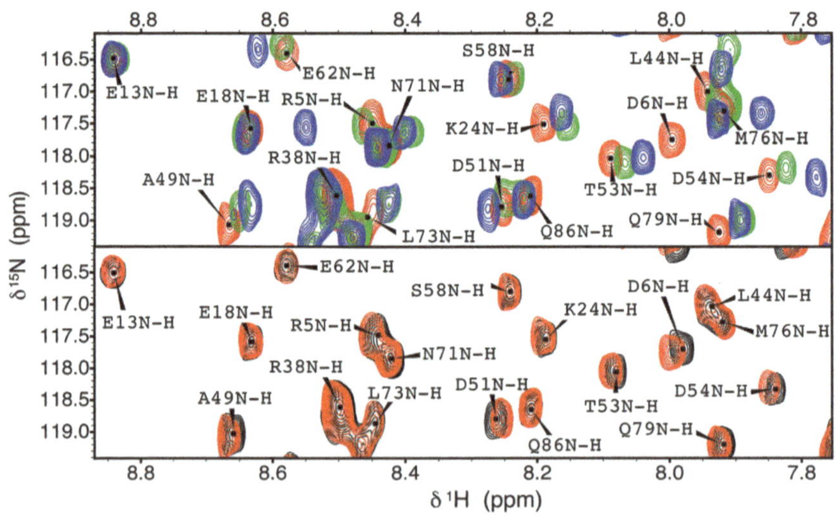

Fig. 4 ASC^{PYD} self-association in the presence of NaCl. Top: region of sofast-HMQC spectra of ASC^{PYD} at different concentrations showing typical shifts upon ASC-ASC binding in the absence and presence of NaCl. Top: Blue: 50 μM ^{15}N-labeled ASC^{PYD}; Green: 50 μM ^{15}N-labeled ASC^{PYD} + 500 μM unlabeled ASC^{PYD}. Red: 50 μM ^{15}N-labeled ASC^{PYD}/100 mM NaCl. Bottom: Red: 50 μM ^{15}N-labeled ASC^{PYD}/100 mM NaCl (same as in top). Black: 50 μM ^{15}N-labeled ASC^{PYD} + 250 μM unlabeled ASC^{PYD} ASC/100 mM NaCl.

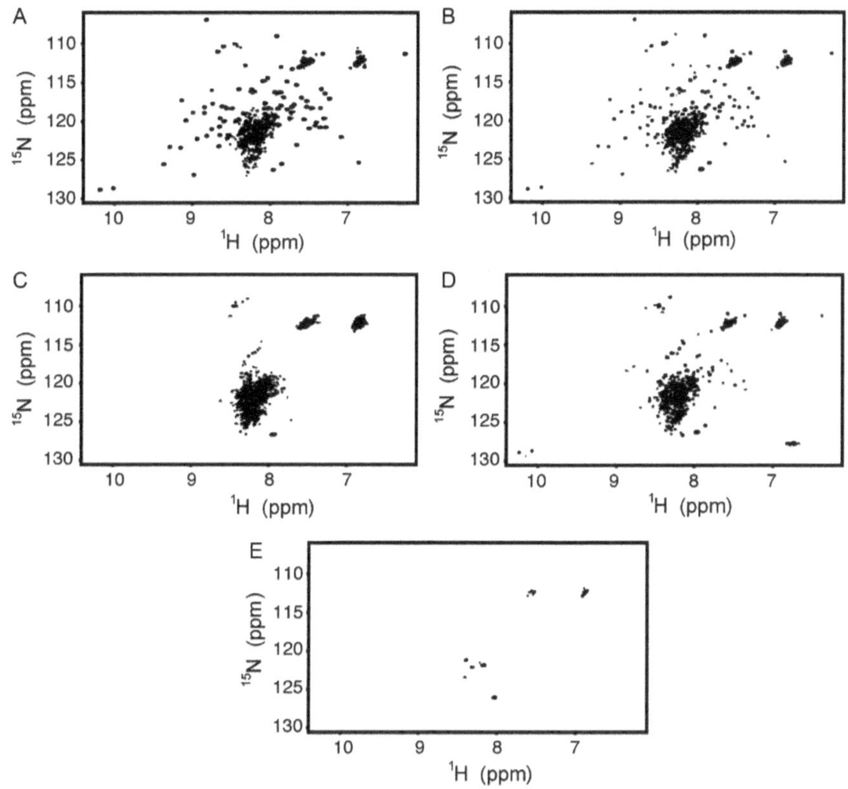

Fig. 5 Factors affecting NLRP3^{PYD} self-association: sample incubation time, protein concentration, temperature and salt concentration. [^{1}H–^{15}N]-HSQC spectra of NLRP3 at pH 3.6. (A) 200 μM protein concentration at 303 K; (B) Same sample used in (A) after overnight incubation at 303 K. (C) Protein concentration of 650 μM, 303 K. (D) Protein concentration of 200 μM, 310 K. (E) Protein concentration of 200 μM, 303 K and 100 mM NaCl.

oligomerization of other Death Domains such as NLRP3^{PYD} (Fig. 5). In the presence of 100 mM of NaCl, the [^{1}H–^{15}N]-HMQC (Cavanagh, Fairbrother, Palmer, Rance, & Skelton, 2007) spectrum of ASC^{PYD} shows a characteristic pattern corresponding to the oligomerized form (Oroz et al., 2016). Different temperature conditions have also been tested by NMR to better understand Death Domain oligomerization for NLRP3^{PYD} (Oroz et al., 2016). The results show that NMR line broadening in [^{1}H–^{15}N]-HSQC (Cavanagh et al., 2007) spectra resulting from NLRP3^{PYD} self-association is more prominent in the presence of salt, after sample incubation, and at higher temperatures (Fig. 5). This latter result points to an important role of hydrophobic interactions in Death Domain self-association (Oroz et al., 2016).

The previous analysis allows the identification of optimum conditions for NMR studies of full-length ASC and its PYD and CARD domains. Three-dimensional NMR experiments at protein concentration close to 200 μM are typically required to obtain the chemical shift assignment of all ^1H, ^{13}C, and ^{15}N nuclei in the protein. Once the chemical shift assignment of the amide ^1H, ^{15}N pairs is known, protein concentration of 10–50 μM is sufficient for fast acquisition NMR experiments aimed at monitoring chemical shift changes to study protein binding. Other conditions for ASC solution NMR samples include: 5 *d*mM$_{15}$-TCEP, 0.1 mM NaN_3, pH 3.8, 5% D_2O/H_2O.

3.2 Protein NMR chemical shift assignment for binding studies: Standard 2D and 3D protein NMR experiments and fast acquisition techniques

After NMR sample preparation, [^1H–^{15}N]-HSQC NMR experiments are typically acquired to obtain what is known as the protein "fingerprint." The [^1H–^{15}N]-HSQC spectrum correlates the nitrogen and the amide proton chemical shifts in two dimensions, producing observable signals for the different amino acids except for proline, which lacks a ^1H–^{15}N amide pair. [^1H–^{15}N]-HSQC spectra also provide information on the state of the protein sample, for instance; whether the sample contains impurities, whether the protein is folded or unfolded and information on protein oligomerization. Correctly folded proteins give rise to well dispersed signals in the spectrum, whereas signals from unfolded proteins are typically clustered in a small region of the spectra, particularly between amide ^1H chemical shifts of 8.0 and 8.5 ppm (Wishart, Bigam, Holm, Hodges, & Sykes, 1995). The [^1H–^{15}N]-HSQC experiment is commonly used to optimize sample conditions prior to the acquisition of lengthy triple resonance 3D experiments for complete chemical shift assignment. The combination of the [^1H–^{15}N]-HSQC with 3D experiments, typically CBCACONH and HNCACB (Grzesiek & Bax, 1992; Wittekind & Mueller, 1993), allow the sequential assignment of all backbone resonances (except the carbonyl ^{13}C) and the C_β of the amino acid side chain. In particular, the HNCACB and CBCACONH provide information on the chemical shifts of the C_α and C_β corresponding to each amide NH group of the same and the following residue, respectively (Grzesiek & Bax, 1992; Wittekind & Mueller, 1993). This combined analysis allows for the determination of the chemical shifts from all amide pairs in the protein, and thus all cross-peaks present in the

[^{1}H–^{15}N]-HSQC can be ascribed to a particular amino acid in the protein sequence (Fig. 6). This spectrum will be used as reference to monitor chemical shift changes upon binding in subsequent NMR titrations experiments.

Because of the low sensitivity of NMR, submillimolar concentrations are typically necessary for standard 2D and 3D experiments using a magnet equipped with a cryoprobe. However, lower protein concentrations (10–50 μM) are necessary for ASC binding studies to ensure that self-association is negligible. NMR fast acquisition techniques (Schanda & Brutscher, 2005) are required at these low concentration values to perform NMR titration experiments within a reasonable time, and simultaneously obtain sufficient signal to noise ratio in the resulting spectra. Fast acquisition techniques use special selective pulses that reduce the recycling time (time between NMR signal accumulation for magnetization recovery) typically by an order of magnitude.

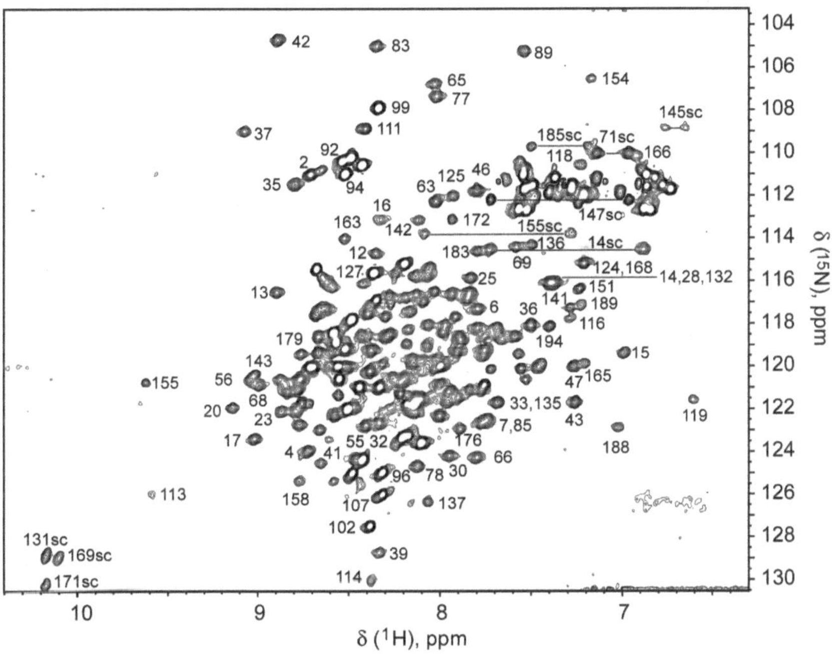

Fig. 6 ^{1}H–^{15}N backbone amide assignment of full-length ASC. [^{1}H–^{15}N]-HSQC spectrum of ^{15}N-, ^{13}C-labeled ASC at pH 3.8. Residue numbers are assigned to the corresponding signals. Assignment of the central region of the spectrum is not shown for clarity. "sc" denotes side chain. ASC is properly folded at pH 3.8.

4. NMR titration experiments

NMR chemical shifts are exquisitely sensitive to slight modifications in the chemical environment of magnetically active nuclei. Upon protein self-association, protein-protein interaction or ligand binding, the chemical environment changes for nuclei located in the specific contact surfaces relative to the unbound state. To understand the effect of complex formation on chemical shifts, it is convenient to briefly define the rate constants playing a role in the binding process:

For a 1:1 interaction (Eq. 1), the dissociation constant of the complex (K_D) is defined as the ratio between the unbinding and binding rate constants (k_{off} and k_{on}, respectively, Eq. 2).

$$P_l + P_u \rightleftharpoons C \tag{1}$$

where P_l and P_u are two proteins, respectively labeled and unlabeled with spin-1/2 magnetically active isotopes, that bind to form complex C.

$$K_D = k_{off} / k_{on} \tag{2}$$

The resulting effect on the chemical shifts of the bound and unbound forms of P_l depends on both the exchange rate for the association ($k_{ex} = [P_u] \cdot k_{on} + k_{off}$) and the difference in the Larmor frequency of the pertinent nucleus in the bound and unbound states. When the exchange rate is substantially smaller than the frequency difference, ($k_{ex} \ll \Delta\omega$), the NMR spectrum will show signals from both the unbound and bound species with intensities typically reflecting the population of each one. In contrast, if the exchange rate is significantly larger than the frequency difference ($k_{ex} \gg \Delta\omega$), the resulting spectrum will be a time average with weighted averaged signals of the unbound and the complexed forms (Eq. 3).

$$\delta_{obs} = Y_{P_l} \cdot \delta_{p_l} + Y_C \cdot \delta_C \tag{3}$$

where δ_{obs} is the observed chemical shift; δ_{Pl} and δ_C are the chemical shifts of P_l when free and complexed, respectively. Y_{pl} and Y_C are the molar fractions of the unbound and the bound protein. For a non-restricted interacting site, it is possible to assume that k_{on} is diffusion controlled and thus, k_{ex} depends only on k_{off}. In a magnet operating at a ^{1}H frequency of 600 MHz, if $k_{ex} \sim 10 \cdot k_{off}$, the system is in the fast exchange limit and a single set of signals will be observed. This situation is typical for processes with dissociation constants in the sub-micromolar range or higher. However, the previous estimation is not valid if k_{on} is not controlled by diffusion.

As mentioned in Section 3, ASC and its two Death Domains are soluble and mostly monomeric at micromolar concentrations at relatively low pH values (pH ~4.0). Under these conditions, it is possible to study ASC^{CARD} and ASC^{PYD} self-association or their binding to other Deaths Domains. The binding is fast relative to the NMR chemical shift time scale ($k_{ex} \gg \Delta\omega$). Thus, one set of signals is monitored during the titration. These signals reflect the binding process according to Eq. 3.

Because the concentration necessary to minimize self-association at the starting point of the titration experiment is very low for solution NMR, it is imperative to use spectrometers equipped with a cryoprobe capable of enhancing the signal to noise ratio relative to regular magnets. Even in the presence of a cryoprobe, the low concentration requires the use of fast-acquisition techniques (Schanda & Brutscher, 2005) to monitor changes in chemical shifts in a reasonable amount of time.

4.1 Equipment

1. Analytical balance
2. Magnetic Spectrometer operating at a ^1H frequency of 600 MHz or higher
3. pH meter with thin electrode
4. UV spectrophotometer

4.2 Chemicals

1. Glycine
2. Deuterated TCEP
3. D_2O
4. Dilute solutions of HCl and NaOH for pH adjustment
5. ^{15}N-labeled and unlabeled protein samples (ASC^{CARD} and ASC^{PYD})
6. HPLC-grade water
7. NaN_3

4.3 Titration protocol

Recalling the chemical equilibrium shown in Eq. (1), P_l in the titration experiment is ^{15}N- or ^{15}N, ^{13}C-labeled protein at a known concentration (10–50 μM). P_u is an unlabeled protein of the same or different species as P_l. Low protein concentration is used as starting point to minimize changes in chemicals shift from potential P_l self-association The isotopically enriched protein is dissolved in the required buffer (20 mM Glycine, 5 mM d15-TCEP, 0.1 mM NaN_3, 5–10% D_2O) and the pH is adjusted to 3.8. Fast acquisition techniques for 2D ^1H–^{15}N correlation spectra (sofast-HMQC; Schanda & Brutscher, 2005) are acquired for this sample, which will provide the chemical shift for the starting point of the titration (no P_u added). The unenriched protein (P_u) is added in lyophilized form directly to the P_l solution, followed by pH adjustment. The changes in chemical shift (δ) are monitored by acquiring 2D [^1H–^{15}N]-sofast-HMQC spectra upon each addition of P_u (Fig. 7A). Protein concentration in the mixtures is randomly checked by UV Absorbance, which does not deviate >10% of the protein concentration measured by weight.

4.4 Identification of chemical shift perturbations and mapping onto the protein three-dimensional structure

Chemical shifts of ^1H and ^{15}N amide bonds are good reporters of perturbations that result from direct protein-protein interactions involving the same residues or adjacent ones. However, chemical shifts may also vary from indirect effects when protein interactions lead to significant conformational changes, which could be understood as a form of allosterism. Examples of the latter case are less common and certain precautions can be taken to identify changes resulting from conformational modifications. In the particular case of ASC, cryo-EM and solid-state NMR data on filaments formed by its CARD and PYD domains indicate that there are no rearrangements of the secondary structural elements upon polymerization (Lu et al., 2014; Sborgi et al., 2015). The Death Domain fold is basically unperturbed, except for some residues located in the contact interfaces.

Because changes in both amide ^1H and ^{15}N chemical shifts are monitored, the data obtained are typically combined by calculating the Euclidean distance for the movement in a plane of the NMR signal as show in Eq. (4).

$$\delta_{obs} = \sqrt{\sum_{i=1}^{N} (\varepsilon_i \cdot \Delta\delta_i)^2} \tag{4}$$

Where $\Delta\delta_i$ is the difference between the chemical shifts in the bound and the free forms. The former are the chemical shifts obtained at each point of the titration and the latter are the chemical shifts before the addition of P_u (the starting point of the titration). The different types of nuclei are indicated by "i." The constant ε is a correction factor that takes into account the different spectral widths for chemical shift variation of the different nuclei. The value

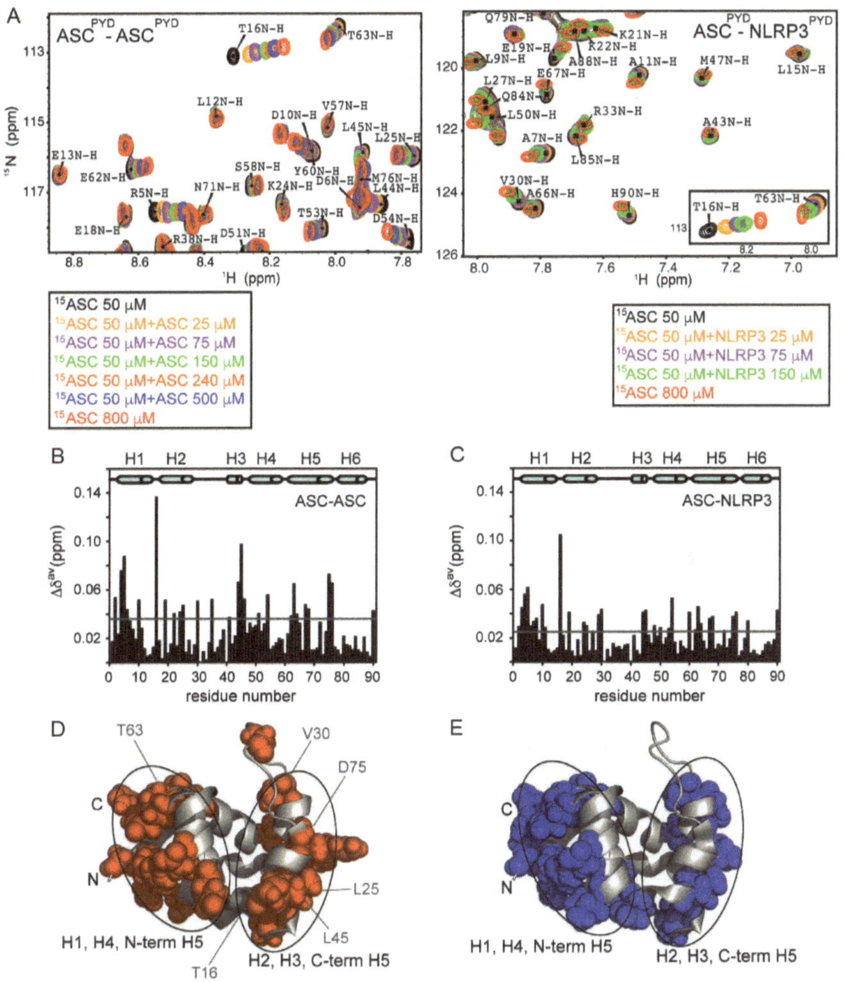

Fig. 7 ASC^{PYD} self-associates and interacts with other NLRP3^{PYD} through two equivalent interfaces. (A) Superimposed sofast-HMQC spectra resulting from the titration of ^{15}N-labeled ASC^{PYD} (50 μM) with unlabeled ASC^{PYD} (increasing concentration). Some NMR signals remain unmodified, whereas others show clear changes in δ upon binding. Similar shifts were observed upon titration with unlabeled NLRP3^{PYD} (right). For comparison, the inset shows a representative region of the spectra acquired for the ASC^{PYD}-NLRP3^{PYD} titration where the observed shifts upon binding are equivalent to those found in ASC^{PYD}-ASC^{PYD}. (B and C) Chemical shift changes upon binding vs. residue number for ASC^{PYD}-ASC^{PYD} (B) and ASC^{PYD}-NLRP3^{PYD} (C) interaction. Dashed lines represent the threshold value calculated as 1.5 times the SD of the average chemical shift change. (D and E) Mapping the solvent exposed residues with δ changes above the threshold value upon ligand binding for ASC^{PYD}-ASC^{PYD} (D, in red) and ASC^{PYD}-NLRP3^{PYD} (E, in blue) onto the ASC^{PYD} 3D structure.

of this constant may vary for each ^{1}H and ^{15}N amide pair, but typically a value of 0.2 is used for ^{15}N δ changes to consider the ratio between the spectral width of ^{1}H vs. ^{15}N (~5 vs. 25 ppm).

Because of the sensitivity of chemical shifts to minute changes in the chemical environment, it is important to identify the changes that are directly indicative of the interaction. For this purpose, only those residues whose NH chemical shifts show perturbations larger than one to two times the standard deviation are selected to represent the binding interface (Fig. 7B and C). The selection of this number is always a compromise because the larger the cutoff, the lower the sensitivity for identifying the binding, and thus potential interacting areas might be overlooked if the associated chemical shift perturbations are small. In addition, it is logical to assume that only residues in the protein surface are available for biding. Therefore, an additional cutoff is considered that selects amino acids with chemical shift changes upon binding that are also sufficiently solvent exposed. A user friendly and popular program to calculate solvent exposed areas of amino acids in protein structures is GETAREA (Fraczkiewicz & Braun, 1998), and a common value used for solvent exposure is >30%. Finally, the interacting surface is defined by mapping onto the known 3D structure of the protein (Fig. 7D), the amino acids that satisfy the conditions: (1) show chemical shift perturbations larger that one to two times the standard deviation and (2) are >30% solvent exposed. The method explained here monitors chemical shift changes of the amide ^{1}H and ^{15}N. These changes are directly related to the interaction, provided that no significant conformational rearrangements occur. Most likely, the amino acid side chain is involved in the interaction, thus it is recommended to depict the whole amino acid on the protein surface to obtain unbiased representation of the interacting surface (Williamson, 2013).

4.5 Determination of the apparent dissociation constant for protein interaction or self-assembly

The difference between the δ_{obs} calculated as shown in Eq. (4) for each titration point and the starting point ($\Delta\delta_{obs}$), only for amino acids satisfying the two criteria specified in the previous section, are plotted against the concentration values of the unlabeled protein ($[P_u]$) used during the titration to create the corresponding binding curves (Fig. 8). The dissociation constants (K_D) for the interaction can be obtained by fitting the experimental data by non-linear least-squares methods to Eq. (5) (Williamson, 2013).

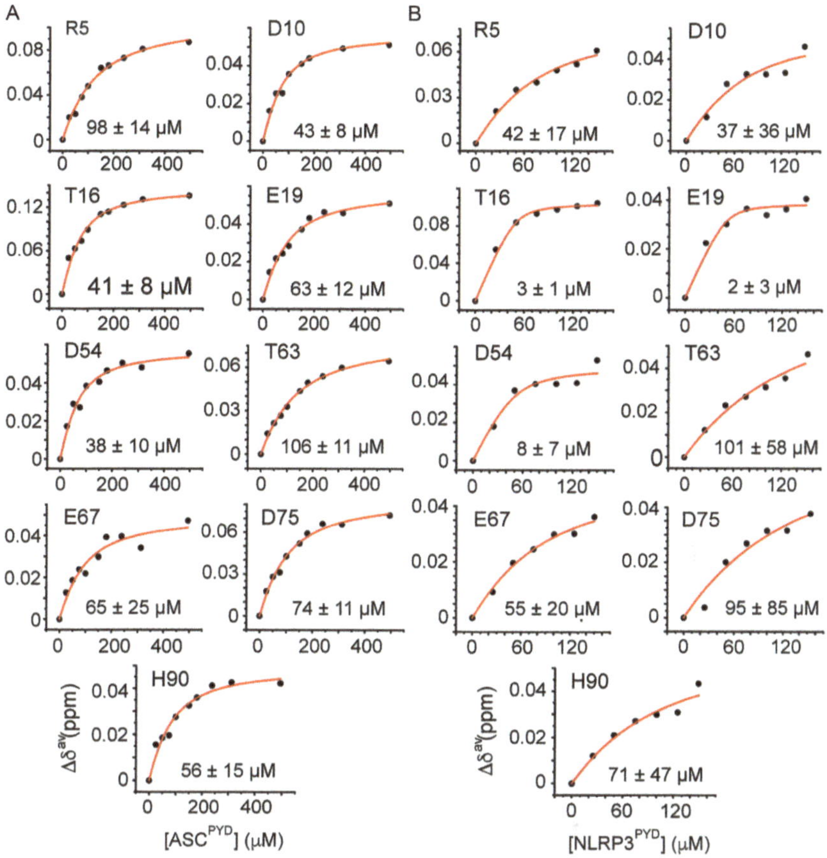

Fig. 8 Apparent dissociation constants for ASC^{PYD}-ASC^{PYD} and ASC^{PYD}-$NLRP3^{PYD}$ interactions. Changes in chemical shift vs. protein concentration for the ASC^{PYD} self-association (A) and the ASC^{PYD}-$NLRP3^{PYD}$ interaction (B). K_D values are shown in each graph. The heteromolecular interaction shows higher affinity.

$$\Delta\delta_{obs} = (\Delta\delta_{max}/2[P_l])\left\{(K_D + [P_u] + [P_l]) - \left((K_D + [P_u] + [P_l])^2 - 4[P_l]\,[P_u]\right)^{1/2}\right\} \tag{5}$$

where, $\Delta\delta_{max}$ is the maximum change in chemical shift obtained by subtracting the value at the starting point from the value when saturation is reached. $[P_l]$ is the starting and total concentration of labeled protein (10–50 μM), $[P_u]$ is total concentration of unlabeled protein at any given point in the titration, and K_D is the dissociation constant. The maximum value of chemical shift change cannot be measured experimentally and may be obtained directly from the curve fitting.

NMR renders multiple probes to monitor binding and thus multiple fittings provide different values of the K_D that are typically within the same order of magnitude. This heterogeneity is due to each probe's specific sensitivity to binding. Therefore, there are two possible options to report the measured K_D values, either by fitting all meaningful binding curves simultaneously to obtain a single value of K_D (Williamson, 2013) or to report the range and the average values from the different probes (Nambayan et al., 2019; Oroz et al., 2016).

The shape of the binding curve also provides information on the presence of single or multiple binding sites when the binding events have different affinities. In the former case, binding isotherms are hyperbolic, whereas they deviate from this behavior to show more complicated curves in the latter. The PYD domain of ASC shows at least two binding regions that correspond to the type I interaction (Oroz et al., 2016). Despite the presences of these two interacting areas, the binding isotherms are hyperbolic most likely due to the equivalent affinity for both sites (Fig. 8). In contrast, ASC^{CARD} titration curves indicate the presence of a binding event happening at significantly high concentration of unlabeled protein added (P_u), reported by the sudden increase of chemical shift perturbation after reaching a plateau (Fig. 9). A possible explanation for this behavior is that the presence of high protein concentration favors the formation of higher order oligomeric species. Apparent K_D values for ASC^{CARD}-ASC^{CARD} self-association were determined excluding the last titration point by fitting the binding isotherms (Fig. 9) to Eq. (5). In addition, the amino acids responsible for the interaction, including their side chains, are mapped onto ASC^{CARD} 3D NMR structure to identify the binding interfaces (Fig. 10).

5. Molecular docking using chemical shift perturbation data

Molecular docking protocols are fast and accurate for defining the relative position of the components of a protein complex if few known restraints can guide the calculations. These restraints may be derived from chemical shift perturbation data, which basically indicates that certain identified amino acids are part of the interacting surface.

One of the most commonly used programs for molecular docking is HADDOCK (http://milou.science.uu.nl/services/HADDOCK2.2) (Van Zundert et al., 2016). This docking protocol defines as active residues those amino acids identified as part of the interacting surface (typically residues

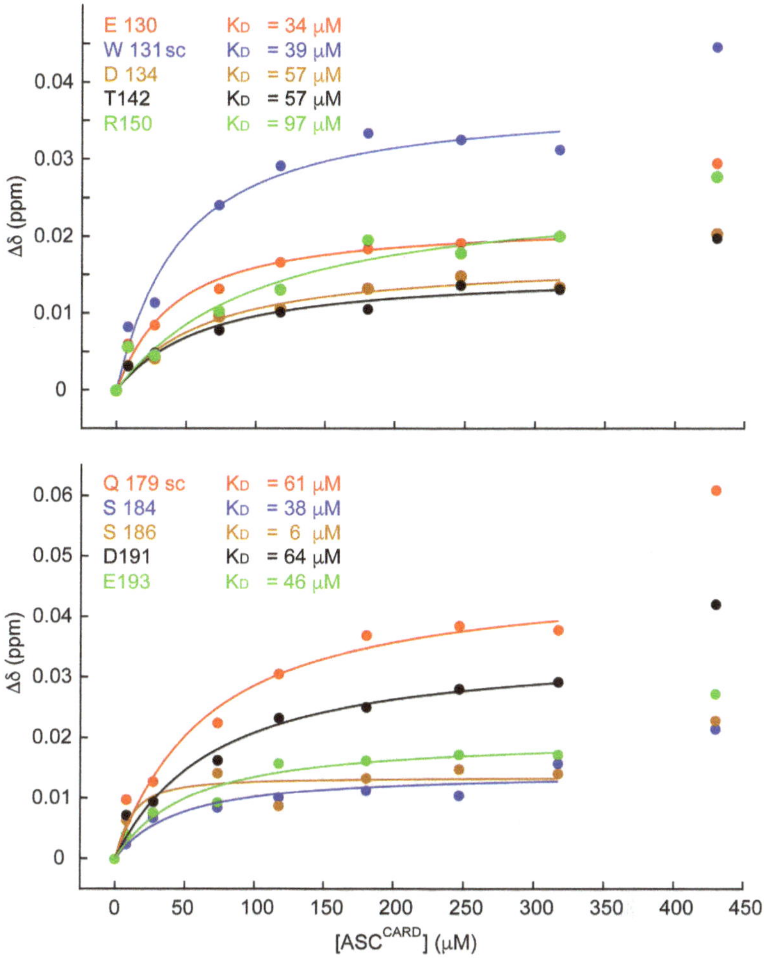

Fig. 9 ASC^{CARD} self-association does not follow a one to one binding mechanism. Chemical shift perturbations of ^{13}C, ^{15}N-labeled ASC^{CARD} vs. unlabeled ASC^{CARD} concentration. Representative solvent-exposed residues with largest perturbations are shown, together with the measured K_D values from the fitting to the hyperbolic section of the curve. The resulting average K_D is ~50 μM.

with large chemical shift perturbations). These amino acids are used to create a list of ambiguous interaction restraints for the docking calculation. In addition, HADDOCK also defines passive residues as those amino acids that are in the vicinity of the active residues, and may or may not participate in the docking protocol. Fig. 11 shows the 3D structure of the ASC dimer calculated by protein-protein docking using experimental NMR restraints

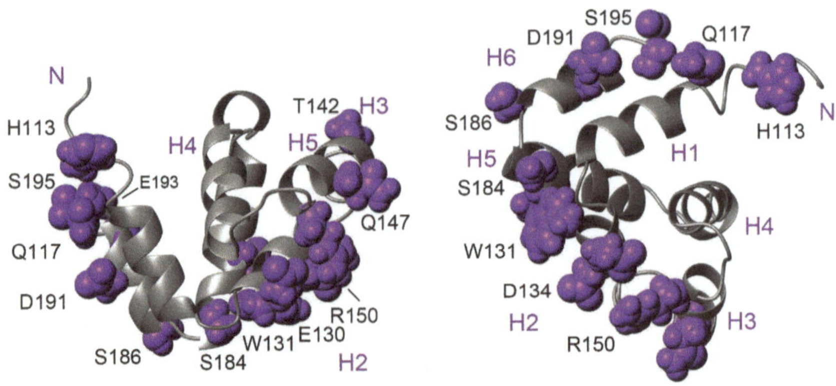

Fig. 10 Interacting surfaces in ASC^{CARD} self-association. ASC^{CARD} structure (De Alba, 2009) is shown as ribbon diagram in two different views. Solvent-exposed amino acids with the largest chemical shift perturbations are depicted in purple and labeled in black. Helices are numbered. Figure was created with MOLMOL (Koradi et al., 1996).

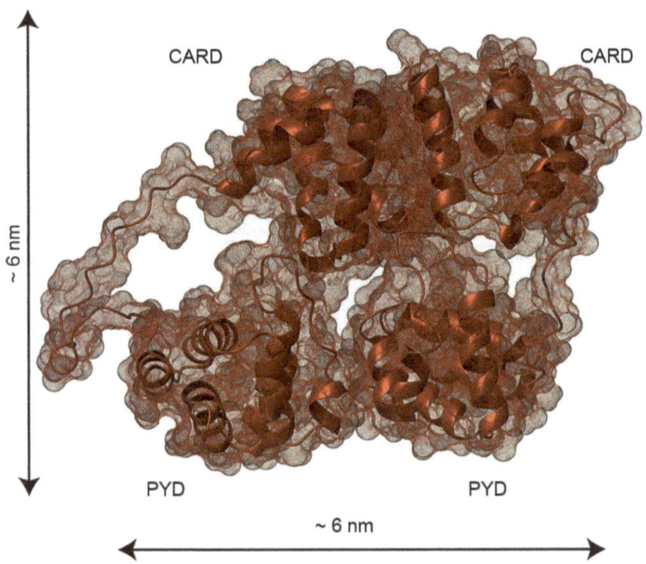

Fig. 11 Dimensions of the ASC dimer. Contact surface of the ASC dimer structure derived from NMR-based protein docking. The PYD and CARD domains are labeled. Dimensions were calculated with MOLMOL (Koradi et al., 1996).

derived from chemical shift perturbation data (Oroz et al., 2016; Nambayan et al., 2019). The dimensions of the dimer match the size of the rings observed by Transmission Electron Microscopy of ASC filaments (Fig. 1), indicating that polymerization of full-length ASC is mediated by both the PYD and CARD domains, and not exclusively by the PYD domain.

5.1 Steps for ASC molecular docking using chemical shift perturbation data

1. Two of the lowest energy structures representative of the solution structure of full-length human ASC (PDB code 2KN6, De Alba, 2009) are used as protomers for the docking protocol.
2. Residues with chemical shift perturbations that are larger than 1.5 times the standard deviation and with solvent exposed areas larger than 30% as calculated by GETAREA (Fraczkiewicz & Braun, 1998) are defined as active for the interaction. Specifically for ASC, the active residues used for the docking calculations were those formed by the α-helices H1, H4 and the N-terminus of H5 (residues G2, R3, D6, D10, T16, L50, D54, E62, T63, E67 and H90), and by helices H2, H3 and the C-terminus of H5 (residues E19, K22, K24, L25, S29, V30, L45 and D75) from ASC^{PYD} (Oroz et al., 2016), and residues H113, Q117, E130, W131, D134, T142, Q147 side chain, R150, Q179 side chain, S184, S186, D191, E193, S195 from ASC^{CARD} (Nambayan et al., 2019).
3. Passive residues were autoselected by the docking protocol around those defined as active.
4. Because the CARD and PYD domains of ASC are connected by a semi-flexible linker, it is recommended to use the HADDOCK protocol interface that allows for backbone and side chain of semi-flexible segments to move during refinement. The semi-flexible segment was defined from residues 90 to 112 (ASC inter-domain linker (De Alba, 2009)).
5. The N and C termini were considered charged.
6. Solvated docking was used with water as solvent for the iterations.
7. The number of starting structures is 1000, out of which 200 are considered for semi-flexible refinement.
8. The top five clusters, each containing five structures, are analyzed and inspected with molecular visualization programs such as MOLMOL (Koradi, Billeter, & Wüthrich, 1996).
9. The majority of the dimers with largest buried surface area and lower overall energy showed the experimentally observed Type I interaction for the PYD-PYD inter-molecular association and the homotypic interaction between the CARDs (Nambayan et al., 2019).

6. Summary

We have described in this chapter some of the most important challenges that researchers may find when attempting to study the structure and function of the inflammasome adapter ASC and its two Death Domains, PYD and CARD, by solution NMR techniques. These challenges start with the overexpression of the protein and protein constructs by recombinant technology and their purification using liquid chromatography. The strong tendency of ASC and its two domains to self-associate and to participate in protein-protein interaction events pose significant difficulties for solution NMR studies. By understanding and describing the different factors playing a critical role in the solubility of these proteins, the researcher is in a better position to overcome these complications. In addition, the specific NMR experiments and protocols used to probe the ASC interaction herein explained, may be applied to other Death Domains and protein interactions of the different inflammasome components.

Acknowledgments

This work was in part supported by NSF-CREST: Center for Cellular and Biomolecular Machines at the University of California, Merced (NSF-HRD-1547848) to P. Diaz-Parga. We are grateful to UC Merced for start-up funds.

References

Arai, M., Ferreon, J. C., & Wright, P. E. (2012). Quantitative analysis of multisite protein–ligand interactions by NMR: Binding of intrinsically disordered p53 transactivation subdomains with the TAZ2 domain of CBP. *Journal of the American Chemical Society*, *134*(8), 3792–3803. https://doi.org/10.1021/ja209936u.

Barbato, G., Ikura, M., Kay, L. E., Pastor, R. W., & Bax, A. (1992). Backbone dynamics of calmodulin studied by nitrogen-15 relaxation using inverse detected two-dimensional NMR spectroscopy: The central helix is flexible. *Biochemistry*, *31*(23), 5269–5278. https://doi.org/10.1021/bi00138a005.

Bertin, J., & DiStefano, P. S. (2000). The PYRIN domain: A novel motif found in apoptosis and inflammation proteins. *Cell Death & Differentiation*, 7(12), 1273–1274. https://doi.org/10.1038/sj.cdd.4400774.

Broz, P., & Dixit, V. M. (2016). Inflammasomes: Mechanism of assembly, regulation and signaling. *Nature Reviews Immunology*, *16*(7), 407–420. https://doi.org/10.1038/nri.2016.58.

Bürckstümmer, T., Baumann, C., Blüml, S., Dixit, E., Dürnberger, G., Jahn, H., et al. (2009). An orthogonal proteomic-genomic screen identifies AIM2 as a cytoplasmic DNA sensor for the inflammasome. *Nature Immunology*, *10*(3), 266–272. https://doi.org/10.1038/ni.1702.

Cavanagh, J., Fairbrother, W. J., Palmer, A. G., Rance, M., & Skelton, N. J. (2007). Multidimensional NMR spectroscopy. *Protein NMR Spectroscopy*, 271–332. https://doi.org/10.1016/b978-012164491-8/50006-3.

Chinnaiyan, A. M., O'Rourke, K., Tewari, M., & Dixit, V. M. (1995). FADD, a novel death domain-containing protein, interacts with the death domain of fas and initiates apoptosis. *Cell*, *81*(4), 505–512. https://doi.org/10.1016/0092-8674(95)90071-3.

Chou, J. J., Matsuo, H., Duan, H., & Wagner, G. (1998). Solution structure of the RAIDD CARD and model for CARD/CARD interaction in Caspase-2 and Caspase-9 recruitment. *Cell*, *94*(2), 171–180. https://doi.org/10.1016/s0092-8674(00)81417-8.

Clore, G. M., & Iwahara, J. (2009). Theory, practice, and applications of paramagnetic relaxation enhancement for the characterization of transient low-population states of biological macromolecules and their complexes. *Chemical Reviews*, *109*(9), 4108–4139. https://doi.org/10.1021/cr900033p.

De Alba, E. (2009). Structure and Interdomain dynamics of apoptosis-associated speck-like protein containing a CARD (ASC). *Journal of Biological Chemistry*, *284*(47), 32932–32941. https://doi.org/10.1074/jbc.m109.024273.

De Alba, E., Baber, J. L., & Tjandra, N. (1999). The use of residual dipolar coupling in concert with backbone relaxation rates to identify conformational exchange by NMR. *Journal of the American Chemical Society*, *121*(17), 4282–4283. https://doi.org/10.1021/ja990062t.

De Alba, E., & Tjandra, N. (2000). Protein backbone 15N relaxation rates as a tool for the diagnosis of structure quality. *Journal of Magnetic Resonance*, *144*(2), 367–371. https://doi.org/10.1006/jmre.2000.2063.

De Alba, E., & Tjandra, N. (2002). NMR dipolar couplings for the structure determination of biopolymers in solution. *Progress in Nuclear Magnetic Resonance Spectroscopy*, *40*(2), 175–197. https://doi.org/10.1016/s0079-6565(01)00042-5.

De Alba, E., & Tjandra, N. (2003). Orientational restraints. *BioNMR in Drug Research*, *16*, 179–205. https://doi.org/10.1002/3527600663.ch8.

Dinarello, C. A. (2011). Interleukin-1 in the pathogenesis and treatment of inflammatory diseases. *Blood*, *117*(14), 3720–3732. https://doi.org/10.1182/blood-2010-07-273417.

Ding, J., Wang, K., Liu, W., She, Y., Sun, Q., Shi, J., et al. (2016). Pore-forming activity and structural autoinhibition of the gasdermin family. *Nature*, *535*(7610), 111–116. https://doi.org/10.1038/nature18590.

Dorfleutner, A., Chu, L., & Stehlik, C. (2015). Inhibiting the inflammasome: One domain at a time. *Immunological Reviews*, *265*(1), 205–216. https://doi.org/10.1111/imr.12290.

Douzi, B. (2017). Protein–protein interactions: surface Plasmon resonance. *Methods in Molecular Biology*, *1615*, 257–275. https://doi.org/10.1007/978-1-4939-7033-9_21.

Doyle, M. L. (1997). Characterization of binding interactions by isothermal titration calorimetry. *Current Opinion in Biotechnology*, *8*(1), 31–35. https://doi.org/10.1016/s0958-1669(97)80154-1.

Duan, H., & Dixit, V. M. (1997). RAIDD is a new 'death' adaptor molecule. *Nature*, *385*(6611), 86–89. https://doi.org/10.1038/385086a0.

Fairbrother, W. J., Gordon, N. C., Humke, E. W., O'Rourke, K. M., Starovasnik, M. A., Yin, J., et al. (2001). The PYRIN domain: A member of the death domain-fold superfamily. *Protein Science*, *10*(9), 1911–1918. https://doi.org/10.1110/ps.13801.

Faustin, B., Lartigue, L., Bruey, J., Luciano, F., Sergienko, E., Bailly-Maitre, B., et al. (2007). Reconstituted NALP1 Inflammasome reveals two-step mechanism of Caspase-1 activation. *Molecular Cell*, *25*(5), 713–724. https://doi.org/10.1016/j.molcel.2007.01.032.

Fernandes-Alnemri, T., Yu, J., Datta, P., Wu, J., & Alnemri, E. S. (2009). AIM2 activates the inflammasome and cell death in response to cytoplasmic DNA. *Nature*, *458*(7237), 509–513. https://doi.org/10.1038/nature07710.

Fielding, L. (2003). NMR methods for the determination of protein- ligand dissociation constants. *Current Topics in Medicinal Chemistry*, *3*(1), 39–53. https://doi.org/10.2174/1568026033392705.

Fink, S. L., & Cookson, B. T. (2005). Apoptosis, pyroptosis, and necrosis: Mechanistic description of dead and dying eukaryotic cells. *Infection and Immunity*, *73*(4), 1907–1916. https://doi.org/10.1128/iai.73.4.1907-1916.2005.

Fink, S. L., & Cookson, B. T. (2006). Caspase-1-dependent pore formation during pyroptosis leads to osmotic lysis of infected host macrophages. *Cellular Microbiology*, *8*(11), 1812–1825. https://doi.org/10.1111/j.1462-5822.2006.00751.x.

Fraczkiewicz, R., & Braun, W. (1998). Exact and efficient analytical calculation of the accessible surface areas and their gradients for macromolecules. *Journal of Computational Chemistry*, *19*(3), 319. https://doi.org/10.1002/(sici)1096-987x(199802)19:3<319::aid-jcc6>3.3.co;2-3.

Fritz, J. H., Ferrero, R. L., Philpott, D. J., & Girardin, S. E. (2006). Nod-like proteins in immunity, inflammation and disease. *Nature Immunology*, 7(12), 1250–1257. https://doi.org/10.1038/ni1412.

Gross, O., Poeck, H., Bscheider, M., Dostert, C., Hannesschläger, N., Endres, S., et al. (2009). Syk kinase signaling couples to the Nlrp3 inflammasome for anti-fungal host defense. *Nature*, *459*(7245), 433–436. https://doi.org/10.1038/nature07965.

Grzesiek, S., & Bax, A. (1992). Correlating backbone amide and side chain resonances in larger proteins by multiple relayed triple resonance NMR. *Journal of the American Chemical Society*, *114*(16), 6291–6293. https://doi.org/10.1021/ja00042a003.

Guo, H., Callaway, J. B., & Ting, J. P. (2015). Inflammasomes: Mechanism of action, role in disease, and therapeutics. *Nature Medicine*, *21*(7), 677–687. https://doi.org/10.1038/nm.3893.

He, Y., Hara, H., & Núñez, G. (2016). Mechanism and regulation of NLRP3 Inflammasome activation. *Trends in Biochemical Sciences*, *41*(12), 1012–1021. https://doi.org/10.1016/j.tibs.2016.09.002.

He, W., Wan, H., Hu, L., Chen, P., Wang, X., Huang, Z., et al. (2015). Gasdermin D is an executor of pyroptosis and required for interleukin-1β secretion. *Cell Research*, *25*(12), 1285–1298. https://doi.org/10.1038/cr.2015.139.

Heck, A. J. (2008). Native mass spectrometry: A bridge between interactomics and structural biology. *Nature Methods*, *5*(11), 927–933. https://doi.org/10.1038/nmeth.1265.

Hofmann, K. (1997). The CARD domain: A new apoptotic signaling motif. *Trends in Biochemical Sciences*, *22*(5), 155–156. https://doi.org/10.1016/s0968-0004(97)01043-8.

Ichinohe, T., Lee, H. K., Ogura, Y., Flavell, R., & Iwasaki, A. (2009). Inflammasome recognition of influenza virus is essential for adaptive immune responses. *The Journal of Experimental Medicine*, *206*(1), 79–87. https://doi.org/10.1084/jem.20081667.

Inohara, N., & Nuñez, G. (2003). NODs: Intracellular proteins involved in inflammation and apoptosis. *Nature Reviews Immunology*, *3*(5), 371–382. https://doi.org/10.1038/nri1086.

Ishima, R., & Torchia, D. A. (2000). Protein dynamics from NMR. *Nature Structural Biology*, 7(9), 740–743. https://doi.org/10.1038/78963.

Jensen, M. R., Ortega-Roldan, J., Salmon, L., Van Nuland, N., & Blackledge, M. (2011). Characterizing weak protein–protein complexes by NMR residual dipolar couplings. *European Biophysics Journal*, *40*(12), 1371–1381. https://doi.org/10.1007/s00249-011-0720-5.

Jin, T., & Xiao, T. S. (2015). Activation and assembly of the inflammasomes through conserved protein domain families. *Apoptosis*, *20*(2), 151–156. https://doi.org/10.1007/s10495-014-1053-5.

Joly, S., Ma, N., Sadler, J. J., Soll, D. R., Cassel, S. L., & Sutterwala, F. S. (2009). Cutting edge: Candida albicans hyphae formation triggers activation of the Nlrp3 inflammasome. *The Journal of Immunology*, *183*(6), 3578–3581. https://doi.org/10.4049/jimmunol.0901323.

Kanneganti, T., Body-Malapel, M., Amer, A., Park, J., Whitfield, J., Franchi, L., et al. (2006). Critical role for cryopyrin/Nalp3 in activation of Caspase-1 in response to viral infection and double-stranded RNA. *Journal of Biological Chemistry*, *281*(48), 36560–36568. https://doi.org/10.1074/jbc.m607594200.

Kersse, K., Verspurten, J., Berghe, T. V., & Vandenabeele, P. (2011). The death-fold superfamily of homotypic interaction motifs. *Trends in Biochemical Sciences*, *36*(10), 541–552. https://doi.org/10.1016/j.tibs.2011.06.006.

Koradi, R., Billeter, M., & Wüthrich, K. (1996). MOLMOL: A program for display and analysis of macromolecular structures. *Journal of Molecular Graphics*, *14*(1), 51–55. https://doi.org/10.1016/0263-7855(96)00009-4.

Kwon, D., Yoon, J. H., Shin, S., Jang, T., Kim, H., So, I., et al. (2012). A comprehensive manually curated protein–protein interaction database for the death domain superfamily. *Nucleic Acids Research*, *40*(D1), D331–D336. https://doi.org/10.1093/nar/gkr1149.

Leavitt, S., & Freire, E. (2001). Direct measurement of protein binding energetics by isothermal titration calorimetry. *Current Opinion in Structural Biology*, *11*(5), 560–566. https://doi.org/10.1016/s0959-440x(00)00248-7.

Leney, A. C., & Heck, A. J. (2017). Native mass spectrometry: What is in the name? *Journal of the American Society for Mass Spectrometry*, *28*(1), 5–13. https://doi.org/10.1007/s13361-016-1545-3.

Li, Y., Fu, T., Lu, A., Witt, K., Ruan, J., Shen, C., et al. (2018). Cryo-EM structures of ASC and NLRC4 CARD filaments reveal a unified mechanism of nucleation and activation of caspase-1. *Proceedings of the National Academy of Sciences*, *115*(43), 10845–10852. https://doi.org/10.1073/pnas.1810524115.

Liu, X., Zhang, Z., Ruan, J., Pan, Y., Magupalli, V. G., Wu, H., et al. (2016). Inflammasome-activated gasdermin D causes pyroptosis by forming membrane pores. *Nature*, *535*(7610), 153–158. https://doi.org/10.1038/nature18629.

Lu, A., Magupalli, V., Ruan, J., Yin, Q., Atianand, M., Vos, M. R., et al. (2014). Unified polymerization mechanism for the assembly of ASC-dependent inflammasomes. *Cell*, *156*(6), 1193–1206. https://doi.org/10.1016/j.cell.2014.02.008.

Malik, A., & Kanneganti, T. (2017). Inflammasome activation and assembly at a glance. *Journal of Cell Science*, *130*(23), 3955–3963. https://doi.org/10.1242/jcs.207365.

Martinon, F., Burns, K., & Tschopp, J. (2002). The inflammasome. *Molecular Cell*, *10*(2), 417–426. https://doi.org/10.1016/s1097-2765(02)00599-3.

Martinon, F., & Tschopp, J. (2005). NLRs join TLRs as innate sensors of pathogens. *Trends in Immunology*, *26*(8), 447–454. https://doi.org/10.1016/j.it.2005.06.004.

Masumoto, J., Taniguchi, S., Ayukawa, K., Sarvotham, H., Kishino, T., Niikawa, N., et al. (1999). ASC, a novel 22-kDa protein, aggregates during apoptosis of human promyelocytic leukemia HL-60 cells. *Journal of Biological Chemistry*, *274*(48), 33835–33838. https://doi.org/10.1074/jbc.274.48.33835.

Masumoto, J., Taniguchi, S., Nakayama, K., Ayukawa, K., & Sagara, J. (2001). Murine Ortholog of ASC, a CARD-containing protein, self-associates and exhibits restricted distribution in developing mouse embryos. *Experimental Cell Research*, *262*(2), 128–133. https://doi.org/10.1006/excr.2000.5078.

Matyszewski, M., Morrone, S. R., & Sohn, J. (2018). Digital signaling network drives the assembly of the AIM2-ASC inflammasome. *Proceedings of the National Academy of Sciences*, *115*(9), E1963–E1972. https://doi.org/10.1073/pnas.1712860115.

Meylan, E., Tschopp, J., & Karin, M. (2006). Intracellular pattern recognition receptors in the host response. *Nature*, *442*(7098), 39–44. https://doi.org/10.1038/nature04946.

Mittermaier, A., & Kay, L. E. (2006). New tools provide new insights in NMR studies of protein dynamics. *Science*, *312*(5771), 224–228. https://doi.org/10.1126/science.1124964.

Nambayan, R. J. T., Sandin, S. I., Quint, D. A., Satyadi, D. M., & De Alba, E. (2019). The inflammasome adapter ASC assembles into filaments with integral participation of its two death domains, PYD and CARD. *Journal of Biological Chemistry*, *294*(2), 439–452. https://doi.org/10.1074/jbc.ra118.004407.

Oroz, J., Barrera-Vilarmau, S., Alfonso, C., Rivas, G., & De Alba, E. (2016). ASC pyrin domain self-associates and binds NLRP3 protein using equivalent binding interfaces. *Journal of Biological Chemistry*, *291*(37), 19487–19501. https://doi.org/10.1074/jbc.m116.741082.

Park, H. H., Lo, Y., Lin, S., Wang, L., Yang, J. K., & Wu, H. (2007). The death domain superfamily in intracellular signaling of apoptosis and inflammation. *Annual Review of Immunology, 25*(1), 561–586. https://doi.org/10.1146/annurev.immunol.25.022106.141656.

Park, H. H., Logette, E., Raunser, S., Cuenin, S., Walz, T., Tschopp, J., et al. (2007). Death domain assembly mechanism revealed by crystal structure of the oligomeric PIDDosome core complex. *Cell, 128*(3), 533–546. https://doi.org/10.1016/j.cell.2007.01.019.

Phizicky, E. M., & Fields, S. (1995). Protein-protein interactions: Methods for detection and analysis. *Microbiological Reviews, 59*, 94–123.

Platnich, J. M., Chung, H., Lau, A., Sandall, C. F., Bondzi-Simpson, A., Chen, H., et al. (2018). Shiga toxin/lipopolysaccharide activates Caspase-4 and Gasdermin D to trigger mitochondrial reactive oxygen species upstream of the NLRP3 inflammasome. *Cell Reports, 25*(6), 1525–1536. e7. https://doi.org/10.1016/j.celrep.2018.09.071.

Proell, M., Riedl, S. J., Fritz, J. H., Rojas, A. M., & Schwarzenbacher, R. (2008). The nod-like receptor (NLR) family: A tale of similarities and differences. *PLoS One, 3*(4). e2119. https://doi.org/10.1371/journal.pone.0002119.

Raines, R. T. (2015). Fluorescence polarization assay to quantify protein-protein interactions: An update. *Methods in Molecular Biology*, 323–327. https://doi.org/10.1007/978-1-4939-2425-7_19.

Rao, V. S., Srinivas, K., Sujini, G. N., & Kumar, G. N. (2014). Protein-protein interaction detection: Methods and analysis. *International Journal of Proteomics, 2014*, 1–12. https://doi.org/10.1155/2014/147648.

Rathinam, V. A., Jiang, Z., Waggoner, S. N., Sharma, S., Cole, L. E., Waggoner, L., et al. (2010). The AIM2 inflammasome is essential for host defense against cytosolic bacteria and DNA viruses. *Nature Immunology, 11*(5), 395–402. https://doi.org/10.1038/ni.1864.

Rathkey, J. K., Benson, B. L., Chirieleison, S. M., Yang, J., Xiao, T. S., Dubyak, G. R., et al. (2017). Live-cell visualization of gasdermin D-driven pyroptotic cell death. *Journal of Biological Chemistry, 292*(35), 14649–14658. https://doi.org/10.1074/jbc.m117.797217.

Sborgi, L., Ravotti, F., Dandey, V., Dick, M., Mazur, A., Reckel, S., et al. (2015). Structure and assembly of the mouse ASC filament by combined NMR spectroscopy and cryo-electron microscopy. *Proceedings of the National Academy of Sciences of the United States of America, 112*, 13237–13242. https://doi.org/10.2210/pdb2n1f/pdb.

Schanda, P., & Brutscher, B. (2005). Very fast two-dimensional NMR spectroscopy for real-time investigation of dynamic events in proteins on the time scale of seconds. *Journal of the American Chemical Society, 127*(22), 8014–8015. https://doi.org/10.1021/ja051306e.

Schroder, K., & Tschopp, J. (2010). The inflammasomes. *Cell, 140*(6), 821–832. https://doi.org/10.1016/j.cell.2010.01.040.

Shi, J., Zhao, Y., Wang, K., Shi, X., Wang, Y., Huang, H., et al. (2015). Cleavage of GSDMD by inflammatory caspases determines pyroptotic cell death. *Nature, 526*(7575), 660–665. https://doi.org/10.1038/nature15514.

Smits, A. H., & Vermeulen, M. (2016). Characterizing protein–protein interactions using mass spectrometry: Challenges and opportunities. *Trends in Biotechnology, 34*(10), 825–834. https://doi.org/10.1016/j.tibtech.2016.02.014.

Sohn, J., & Hur, S. (2016). Filament assemblies in foreign nucleic acid sensors. *Current Opinion in Structural Biology, 37*, 134–144. https://doi.org/10.1016/j.sbi.2016.01.011.

Srinivasula, S. M., Poyet, J., Razmara, M., Datta, P., Zhang, Z., & Alnemri, E. S. (2002). The PYRIN-CARD protein ASC is an activating adaptor for caspase-1. *Journal of Biological Chemistry, 277*(24), 21119–21122. https://doi.org/10.1074/jbc.c200179200.

Stehlik, C., Lee, S. H., Dorfleutner, A., Stassinopoulos, A., Sagara, J., & Reed, J. C. (2003). Apoptosis-associated speck-like protein containing a caspase recruitment domain is a regulator of procaspase-1 activation. *The Journal of Immunology, 171*(11), 6154–6163. https://doi.org/10.4049/jimmunol.171.11.6154.

Stehlik, C., & Reed, J. C. (2004). The PYRIN connection. *The Journal of Experimental Medicine, 200*(5), 551–558. https://doi.org/10.1084/jem.20032234.

Stratmann, S. A., Morrone, S. R., Van Oijen, A. M., & Sohn, J. (2015). The innate immune sensor IFI16 recognizes foreign DNA in the nucleus by scanning along the duplex. *eLife, 4*, e11721. https://doi.org/10.7554/elife.11721.

Thornberry, N. A., Bull, H. G., Calaycay, J. R., Chapman, K. T., Howard, A. D., Kostura, M. J., et al. (1992). A novel heterodimeric cysteine protease is required for interleukin-1β processing in monocytes. *Nature, 356*(6372), 768–774. https://doi.org/10.1038/356768a0.

Tjandra, N., & Bax, A. (1997). Direct measurement of distances and angles in biomolecules by NMR in a dilute liquid crystalline medium. *Science, 278*(5340), 1111–1114. https://doi.org/10.1126/science.278.5340.1111.

Tjandra, N., Feller, S. E., Pastor, R. W., & Bax, A. (1995). Rotational diffusion anisotropy of human ubiquitin from 15N NMR relaxation. *Journal of the American Chemical Society, 117*(50), 12562–12566. https://doi.org/10.1021/ja00155a020.

Tominaga, K., Yoshimoto, T., Torigoe, K., Kurimoto, M., Matsui, K., Hada, T., et al. (2000). IL-12 synergizes with IL-18 or IL-1β for IFN-γ production from human T cells. *International Immunology, 12*(2), 151–160. https://doi.org/10.1093/intimm/12.2.151.

Van Damme, J., De Ley, M., Opdenakker, G., Billiau, A., De Somer, P., & Beeumen, J. V. (1985). Homogeneous interferon-inducing 22K factor is related to endogenous pyrogen and interleukin-1. *Nature, 314*(6008), 266–268. https://doi.org/10.1038/314266a0.

Van Zundert, G., Rodrigues, J., Trellet, M., Schmitz, C., Kastritis, P., Karaca, E., et al. (2016). The HADDOCK2.2 web server: User-friendly integrative modeling of biomolecular complexes. *Journal of Molecular Biology, 428*(4), 720–725. https://doi.org/10.1016/j.jmb.2015.09.014.

Wagner, G. (1993). NMR relaxation and protein mobility. *Current Opinion in Structural Biology, 3*(5), 748–754. https://doi.org/10.1016/0959-440x(93)90059-t.

Weber, C. H., & Vincenz, C. (2001). The death domain superfamily: A tale of two interfaces? *Trends in Biochemical Sciences, 26*(8), 475–481. https://doi.org/10.1016/s0968-0004(01)01905-3.

Williamson, M. P. (2013). Using chemical shift perturbation to characterize ligand binding. *Progress in Nuclear Magnetic Resonance Spectroscopy, 73*, 1–16. https://doi.org/10.1016/j.pnmrs.2013.02.001.

Wishart, D. S., Bigam, C. G., Holm, A., Hodges, R. S., & Sykes, B. D. (1995). 1H, 13C and 15N random coil NMR chemical shifts of the common amino acids. I. Investigations of nearest-neighbor effects. *Journal of Biomolecular NMR, 5*(1), 67–81. https://doi.org/10.1007/bf00227471.

Wittekind, M., & Mueller, L. (1993). HNCACB, a high-sensitivity 3D NMR experiment to correlate amide-proton and nitrogen resonances with the alpha- and beta-carbon resonances in proteins. *Journal of Magnetic Resonance, Series B, 101*(2), 201–205. https://doi.org/10.1006/jmrb.1993.1033.

Xu, H., Yang, J., Gao, W., Li, L., Li, P., Zhang, L., et al. (2014). Innate immune sensing of bacterial modifications of rho GTPases by the pyrin inflammasome. *Nature, 513*(7517), 237–241. https://doi.org/10.1038/nature13449.

Further reading

Inohara, N., Chamaillard, M., McDonald, C., & Nuñez, G. (2005). NOD-LRR PROTEINS: Role in host-microbial interactions and inflammatory disease. *Annual Review of Biochemistry, 74*(1), 355–383. https://doi.org/10.1146/annurev.biochem.74.082803.133347.

Mittermaier, A. (2006). New tools provide new insights in NMR studies of protein dynamics. *Science*, *312*(5771), 224–228. https://doi.org/10.1126/science.1124964.

Palmer, A. G. (2004). NMR characterization of the dynamics of biomacromolecules. *Chemical Reviews*, *104*(8), 3623–3640. https://doi.org/10.1021/cr030413t.

Ting, J. P., Kastner, D. L., & Hoffman, H. M. (2006). CATERPILLERs, pyrin and hereditary immunological disorders. *Nature Reviews Immunology*, *6*(3), 183–195. https://doi.org/10.1038/nri1788.

CHAPTER FIFTEEN

Tracking DNA sensor and inflammasome complexes in cells using confocal microscopy

Brendan Antiochos*
Johns Hopkins Division of Rheumatology, Baltimore, MD, United States
*Corresponding author: e-mail address: bantioc1@jhmi.edu

Contents

Abstract

The innate immune system provides host defense against a variety of threats, including microbial infection and cellular damage. Pattern recognition receptors identify both pathogen and damage associated molecular motifs associated with these events. Nucleic acid is one such danger signal that is detected by a set of cytoplasmic sensors. Oligomerization is a key mode of activation shared by many of these sensors, who form large multimeric structures when activated. Microscopy is one of several techniques that facilitates study of these molecules, permitting visualization of oligomeric forms, and tracking of relevant signaling partners. In this chapter, we will discuss methods of generating and imaging cytoplasmic DNA sensors and inflammasome complexes by fluorescence microscopy.

The innate immune system provides the host with the ability to identify and respond to a variety of intracellular threats, including cellular injury, genomic instability, and infection (Fischer, 2019; Li & Chen, 2018).

Methods in Enzymology, Volume 625
ISSN 0076-6879
https://doi.org/10.1016/bs.mie.2019.05.005

Both pathogen associated molecular patterns (PAMPs) and damage associated molecular patterns (DAMPs) can elicit inflammatory signaling cascades through activation of innate cellular sensors (Palmai-Pallag & Bachrati, 2014). Innate immune sensors are of interest to the fields of immunology, microbiology (Crane, Lee, FitzGerald, & Jamieson, 2018), autoimmunity (Caso et al., 2018) and cancer research (Li & Chen, 2018), and numerous techniques can be employed in their study.

Double-stranded DNA (dsDNA) is an immunostimulatory molecular pattern that activates numerous intracellular sensors when mislocalized within the cell (Fischer, 2019; Unterholzner, 2013). Cytoplasmic dsDNA can indicate the presence of microbial infection as well as various forms of cellular injury that result in damaged endogenous nucleic acid (Dhanwani, Takahashi, & Sharma, 2018). These cellular events indicate great potential harm to the cell, and are therefore monitored for by an array of dsDNA sensors, which include cyclic GMP-AMP synthase (cGAS) (Sun, Wu, Du, Chen, & Chen, 2013), the AIM-like receptors (AIM2, IFI16, IFIX and MNDA), DAI, RNA Polymerase III, DDX41, and others (Unterholzner, 2013). Two main signaling pathways have been identified downstream of these sensors: the Type I IFN response and the pro-caspase-1 inflammasome. Through these pathways, identification of dsDNA in the cytosol elicits an immune response that ranges from antiviral IFN activation (Unterholzner, 2013) to pyroptosis via IL-1 (Hornung et al., 2009).

Numerous innate signaling pathways are activated by the assembly of oligomeric molecular structures. This mode of activation through oligomerization is shared by many innate pattern recognition receptors, including the Toll-like receptors (TLRs), NOD-like receptors (NLRs), the cytoplasmic nucleic acid sensors of RNA (RLRs) and DNA (ALRs), and components of the inflammasome (Nanson, Rahaman, Ve, & Kobe, 2018; Vajjhala, Ve, Bentham, Stacey, & Kobe, 2017). The ALRs AIM2 and IFI16 generate large filamentous oligomers when activated; these filaments can be imaged by electron microscopy as well as immunofluorescence imaging (Antiochos, Matyszewski, Sohn, Casciola-Rosen, & Rosen, 2018; Morrone et al., 2015, 2014; Sohn & Hur, 2016). Both inflammasome proteins ASC and pro-caspase 1 form filaments when activated, and have a similar appearance to ALR filaments when viewed by EM (Franklin et al., 2014; Lu et al., 2014). When imaged within cells, inflammasome activation leads to the generation of characteristic "ASC specks," representing an oligomerized state of ASC subunits (Franklin, Latz, & Schmidt, 2018). cGAS and STING both dimerize when activated

(Hooy & Sohn, 2018; Ouyang et al., 2012), and STING has been visualized in perinuclear clusters in its activated state (Ablasser et al., 2013).

Several techniques can be employed to measure activation of the dsDNA sensing pathways. Biochemical or molecular analysis of the downstream pathway activation state is one option—this could include Western blotting of mature caspase 1 or PCR analysis of IFN expression (Hornung et al., 2009; Unterholzner et al., 2010). This read-out can be coupled to knock-down or knock-out of the sensor of interest to implicate its role in activating the relevant pathway. Semi-denaturing or non-denaturing gel electrophoresis can be employed to demonstrate oligomerization of activated proteins (Cai et al., 2014; Motani, Ito, & Nagata, 2015) when coupled to Western blotting. Activation of many dsDNA sensors and inflammasome components can also be directly visualized by immunofluorescence imaging (Hornung et al., 2009; Sun et al., 2013). Given the considerable conformational changes that occur with activation of many of these molecules, cellular imaging affords the ability to specifically identify cells in which dsDNA sensing machinery has encountered its ligand and assumed an activated state. Immunofluorescent imaging also permits the simultaneous tracking of interactions with additional proteins of interest found in the multi-molecular complexes that are integral to these pathways.

Imaging of dsDNA sensors can be conducted in cell culture as well as ex vivo in tissues from animal models or human subjects. The following protocols will provide guidance for typical experimental procedures that permit imaging of dsDNA sensors and inflammasome complexes in cultured cells. Several general considerations need to be addressed in designing such an experiment.

1. Endogenous versus exogenous protein

Two sources of dsDNA sensor/inflammasome protein can be studied in a cell culture system: (1) native endogenous proteins expressed by the cell line of interest, or (2) exogenous protein expressed by either transient transfection or genetic editing. The latter category includes genes incorporated via CRISPR-Cas9 system, Tetracycline-inducible expression systems, or transient plasmid-based DNA transfection systems.

Activation of endogenous protein may represent a more physiologically relevant stimulus for study of downstream cellular events, and several well-characterized ligands for dsDNA sensors and inflammasome complexes exist to permit this approach. Drawbacks of studying endogenous protein include the variable expression levels of these sensors in several cell lines, including the potential for mutational loss in some cancer cell lines. For example, the widely used HEK 293T cell line is deficient in nearly all proteins within this group, including AIM2, IFI16 and ASC. It is therefore critical to determine the expression levels of target proteins in any cell line of interest: this is best done at the protein (rather than RNA) level and can be performed using either Western blotting of cell lysates or immunofluorescence of fixed cells. Among the more widely used immortalized cell lines, THP1 cells are of particular utility given their expression of numerous dsDNA sensors and inflammasome components, including cGAS/STING, IFI16, AIM2, ASC and pro-caspase1 (Paijo et al., 2016; Veeranki, Duan, Panchanathan, Liu, & Choubey, 2011).

In situations where the desired cell line does not express the protein of interest, or where experimental manipulation of the protein is desired (such as the generation of mutations or genetic protein tags), the use of exogenous protein may be beneficial. A simple means of introducing exogenous protein is via transient DNA transfection. This method allows for high level protein expression in suitable cells, but has the drawback of introducing dsDNA (the plasmid itself) which will lead to the activation of other endogenous dsDNA sensors present in the cell. In this light, the HEK 293T cell line is of particular utility, as it lacks inflammasome components (ASC), as well as IFI16, AIM2, cGAS and STING, and can therefore be used to reconstitute these dsDNA sensing pathways by transfection of the genes of interest without concern for activating endogenous elements of the same pathway by the transfection process alone (Fernandes-Alnemri, Yu, Datta, Wu, & Alnemri, 2009; Hornung et al., 2009; Zhang et al., 2014).

An alternative to DNA transfection is RNA transfection, which has been used in the study of the ALR IFI16 (Almine et al., 2017). With this technique, RNA is transcribed using an in vitro system, generating mRNA suitable for expression in mammalian cells. The parent plasmid DNA is destroyed with DNase, and the mRNA alone is transfected into the cell for expression. This approach avoids the issue of activating endogenous dsDNA sensors with the transfection process alone (though RNA sensors may be activated). While RNA transfection typically results in lower expression levels than DNA transfection, this system can be optimized to yield expression of sufficient protein for visualization of exogenous protein when DNA transfection is not feasible.

2. Sensor ligand

The cytoplasmic dsDNA sensors interact with nucleic acid in a sequence independent manner. Accordingly, a wide variety of DNA sources can be utilized to activate the sensors in a target cell, including plasmid DNA, synthetic DNA (Poly (dA:dT)), or PCR-generated DNA fragments. PCR can be used to create relatively short dsDNA fragments, ranging in size from a few hundred to a few thousand base pair (Luecke et al., 2017). An important consideration when selecting a source of dsDNA is the means by which the DNA will be visualized (if necessary). Intercalating dye such as DAPI can be used to visualize all dsDNA within the cell, including the transfected dsDNA, as well as the host cell's nucleus. DAPI has the advantages of being inexpensive, readily available, and an easily detected fluorescent dye. However, other options are available for specific labeling of the transfected dsDNA, which may be of use to discriminate exogenous dsDNA from the cell's endogenous nucleic acid. Rhodamine-labeled Poly(dA:dT) is commercially available (InvivoGen) and can be readily visualized after transfection with no additional labeling. An alternative approach is biotinylation of unlabeled DNA (plasmid or other), followed by fluorescent labeling with Streptavidin-coupled fluorophores. The PHOTOPROBE biotin kit (Vector Labs) serves well for this purpose. DNA can also be directly labeled with fluorescent dyes such as FITC or Cy3 using a kit from Mirus Bio (Fernandes-Alnemri et al., 2009). It is important to note that this sort of modification of the target dsDNA can alter the sensor's ability to recognize the ligand, and may alter the sensor's activation properties. One should therefore select minimally sized DNA tags, such as short arm (versus long arm) biotin when labeling plasmid.

An additional means of labeling DNA within cells is the use of anti-dsDNA antibodies, which can be used in co-staining experiments when the target protein is identified by immunolabeling. Monoclonal antibodies against dsDNA are commercially available, including clone AC-30-10 (Millipore Sigma). When antibody labeling of exogenous dsDNA is employed, an unmodified DNA ligand can be utilized.

3. Inflammasome inducers

Numerous compounds have been used to experimentally induce inflammasome assembly, reflecting the wide spectrum of cellular stimuli that are capable of activating this pathway (Latz, Xiao, & Stutz, 2013). While the AIM2 inflammasome in particular can be induced with dsDNA, a variety of other compounds, including toxins, bacterial motifs, uric acid, and other molecules, can activate other inflammasomes, in particular the NLRP3 inflammasome (Sidiropoulos, Goulielmos, Voloudakis, Petraki, & Boumpas, 2008). These compounds are available from a variety of sources, including the vendor InvivoGen.

4. Immunofluorescence cell imaging protocols

Protocol 1 explains the approach to imaging an exogenous DNA sensor with a fluorescent tag via plasmid-based transfection. Protocol 2 explains immunofluorescent imaging of an endogenous DNA sensor.

5. Reagents/tools

Forceps
30 mL glass beakers
12-Well plate
24-Well plate
12 mm round glass coverslips
4% PFA solution (in 1 × PBS)
100% Acetone
1 × PBS
5% BSA in 1 × PBS
1% BSA in 1 × PBS
ProLong Gold DAPI mounting medium (ThermoFisher)
Microscope slides
20–200 μL pipette
Adhesion reagent: (fibronectin solution or HistoGrip)
1.5 mL Eppendorf tubes (clear and amber)
Incubation chamber
Aluminum foil
Kimwipes
Clear nail polish

6. Example protocol 1

Imaging plasmid-based expression of fluorescently labeled AIM2 in HEK 293T cells.

For this example, we will transfect HEK 293T cells with a plasmid that expresses GFP-tagged AIM2.

6.1 Day 1

- Autoclave coverslips and forceps prior to use
- Treat coverslips with an agent to enhance cellular adhesion (HEK 293T are weakly adherent). Options for this include:
  - HistoGrip (ThermoFisher)
  - Fibronectin (Sigma)
- After treatment, wash coverslips once in 1 × PBS
- Place coverslips into wells of a 12-well plate (1 coverslip per well for each treatment condition needed)
- Seed cells into each well in ~1 mL media

6.2 Day 2

- When cells have reached appropriate confluence (75–90%), transfect cells with plasmid of interest. For this example, HEK 293T cells are transfected with an AIM2-GFP construct. The quantity of plasmid DNA required will vary based on the specific plasmid and cell line. HEK 293T express most suitable vectors quite well, and relatively low concentrations of DNA can be used (0.1–1.0 μg/mL).
- Allow expression for 8–12 h.
- *Note*: From this point forward, cells should be shielded from light to protect the fluorescent protein of interest.
- Gently remove media and wash coverslips with 1 mL 1 × PBS. Remove and apply solutions gently to avoid dislodging cells from coverslips.
- Apply 4% PFA to each coverslip to fix cells.
- *Note*: PFA is toxic and appropriate precautions (such as working in fume hood) should be followed.
- Allow cells to sit covered and shielded from light at room temperature in PFA for 10 min.
- *Note*: an alternative fixation technique is the application of ice cold 100% methanol for 20 min on ice.
- Remove PFA for appropriate disposal, wash cells in 1 × PBS in 12-well plate.
- Fill one glass beaker with 100% acetone
- Fill a second beaker with 1 × PBS
- Fill a third beaker with dH_2O
- Prior to processing each coverslip, apply 12 μL of mounting medium on a glass slide. If air bubbles are introduced into the drop of mounting

medium, remove them with a pipette tip prior to mounting the cells. Cover the DAPI containing medium and slide with aluminum foil to protect the fluorescent dye.

- One at a time, remove each coverslip from the 12-well plate using the forceps, and immediately submerge in acetone to permeabilize the cell membrane.
- *Note*: When lifting coverslips from the plate, be sure to touch only the edges of the coverslip, rather than the middle. This will prevent damage to cells in the middle of the coverslip. Do not image portions of the coverslip at the margins where damage may have occurred through manipulating the coverslip with forceps. Use a consistent approach to holding coverslips with the forceps, such as always keeping the thumb oriented in parallel with the surface of the coverslip that contains the cells. This will prevent confusion regarding which side of the coverslip is "up" versus "down."
- Hold the coverslip submerged in acetone for 30 s.
- Move the coverslip to the second beaker (1 × PBS) and gently dip the coverslip in and out of the solution 10–20 times. Perform this washing gently to avoid dislodging cells from the surface of the coverslip.
- Next, dip the coverslip 10–20 times in the third beaker (dH_2O).
- Gently, wipe the back of the coverslip (the side with no cells) using a kimwipe.
- Place the coverslip face down on the 12 μL drop of mounting medium on the slide.
- Once all coverslips have been processed, incubate the slides at 4 °C overnight, protected from light.
- The next day, seal the coverslips using clear nail polish around the margins of the coverslip. Using the minimal amount of sealant needed to cover the edges of the coverslip.
- The samples can now be imaged by microscopy Fig. 1.

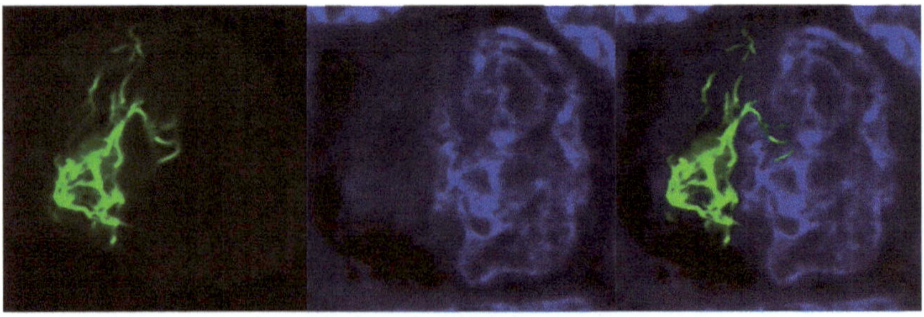

Fig. 1 HEK 293T cell transfected with AIM2-GFP imaged by confocal microscopy. Left panel: GFP. Middle panel: DAPI. Right panel: merged image.

7. Example protocol 2

Imaging endogenous IFI16 filaments in adherent cells.

For this example, we will activate endogenous IFI16 in an epithelial cell line (HSG) using Poly(dA:dT).

7.1 Day 1

- Autoclave coverslips and forceps prior to use.
- If desired, coverslips can be treated with HistoGrip or Fibronectin to promote cellular adhesion. However, in contrast to HEK 293T, many other epithelial cell lines such as HeLa and primary keratinocytes adhere quite well to untreated glass coverslips.
- Place coverslips into well of a 12-well plate (1 coverslip per well for each treatment condition needed).
- Seed cells into each well in ~1 mL media.

7.2 Day 2

- When cells have reached appropriate confluence (75–90%), transfect cells with Poly(dA:dT). Transfection conditions will vary on the regent used and cell type. For the HSG cell line, cells can be transfected as follows:
  - Combine 100 μL OMEM with 4 μL Lipofectamine 2000 and 1 μg Poly(dA:dT).
  - Incubate 10 min at room temperature protected from light.
  - Apply the OMEM/Lipofectamine/Poly(dA:dT) solution to cells in 1 mL of media.
- Allow incubation for 8–24 h.
- Gently remove media and wash coverslips with 1 mL 1 × PBS. Remove and apply solutions gently to avoid dislodging cells from coverslips.
- Apply 4% PFA to each coverslip to fix cells.
- *Note*: PFA is toxic and appropriate precautions (such as working in fume hood) should be followed.
- Allow cells to sit covered and shielded from light at room temperature in PFA for 10 min.
- *Note*: an alternative fixation technique is the application of ice cold 100% methanol for 20 min on ice.
- Remove PFA for appropriate disposal, wash coverslips in 1 × PBS in 12-well plate.
- Potential stop point: Once the cells have been fixed with PFA, they can be stored in 1 × PBS at 4 °C for future labeling (if desired).

- Fill one 30 mL glass beaker with 100% acetone
- Fill a second 30 mL beaker with 1 × PBS.
- Place the lid of a 24-well plate on the bench top, face up, for processing the coverslips.
- Remove each coverslip, one at a time, from the 12-well plate, using forceps, and submerge in acetone for 30 s to permeabilize the cell membrane.
- *Note*: When lifting coverslips from the plate, be sure to touch only the edges of the coverslip, rather than the middle. This will prevent damage to cells in the middle of the coverslip. Do not image portions of the coverslip at the margins where damage may have occurred through manipulating the coverslip with forceps. Use a consistent approach to holding coverslips with the forceps, such as always keeping the thumb oriented in parallel with the surface of the coverslip that contains the cells. This will prevent confusion regarding which side of the coverslip is "up" versus "down."
- After permeabilizing in acetone, move the coverslip directly into the second beaker (1 × PBS) and gently dip the coverslip in and out of the solution 10–20 times. Perform this washing gently to avoid dislodging cells from the surface of the coverslip.
- Gently, wipe the back of the coverslip (the side with no cells) using a kimwipe.
- Place the coverslip face up (cells up) in one of the positions on the lid of the 24-well plate. If multiple conditions and labeling combinations are being processed simultaneously, it may be of use to make a key indicating the treatment/staining conditions for each sample on the plate.
- After placing the coverslip on the plate, apply 50 μL of 5% BSA in PBS (blocking solution) to the surface of the coverslip, taking care not to scrape the surface of the coverslip with the pipette tip.
- Process all coverslips similarly.
- Incubate the cells in blocking solution, protected from light, for 30 min at room temperature. This step can also be performed at 4 °C overnight, if desired.
- Dilute the primary antibody in 1% BSA in PBS, and spin in desktop centrifuge at maximum speed for 3 min. Transfer the majority of the diluted antibody to a fresh tube, leaving ~10–20 μL at the bottom with any precipitated material. For this example, we have used anti-IFI16 mouse monoclonal (Sigma, Clone IFI-230) at 1:100 dilution.
- After blocking, lift the first coverslip from the plate and wash in 1 × PBS.

- *Note*: if lifting the coverslip is difficult, apply PBS to the edge of the coverslip to help lift it from the surface of the plate.
- Gently, wipe the back of the coverslip with a kimwipe.
- Place the coverslip face up on a dry plate. Apply 50 μL of diluted primary antibody.
- Process all remaining coverslips similarly.
- Cover the coverslips and incubate for 1 h at room temperature. Overnight incubation at 4 °C is an alternative.
- Dilute the appropriate fluorescently labeled secondary antibody in 1% BSA in PBS in an amber tube (to protect the fluorophore from light). In this example, we are using Goat anti-Mouse Alexa Fluor 488 (ThermoFisher A-11001) diluted 1:200.
- Centrifuge the diluted antibody at maximum speed × 3 min. Transfer the solution to a second tube, leavening 10–20 μL in the bottom of the tube to avoid any precipitated material.
- Fill three beakers with 1 × PBS.
- Lift the first coverslip from the plate and tap the side gently to a kimwipe to remove the diluted primary antibody. Next, dip the coverslip 10–20 times in each PBS wash beaker to thoroughly wash the primary antibody.
- Gently wipe the back surface of the coverslip with a kimwipe, then place on a dry plate surface.
- Apply 50 μL of the diluted secondary antibody to the surface of the coverslip.
- Process all remaining coverslips similarly.
- Incubate the coverslips 30 min at room temperature, protected from light.
- Fill two beakers with 1 × PBS and third beaker with dH_2O.
- Arrange slides to receive coverslips.
- Prior to processing each coverslip, apply 12 μL of mounting medium to the surface of a slide. If air bubbles are introduced into the drop of mounting medium, remove them with a pipette tip prior to mounting the cells. Cover the slide with foil while processing the coverslip.
- Lift the first coverslip from the surface of the plate using forceps.
- Wash the coverslip by dipping in each PBS beaker 10–20 times, followed by final wash in dH_2O.
- Gently wipe the back of the coverslip with a kimwipe, then place face *down* on the mounting medium.
- Process all remaining samples similarly.
- Allow the samples to sit at 4 °C overnight.

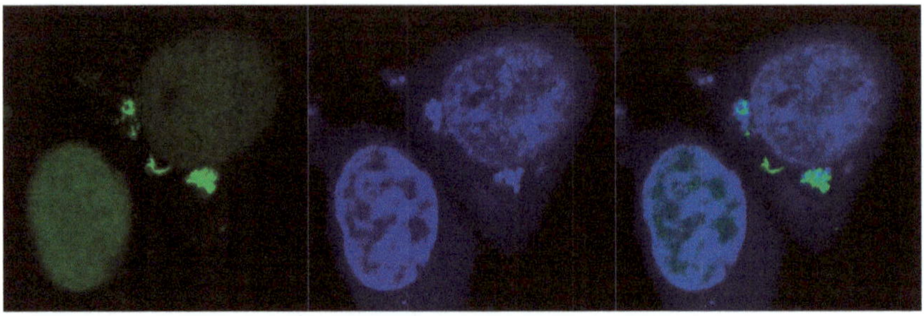

Fig. 2 HSG cell transfected with Poly(dA:dT) and immunostained for IFI16. Left panel: IFI16. Middle panel: DAPI. Right panel: merged image.

- The next day, seal the coverslips using clear nail polish around the margins of the coverslip. Using the minimal amount of sealant needed to cover the edges of the coverslip.
- The samples can now be imaged by microscopy Fig. 2.

8. Handling coverslips: Do's and don'ts

Use a consistent technique when lifting and manipulating coverslips that allows you to be certain which side of the coverslip is "up" and which is "down." One approach is to always lift the coverslip with the cell side up, and keep the thumb facing upward along with the cells (Fig. 3, left panel). When manipulating coverslips it is important to grip only the edge of the coverslip (Fig. 3, middle panel), rather than the gripping across the middle of the coverslip (Fig. 3, right panel). Cells that have been crushed by forceps are unsuitable for imaging.

When applying antibody dilutions, be sure to avoid touching the cells with the pipette tip (left panel). 50 μL is sufficient volume to cover the surface of a 12 mm round glass coverslip. Coverslips will be easier to lift if placed on a clean, dry surface, such as the lid of a 24-well plate (Fig. 4, left panel). Incubations should be carried out inside a humidified chamber, such as this sealable plastic tray containing a damp paper towel (Fig. 4, middle panel). This will prevent evaporation of the blocking solution or antibody applied to the samples. When mounting the coverslips after staining is completed, place coverslips face down onto a 10–15 μL drop of mounting medium (Fig. 4, right panel). This is the only step in the staining process when the coverslip should be applied face down to any surface! Coverslips should not be repositioned once applied to the mounting medium. In case of a dropped coverslip at any point during this protocol: do not despair, as

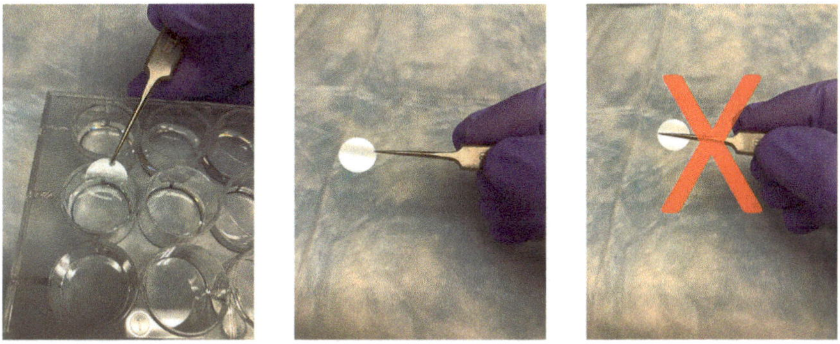

Fig. 3 When manipulating coverslips, maintain consistent orientation with the cell surface always facing up (left panel). Coverslips should be gripped at the edge only (middle panel), to avoid crushing cells in the center of the coverslip that will be used for imaging (right panel).

Fig. 4 The lid of a 24-well cell culture chamber can be used to stain coverslips (left panel). This surface should be kept inside a humidity chamber to avoid drying of the coverslips during incubations (middle panel). At the final stage of the protocol, coverslips are turned face down for mounting on slides (right panel).

dropped coverslips can often be salvaged. In case of a drop, immediately pick up the coverslip and place it in PBS wash solution to prevent drying. Then, lift the coverslip and carefully inspect both surfaces while the PBS evaporates (partially)—it is often possible to see the cells adherent to the "up" surface, as opposed to the smooth, cell free surface of the "down" side.

References

Ablasser, A., Schmid-Burgk, J. L., Hemmerling, I., Horvath, G. L., Schmidt, T., Latz, E., et al. (2013). Cell intrinsic immunity spreads to bystander cells via the intercellular transfer of cGAMP. *Nature*, *503*(7477), 530–534. https://doi.org/10.1038/nature12640.

Almine, J. F., O'Hare, C. A., Dunphy, G., Haga, I. R., Naik, R. J., Atrih, A., et al. (2017). IFI16 and cGAS cooperate in the activation of STING during DNA sensing in human keratinocytes. *Nature Communications, 8*, 14392. https://doi.org/10.1038/ncomms14392.
Antiochos, B., Matyszewski, M., Sohn, J., Casciola-Rosen, L., & Rosen, A. (2018). IFI16 filament formation in salivary epithelial cells shapes the anti-IFI16 immune response in Sjögren's syndrome. *JCI Insight, 3*(18). https://doi.org/10.1172/jci.insight.120179.
Cai, X., Chen, J., Xu, H., Liu, S., Jiang, Q. X., Halfmann, R., et al. (2014). Prion-like polymerization underlies signal transduction in antiviral immune defense and inflammasome activation. *Cell, 156*(6), 1207–1222. https://doi.org/10.1016/j.cell.2014.01.063.
Caso, F., Costa, L., Nucera, V., Barilaro, G., Masala, I. F., Talotta, R., et al. (2018). From autoinflammation to autoimmunity: Old and recent findings. *Clinical Rheumatology, 37*(9), 2305–2321. https://doi.org/10.1007/s10067-018-4209-9.
Crane, M. J., Lee, K. M., FitzGerald, E. S., & Jamieson, A. M. (2018). Surviving deadly lung infections: Innate host tolerance mechanisms in the pulmonary system. *Frontiers in Immunology, 9*, 1421. https://doi.org/10.3389/fimmu.2018.01421.
Dhanwani, R., Takahashi, M., & Sharma, S. (2018). Cytosolic sensing of immunostimulatory DNA, the enemy within. *Current Opinion in Immunology, 50*, 82–87. https://doi.org/10.1016/j.coi.2017.11.004.
Fernandes-Alnemri, T., Yu, J. W., Datta, P., Wu, J., & Alnemri, E. S. (2009). AIM2 activates the inflammasome and cell death in response to cytoplasmic DNA. *Nature, 458*(7237), 509–513. https://doi.org/10.1038/nature07710.
Fischer, S. (2019). Pattern recognition receptors and control of innate immunity: Role of nucleic acids. *Current Pharmaceutical Biotechnology, 19*, 1203–1209. https://doi.org/10.2174/138920112804583087.
Franklin, B. S., Bossaller, L., De Nardo, D., Ratter, J. M., Stutz, A., Engels, G., et al. (2014). The adaptor ASC has extracellular and 'prionoid' activities that propagate inflammation. *Nature Immunology, 15*(8), 727–737. https://doi.org/10.1038/ni.2913.
Franklin, B. S., Latz, E., & Schmidt, F. I. (2018). The intra- and extracellular functions of ASC specks. *Immunological Reviews, 281*(1), 74–87. https://doi.org/10.1111/imr.12611.
Hooy, R. M., & Sohn, J. (2018). The allosteric activation of cGAS underpins its dynamic signaling landscape. *eLife, 7*. https://doi.org/10.7554/eLife.39984.
Hornung, V., Ablasser, A., Charrel-Dennis, M., Bauernfeind, F., Horvath, G., Caffrey, D. R., et al. (2009). AIM2 recognizes cytosolic dsDNA and forms a caspase-1-activating inflammasome with ASC. *Nature, 458*(7237), 514–518. https://doi.org/10.1038/nature07725.
Latz, E., Xiao, T. S., & Stutz, A. (2013). Activation and regulation of the inflammasomes. *Nature Reviews. Immunology, 13*(6), 397–411. https://doi.org/10.1038/nri3452.
Li, T., & Chen, Z. J. (2018). The cGAS-cGAMP-STING pathway connects DNA damage to inflammation, senescence, and cancer. *The Journal of Experimental Medicine, 215*(5), 1287–1299. https://doi.org/10.1084/jem.20180139.
Lu, A., Magupalli, V. G., Ruan, J., Yin, Q., Atianand, M. K., Vos, M. R., et al. (2014). Unified polymerization mechanism for the assembly of ASC-dependent inflammasomes. *Cell, 156*(6), 1193–1206. https://doi.org/10.1016/j.cell.2014.02.008.
Luecke, S., Holleufer, A., Christensen, M. H., Jonsson, K. L., Boni, G. A., Sorensen, L. K., et al. (2017). cGAS is activated by DNA in a length-dependent manner. *EMBO Reports, 18*(10), 1707–1715. https://doi.org/10.15252/embr.201744017.
Morrone, S. R., Matyszewski, M., Yu, X., Delannoy, M., Egelman, E. H., & Sohn, J. (2015). Assembly-driven activation of the AIM2 foreign-dsDNA sensor provides a polymerization template for downstream ASC. *Nature Communications, 6*, 7827. https://doi.org/10.1038/ncomms8827.
Morrone, S. R., Wang, T., Constantoulakis, L. M., Hooy, R. M., Delannoy, M. J., & Sohn, J. (2014). Cooperative assembly of IFI16 filaments on dsDNA provides insights

into host defense strategy. *Proceedings of the National Academy of Sciences of the United States of America*, *111*(1), E62–E71. https://doi.org/10.1073/pnas.1313577111.

Motani, K., Ito, S., & Nagata, S. (2015). DNA-mediated cyclic GMP-AMP synthase-dependent and -independent regulation of innate immune responses. *Journal of Immunology*, *194*(10), 4914–4923. https://doi.org/10.4049/jimmunol.1402705.

Nanson, J. D., Rahaman, M. H., Ve, T., & Kobe, B. (2018). Regulation of signaling by cooperative assembly formation in mammalian innate immunity signalosomes by molecular mimics. *Seminars in Cell & Developmental Biology*, pii: S1084-9521(17)30423-8. https://doi.org/10.1016/j.semcdb.2018.05.002.

Ouyang, S., Song, X., Wang, Y., Ru, H., Shaw, N., Jiang, Y., et al. (2012). Structural analysis of the STING adaptor protein reveals a hydrophobic dimer interface and mode of cyclic di-GMP binding. *Immunity*, *36*(6), 1073–1086. https://doi.org/10.1016/j.immuni.2012.03.019.

Paijo, J., Doring, M., Spanier, J., Grabski, E., Nooruzzaman, M., Schmidt, T., et al. (2016). cGAS senses human cytomegalovirus and induces type I interferon responses in human monocyte-derived cells. *PLoS Pathogens*, *12*(4), e1005546. https://doi.org/10.1371/journal.ppat.1005546.

Palmai-Pallag, T., & Bachrati, C. Z. (2014). Inflammation-induced DNA damage and damage-induced inflammation: A vicious cycle. *Microbes and Infection*, *16*(10), 822–832. https://doi.org/10.1016/j.micinf.2014.10.001.

Sidiropoulos, P. I., Goulielmos, G., Voloudakis, G. K., Petraki, E., & Boumpas, D. T. (2008). Inflammasomes and rheumatic diseases: Evolving concepts. *Annals of the Rheumatic Diseases*, *67*(10), 1382–1389. https://doi.org/10.1136/ard.2007.078014.

Sohn, J., & Hur, S. (2016). Filament assemblies in foreign nucleic acid sensors. *Current Opinion in Structural Biology*, *37*, 134–144. https://doi.org/10.1016/j.sbi.2016.01.011.

Sun, L., Wu, J., Du, F., Chen, X., & Chen, Z. J. (2013). Cyclic GMP-AMP synthase is a cytosolic DNA sensor that activates the type I interferon pathway. *Science*, *339*(6121), 786–791. https://doi.org/10.1126/science.1232458.

Unterholzner, L. (2013). The interferon response to intracellular DNA: Why so many receptors? *Immunobiology*, *218*(11), 1312–1321. https://doi.org/10.1016/j.imbio.2013.07.007.

Unterholzner, L., Keating, S. E., Baran, M., Horan, K. A., Jensen, S. B., Sharma, S., et al. (2010). IFI16 is an innate immune sensor for intracellular DNA. *Nature Immunology*, *11*(11), 997–1004. https://doi.org/10.1038/ni.1932.

Vajjhala, P. R., Ve, T., Bentham, A., Stacey, K. J., & Kobe, B. (2017). The molecular mechanisms of signaling by cooperative assembly formation in innate immunity pathways. *Molecular Immunology*, *86*, 23–37. https://doi.org/10.1016/j.molimm.2017.02.012.

Veeranki, S., Duan, X., Panchanathan, R., Liu, H., & Choubey, D. (2011). IFI16 protein mediates the anti-inflammatory actions of the type-I interferons through suppression of activation of caspase-1 by inflammasomes. *PLoS One*, *6*(10), e27040. https://doi.org/10.1371/journal.pone.0027040.

Zhang, Y., Yeruva, L., Marinov, A., Prantner, D., Wyrick, P. B., Lupashin, V., et al. (2014). The DNA sensor, cyclic GMP-AMP synthase, is essential for induction of IFN-beta during Chlamydia trachomatis infection. *Journal of Immunology*, *193*(5), 2394–2404. https://doi.org/10.4049/jimmunol.1302718.

CHAPTER SIXTEEN

Assessment of inflammasome and type I IFN responses to DNA viruses and DNA PAMPS

Guido Massaccesi, Michael A. Chattergoon*
Johns Hopkins University School of Medicine, Baltimore, MD, United States
*Corresponding author: e-mail address: chattergoon@jhu.edu

Contents

Abstract

The innate immune system is an evolutionarily conserved host defense system and is the first barrier to infection. The system utilizes genetically conserved receptors to identify the presence of microbial structures. Engagement of innate immune receptors by primarily by ligands that discriminate pathogens from the host activates programmed responses that limit pathogen expansion. Despite its ubiquitous nature, surprisingly DNA is a critical structure that triggers innate immune responses. Focusing on structural modifications or aberrant location of DNA, innate immune receptors identify physiologic stress. Inflammasomes and interferons are critical innate immune pathways that are activated by DNA. DNA binding proteins that tie recognition of DNA to both programmed responses have been identified, and their importance demonstrated in

Methods in Enzymology, Volume 625
ISSN 0076-6879
https://doi.org/10.1016/bs.mie.2019.05.008

infection models. In this chapter, we discuss techniques to analyze AIM2 inflammasome and cGAS interferon activation by synthetic DNA and DNA viruses. We also discuss methods to measure the activity of these immune pathways.

1. Introduction

The innate immune system is an ancient evolutionarily conserved host defense system (Buchmann, 2014). The innate immune system in humans is comprised of several haematopoietically derived cells that perform specialized functions. However, virtually all cells have intrinsic immune defenses which act as an early warning system while limiting pathogen spread. The combined actions of these specialized immune cells and tissue resident cells serve as the first line of defense against invading microbial pathogens and other toxic insults.

Activating an innate immune response involves complex interactions between the pathogen and the host. Membrane-localized, endosomal, cytosolic and nuclear germ-line encoded sensors, collectively referred to as pattern recognition receptors (PRR) are critical to recognizing the presence of pathogens and initiating innate immune responses. Innate immune PRRs, prototypically Toll-like receptors (TLR)s (Kawai & Akira, 2010), expressed by tissue-resident or hematopoietic cells, "sense" pathogen-associated molecular patterns (PAMPs) and trigger downstream effector programs. Besides TLRs, other innate PRRs include (1) Nucleotide-binding Oligomerization Domain (NOD)-like receptors, (2) complement receptors, (3) scavenger receptors (e.g., CD36, Scavenger Receptor [SR]-A, Lectin-like oxidized low-density lipoprotein (LDL) receptor-1 [LOX-1], and Drosophila SR-C [dSR—C]), (4) C-type lectin receptors (e.g., mannose-binding proteins, dectin-1, dectin-2, and Dendritic Cell-Specific Intercellular adhesion molecule-3-Grabbing Non-integrin [DC-SIGN]), and (5) intracellular nucleic acid-sensing receptors (e.g., Absent in Melanoma 2 [AIM2], Melanoma Differentiation-Associated protein 5 [MDA-5], Retinoic-Acid Inducible Gene I [RIG-I], and cyclic GMP-AMP synthetase [cGAS]). Triggering of a PRR activates programmed responses designed to limit the expansion of the pathogen and support the development of adaptive immunity and immunologic memory. Inflammasomes and interferon type I (IFN-I) are examples of important programmed responses with antimicrobial effects.

1.1 Induction of interferon type I by DNA

The recognition or "sensing" of nucleic acids is a primary mechanism used by the innate immune system to mount an immune response against microbial pathogens. dsRNA and ssRNA can be recognized in the endosomal system by some TLRs leading to the production of IFN-I and other inflammatory cytokines (Kawai & Akira, 2010). In the cytoplasm, dsRNA is recognized by MDA5 and RIG-I; the latter also sensing 5′ triphosphate RNA. Binding of these RNA species to MDA5 and RIG-I recruits Mitochondrial antiviral-signaling protein (MAVS) leading to downstream activation of IRF3, MAPK, and NF-κB (Kato, Takahasi, & Fujita, 2011).

Like RNA, the presence of dsDNA can also be recognized in the endosome by TLRs leading to production on IFN-I. Several cytosolic DNA sensors including Gamma-interferon-inducible protein-16 (IFI16) and DEAD-Box Helicase 41 (DDX41) have been described. However, the mechanism by which these receptors mediate a DNA-dependent immune response is not entirely understood (Unterholzner et al., 2010; Zhang et al., 2011). Recently, it was recognized that cGAS is a significant driver of IFN-I expression in response to dsDNA in the cytoplasm (Sun, Wu, Du, Chen, & Chen, 2013; Wu et al., 2013). Recognition of dsDNA by cGAS induces enzymatic conversion of ATP and GTP into the second messenger, cyclic di-GMP-AMP (cGAMP) (Gao et al., 2013; Li et al., 2013). cGAMP then binds to Stimulator of interferon genes (STING), also known as transmembrane protein 173 (TMEM173) which resides in the endoplasmic reticulum. cGAMP binding to STING dimers induces a conformational change, and STING then binds to TANK-Binding Kinase 1 (TBK1) and translocates to the perinuclear Golgi region (Ishikawa, Ma, & Barber, 2009). TBK1 phosphorylates STING, allowing the recruitment of IRF3 which, after binding to STING, gets phosphorylated and activated by TBK1 (Liu et al., 2015). Binding of DNA and activation of cGAS is sequence-independent, and a broad spectrum of viral/bacterial pathogens, as well as endogenous DNA, induce IFN-I responses through the cGAS-STING-IRF3 axis. There is emerging evidence from monogenic interferonopathies and related mouse models that DNA sensing by the cGAS-STING pathway may be involved in the pathogenesis of autoinflammatory disorders (Crow & Manel, 2015).

1.2 Interferon type I response to microbial DNA

Infectious diseases as a result of DNA virus infections are a significant health concern worldwide, and a thorough mechanistic understanding of host antiviral responses is central to the development of antivirals and vaccines. Type I interferon production by the host is the frontline anti-viral defense strategy, and it is one of the primary immune responses to DNA viruses.

As previously discussed, IFN-I responses as a result of sensing viral DNA can be initiated in the cytoplasm and endosomal compartments. Many DNA virus are detected in endosomes by TLR9 and to a lesser extent TLR7 (Cerullo et al., 2007; Delale et al., 2005; Krug et al., 2004; Rasmussen et al., 2007; Tabeta et al., 2004). Most DNA viruses access the cytosolic compartment during their life cycle providing an opportunity to the innate immune system for early detection and response. Therefore, the detection of viral products in the cytosol is integral to innate anti-viral defense. One of the most commonly observed virus-associated molecular patterns in the cytosol is viral nucleic acid the detection of which forms the basis of the cytosolic sensing of DNA viruses. Several DNA viruses including herpes simplex virus 1, hepatitis B virus, vaccinia virus, adenovirus, and Kaposi's sarcoma-associated herpesvirus are recognized by the cGAS (Dansako et al., 2016; Lam, Stein, & Falck-Pedersen, 2014; Wu et al., 2015). Additional receptors, such as DAI, RNA polymerase III, IFI16 and others, have been implicated in recognition of DNA after viral infection (Chiu, Macmillan, & Chen, 2009; DeFilippis, Alvarado, Sali, Rothenburg, & Fruh, 2010; Unterholzner et al., 2010). These receptors appear to utilize a common signaling pathway which converges on STING to turn on interferon production at the transcriptional level.

Recognition of bacterial DNA is also used as a mechanism to mount responses to bacteria rapidly. Infection with *Mycobacterium tuberculosis*, *Listeria monocytogenes*, *Chlamydia trachomatis*, and *Francisella tularensis* activates IFN-I expression that is dependent on cGAS activity and bacteria-produced cyclic dinucleotides that act directly on STING (Collins et al., 2015; Hansen et al., 2014; Storek, Gertsvolf, Ohlson, & Monack, 2015; Watson et al., 2015; Woodward, Iavarone, & Portnoy, 2010).

1.3 Inflammasomes in innate immunity

Inflammasomes are macromolecular platforms for the recruitment and activation of inflammatory caspases in the context of stress or danger signals. Inflammasome complexes are composed of (1) a PRR, (2) the adaptor molecule, apoptosis-associated speck-like protein containing a CARD (caspase recruitment domain) (ASC) and (3) the protease pro-caspase-1. Five inflammasomes have been identified in humans; they are named by their core PRR and include (1) the Nod-like receptor (NLR)s (e.g., NLRP1, NLRP3, NLRC4), (2) the PYHIN Pyrin and HIN domain (PYHIN) molecule AIM2 and (3) a third group comprised of the tripartite motif family member pyrin. Regulation of inflammation through the control of IL-1β and IL-18 production appears to be a major function of inflammasomes. IL-1β is a major pro-inflammatory cytokine that initiates inflammatory cascades and mediates immune-cell recruitment and activation (Netea, van de Veerdonk, van der Meer, Dinarello, & Joosten, 2015). IL-18 is a multifunctional cytokine that exerts many context-dependent biological effects including induction of IFN-γ and perforin from NK and T-cells, activation of mast cells and basophils and promoting tissue remodeling and wound healing (Samarani et al., 2016).

Assembly and activation of inflammasomes are tightly regulated and is induced by two signals (Chen & Ichinohe, 2015; He, Hara, & Nunez, 2016). *Signal 1* or *priming* induces deubiquitination and dephosphorylation of NLRP3 as well as the transcription of inflammasome components and pro-cytokines. Signal 1 is mediated by the transcription factor NFκB and can be initiated by intracellular and plasma membrane PRR including TLR, cytokine receptors as well as other NFκB activators. *Signal 2* or *activation* is less defined but involves potassium efflux, calcium mobilization from intracellular store compartments, and the production of ROS leads to assembly of the inflammasome complex. NACHT, LRR and PYD domains-containing protein 3 (NLRP3) is the most studied member of the NLR family. NLRP3 is activated in innate immune cells by a variety of stimuli, including pathogens and danger signals such as monosodium urate and ATP (Yao et al., 2017). Upon stimulation, NLRP3 undergoes conformational changes allowing recruitment and polymerization of the adaptor ASC—which is visualized as an approx. 1 μm speck in the cytoplasm—through PYD–PYD domain association. ASC filaments recruit pro-caspase-1 through CARD–CARD domain interactions to form the signaling complex. Autocleavage and the removal of regulatory domains generate the tetrameric activated caspase-1 which cleaves pro-IL-1β and pro-IL-18 to form mature, active IL-1β and IL-18 (Coll, O'Neill, & Schroder, 2016; Guo, Callaway, & Ting, 2015). Inflammasomes may also mediate a form of cell death called pyroptosis through cleavage of Gasdermin-D which can form pores in the plasma membrane (He et al., 2016).

Diverse microbes activate NLRP3 inflammasomes and mice lacking NLRP3, ASC, or caspase-1 are more susceptible to infection with many bacteria (Ceballos-Olvera, Sahoo, Miller, Del Barrio, & Re, 2011; Costa et al., 2012). The NLRP3 inflammasome is also activated by a number of RNA viruses underscoring the critical role of this innate immune mechanism in viral detection and defense. Prominent human pathogens that are recognized by NLRP3 include Influenza virus, respiratory syncytial virus, hepatitis C virus and HIV (Chattergoon et al., 2014; Kanneganti et al., 2006; Veenhuis et al., 2017). NLRP3 also recognizes enveloped and non-enveloped DNA viruses, such as herpes viruses, MVA and adenovirus (Delaloye et al., 2009; Muruve et al., 2008; Nour et al., 2011). Dysregulated NLRP3 activity is observed with a variety of inflammatory disorders, such as cryopyrin-associated periodic syndromes and contributes to the pathogenesis of metabolic diseases including diabetes (Masters et al., 2010). Of note sustained IL-18 elevation is also a hallmark of HIV-1 and HCV infections but clear mechanistic linkages between NLPR3 and increased rates of coronary artery disease, diabetes, etc. are yet to be established in this setting (Ahmad, Sindhu, Toma, Morisset, & Ahmad, 2002; Chattergoon et al., 2011; Torre et al., 2000; Veenhuis et al., 2017). While recent advances have been made in understanding inflammasome activation, NLPR3 regulation is not well understood and significant ambiguities about activation and inhibition of inflammasome-complex formation remain.

1.4 Activation of AIM2 inflammasomes by microbial DNA

AIM2 is also a key sensor of pathogens that detects the presence of foreign DNA accumulating in the cytosol during the life cycle of intracellular pathogens including viruses, bacteria, and parasites. AIM2 achieves specificity by identifying altered or mislocalized DNA molecules. It can detect damaged DNA and the aberrant presence of DNA within the cytosolic compartment such as genomic DNA released into the cytosol upon loss of nuclear envelope integrity. dsDNA binds to AIM2 (absent in melanoma 2) in the cytoplasm, driving the activation of the inflammasome, the expression of IL-1β and IL-18, and the activation of pyroptosis in a caspase-1-dependent manner (Fernandes-Alnemri et al., 2010; Hornung et al., 2009). AIM2 appears to be able to recognize DNA molecules independently of the sequence and compared to other DNA sensors AIM2 is specialized in inflammasome assembly rather than promoting type I IFN responses.

Transfection of host, bacterial and viral DNA, besides its effects on the transcriptional activation of type-I IFN, has been shown to initiate a cascade leading to the proteolytic maturation of IL-1β. These DNA molecules relied on an ASC-dependent, NLRP3-independent but AIM2 dependent inflammasome (Muruve et al., 2008). During infections with intracellular pathogens, microbial DNA can be released in the cytosol. AIM2 function as a DNA sensor has been extensively described in the context of several microbes and pathogens; many bacterial species have been found to activate AIM2 including *Francisella tularensis*, *Listeria monocytogenes*, *Streptococcus pneumoniae*, species of *Mycobacterium* as well as *Legionella pneumophila* and *Staphylococcus aureus*. These observations have been extensively reviewed recently (Lugrin & Martinon, 2017).

Cytomegalovirus (CMV), Herpes Simplex Virus 1 (HSV-1), vaccinia virus, and human papillomavirus have been shown to induce AIM2 inflammasome (Hornung et al., 2009; Huang et al., 2017; Man et al., 2015; Rathinam et al., 2010; Reinholz et al., 2013). The mechanism by which viral DNA is liberated and exposed to AIM2 in the cytosol is not clearly defined. Viruses have been shown to evade AIM2 sensing possibly by inhibiting the ability of AIM2 to interact with released DNA. A recent study reported that Epstein–Barr virus (EBV) may activate AIM2 in THP1 cells (Torii et al., 2017). However, despite that AIM2 can detect viral DNA its contribution to viral immunity is still unclear.

2. Protocols

In this chapter, we will focus on the induction of type 1 interferons and inflammasomes in human immune cells. The reader should be aware that methods to activate similar pathways in murine cells and non-immune cells in humans differ slightly. The THP-1 cell line is extensively used in studies of innate immunity. When undifferentiated, this cell line mimic human monocytes. Several agents can be used to trigger differentiation to a macrophage phenotype. In this state, type I interferons and inflammasome responses can be activated by DNA oligomers, DNA viruses as well as bacterial and fungal infection.

2.1 Methods to induce interferon type I in innate immune cells with DNA

Type I interferons can be readily induced by DNA in a number of cell lines in which the cGAS-STING signaling system is functional. The monocytic cell line THP-1 can be stimulated to produce IFN-β when challenged with purified dsDNA oligomers >80 bp, plasmids, bacterial DNA and human DNA. We have also used THP-1 reporter cell lines where production of reporter enzymes is a surrogate for activation of IFN-I and inflammasomes. In the case of non-synthesized DNA, care must be taken to remove contaminants that may themselves be PAMPS. The following protocol is provided as a general example of a method that readily targets the cGAS-STING-IRF3 pathway in THP-1 cells. We also use this protocol with minor modifications to induce IFN-I from primary human monocytes, monocyte derived macrophages and plasmacytoid dendritic cells. We have also applied this protocol to non-immune cells with intact cGAS-STING signaling to stimulate interferon β.

Materials

1. THP-1 cell line (ATCC® TIB-202™) or THP-1 Dual (Invivogen NF-κB-SEAP IRF-Luc Reporter Monocytes)
2. PMA (Phorbol 12-myristate 13-acetate) (Millipore Sigma P1585)
3. *Culture Media*:
   a. RPMI 1640 (Gibco #11875093)
   b. 10% Fetal Bovine Serum (Corning)
   c. 1% L-Glutamine (Gibco # 2503008)
   d. 1% Penicillin-Streptomycin (Gibco # 15140122)
   e. 1% HEPES Buffer (Corning # 25-060-CI)
4. Lipofectamine LTX (ThermoFisher Scientific # 15338100)
5. Opti-MEM media (Gibco # 31985088)
6. Synthesized dsDNA oligomers

Protocol

1. Suspend THP-1 monocytic cells at a concentration of 5×10^5 cells/mL in *Culture media* and aliquot 200 μL to each well of a flat-bottom tissue culture treated 96-well plate.
2. Add PMA to a final concentration of 50 nM.
3. Incubate plates at 37 °C, 5% CO_2 for 16 h in a humidified incubator.
4. Aspirate PMA-containing *Culture media* without disturbing the cells (cells should be adherent after the 16 h incubation period).
5. Wash cells with 300 μL/well of 1 × PBS for 5 min at room temperature. Repeat this procedure.
6. Add 200 μL/well of fresh *Culture media* and incubate plates/cells for an additional 48 h at 37 °C, 5% CO_2.
7. Lipofectamine LTX reagent is diluted in Opti-MEM medium at a ratio of 1 μL of Lipofectamine to 24 μL Opti-MEM.
8. dsDNA is diluted to 1–20 μg/mL in Opti-MEM.
9. Mix diluted dsDNA and Lipofectamine LTX reagent in a 1:1 ratio. Incubate this mixture for 30 min at room temperature. Once mixed do not vortex the mixture.
10. Remove media and wash cells as in step 6 then add 180 μL/well of *Culture media.*
11. Add 20 μL of DNA- Lipofectamine LTX lipid complex to the cells.

12. Incubate for 6–24 h at 37°C, 5% CO_2.
13. Remove and store supernatants at −80 °C for future cytokine measurement and analysis.
14. Freeze cells pellets at −80°C for subsequent gene expression analysis.

2.2 Methods to induce interferon type I in innate immune cells with DNA viruses

Macrophages IFN-I is important for control of herpesvirus infections. The THP-1 monocyte-like cell line can be differentiated into a macrophage-like phenotype with PMA, upon which DNA viruses like HSV-1 and CMV induce IFN-β expression and induction of interferon stimulated genes. Several studies have highlighted roles for IFI16 in this process (Horan et al., 2013).

Materials

1. THP-1 cell line (ATCC® TIB-202™) or THP-1 Dual (Invivogen NF-κB-SEAP IRF-Luc Reporter Monocytes)
2. PMA (Phorbol 12-myristate 13-acetate) (Millipore Sigma P1585)
3. *Culture Media*:
   a. RPMI 1640 (Gibco #11875093)
   b. 10% Fetal Bovine Serum (Corning)
   c. 1% L-Glutamine (Gibco # 2503008)
   d. 1% Penicillin-Streptomycin (Gibco # 15140122)
   e. 1% HEPES Buffer (Corning # 25–060-CI)
4. HSV-1 (ATCC® VR-733), Human CMV strain AD169 (ATCC® VR538)

Protocol

1. Suspend THP-1 monocytic cells at a concentration of 5×10^5 cells/mL in *Culture media* and aliquot 200 μL to each well of a flat-bottom tissue culture treated 96-well plate.
2. Add PMA to a final concentration of 50 nM.
3. Incubate plates at 37 °C, 5% CO_2 for 16 h in a humidified incubator.
4. Aspirate PMA-containing *Culture media* without disturbing the cells (cells should be adherent after the 16 h incubation period).
5. Wash cells with 300 μL/well of 1 × PBS for 5 min at room temperature. Repeat this procedure.
6. Add 200 μL/well of fresh *Culture media* and incubate plates/cells for an additional 48 h at 37 °C, 5% CO_2.
7. Add HSV-1 at MOI 3–10 or CMV at MOI 5–10 to the differentiated THP-1.

8. Incubate for 6–24 h at 37 °C, 5% CO_2.
9. Remove and store supernatants at −80 °C for future cytokine measurement and analysis.
10. Freeze cells pellets at −80°C for subsequent gene expression analysis.

2.3 Methods to induce AIM2 inflammasome with DNA

The principal inflammasome cytokines IL-1β and IL-18 can be readily induced in monocytic derived cells like THP-1 by targeting AIM2 using purified dsDNA oligomers >250 bp, plasmids, bacterial DNA and human DNA. The following protocol is provided as a general example of a method that activates the AIM2 inflammasome using poly dA:dT.

Materials

1. THP-1 cell line (ATCC® TIB-202™) or THP-1 Dual (Invivogen NF-κB-SEAP IRF-Luc Reporter Monocytes)
2. PMA (Phorbol 12-myristate 13-acetate) (Millipore Sigma P1585)
3. *Culture Media*:
   a. RPMI 1640 (Gibco #11875093)
   b. 10% Fetal Bovine Serum (Corning)
   c. 1% L-Glutamine (Gibco # 2503008)
   d. 1% Penicillin-Streptomycin (Gibco # 15140122)
   e. 1% HEPES Buffer (Corning # 25-060-CI).
4. Lipofectamine LTX (ThermoFisher Scientific # 15338100)
5. Opti-MEM media (Gibco # 31985088)
6. Poly dA:dT (Invivogen Cat# tlrl-patn)
7. Lipopolysaccharide (LPS, Invivogen LPS-EB Ultrapure)

Protocol

1. Suspend THP-1 monocytic cells at a concentration of 5×10^5 cells/mL in *Culture media* and aliquot 200 μL to each well of a flat-bottom tissue culture treated 96-well plate.
2. Add PMA to a final concentration of 50 nM.
3. Incubate plates at 37 °C, 5% CO_2 for 16 h in a humidified incubator.
4. After 48 h incubation aspirate PMA-containing *Culture media* without disturbing the cells (cells should be adherent after the 16 h incubation period).
5. Wash cells with 300 μL/well of 1 × PBS for 5 min at room temperature. Repeat this procedure.
6. Add 200 μL/well of fresh *Culture media* to each well. Add Lipopolysaccharide (LPS) at a final concentration of 100 ng/mL and return plates to a 37 °C, 5% CO_2 incubator for 4 h.

7. Dilute Lipofectamine LTX reagent in Opti-MEM medium at a ratio of 1 μL of Lipofectamine to 24 μL Opti-MEM.
8. Dilute Poly dA:dT to a concentration of 0.1–1 mg/mL in Opti-MEM medium.
9. Mix diluted poly dA:dT to Lipofectamine LTX reagent mix in a 1:1 ratio. Incubate this new mix for 30 min at room temperature.
10. Following incubation, LPS containing *Culture media* should be removed and the cells washed with 1 × PBS. Add 180 μL of fresh media is added to each well.
11. Poly dA:dT lipofectamine mix is added at 20 μL to each well and cells are incubated at 37 °C, 5% CO_2 for 24 h.
12. Remove supernatants for cytokine measurement and analysis through ELISA/luminometer. Store supernatants at −80 °C until assays can be performed.
13. Freeze cells at −80 °C for subsequent gene expression analysis.

3. Analysis

3.1 Measurement of interferon type I (ELISA, qPCR, luciferase)

Note: Measurement of IFN-I activity can be done through commercial ELISAs, qPCR gene expression assays, and cell-based reporter luciferase assays.

3.2 Measurement of IFN-α by ELISA

Materials

1. Human Interferon alpha-1 SimpleStep ELISA® Kit (Abcam)
2. Spectrophotometer

Protocol

We follow the manufacturer's recommendations which are outlined below with the following modifications.

1. Prepare the appropriate amount of antibody coated well strips. We typically perform measurements on each sample in duplicate or triplicate.
2. Equilibrate all reagents, experimental samples (tissue culture supernatants), and protein standards to room temperature.
3. Add standard or experimental samples to appropriate wells in duplicate.
4. Add capture and detector antibody cocktail to all wells. Incubate at room temperature for 1 h.
5. Aspirate and wash each well with the provided wash buffer three times × 300 μL.

6. Add TMB Substrate to each well and incubate at room temperature.
7. Stop the reaction at the predefined timepoint.
8. Read absorbance of each well in spectrophotometer at 450 nm.

3.2.1 ifnb *mRNA as measured by qRT-PCR gene expression assay*

Materials

1. RNeasy Plus kit (Qiagen)
2. MultiScribe Reverse Transcriptase (Thermo Fisher)
3. Using IDT primers for IFN-β, IFN-α, along with housekeeping genes such as ACTB and RPLP0
4. You will also need a method to quantify RNA concentration. We use the Nanodrop 2000 Spectrophotometer (Thermo Fisher)

Protocol

1. Thaw frozen cell pellets (Sections 2.1–2.3) on ice.
2. Extract total RNA from cells using Qiagen RNeasy Plus kit following the manufacturer's protocol.
3. Measure RNA concentration using Nanodrop. Adjust the concentration of each sample to 25 ng/μL using molecular grade nuclease free water.
4. Perform RT PCR using MultiScribe Reverse Transcriptase kit following the manufacturer's protocol. We recommend using a minimum of 200 ng of RNA.
5. Using probed IDT primers for genes of our choice, along with housekeeping genes such as RPLP0 or ActinB, perform qPCR.
6. Analyze relative differential gene expression using housekeeping genes and control conditions as baseline gene expression patterns.

3.2.2 Luciferase assay for IFN type I activity

Materials

1. THP-1 Dual Cells (Invivogen NF-κB-SEAP IRF-Luc Reporter Monocytes)
2. Quanti-Luc (Invivogen #rep-qlc1)
3. 96-well White Flat Bottom Polystyrene Plates (Corning # 3912)
4. Luminometer

Protocol

1. Dissolve one pouch of Quanti-Luc powder into 25 mL of cell-culture grade water.
2. Add 10 μL of supernatants from experimental set-up (Sections 2.1–2.3) into opaque white flat bottom plate to each well necessary for experiment, making sure that the liquid is at the bottom of the well.

3. Prime luminometer with Quanit-luc solution.
4. Read wells containing supernatants with the following parameters:
   a. 50 μL Quanti-luc reagent injection.
   b. Delay of 4 s per well.
   c. 0.1 s read time per well.
5. Units for values will be in RLU (Relative Light Units) with higher values corresponding to higher IFN-I

3.3 Measurement of IL-18 (ELISA)

Measurement of IL-18 or IL-1β activity can be done through commercial ELISAs or qRT-PCR gene expression assays. In general analysis of pro-IL-1β is performed by qRT-PCR as mRNA for pro-IL-18 is only weakly inducible in activated cells.

Materials

1. MBL Human IL-18 ELISA Kit (MBL)
2. Spectrophotometer

Protocol

1. Remove appropriate number of antibody coated well strips.
2. Equilibrate all reagents to room temperature and prepare all reagents, samples (1:5 dilution with assay diluent) and standards as instructed.
3. Add standard or diluted sample to appropriate well
4. Incubate at room temperature for 2 h.
5. Aspirate and wash each well.
6. Add antibody cocktail to all wells and incubate at room temperature for 1 h.
7. Aspirate and wash each well.
8. Add substrate to each well and incubate at room temperature for 30 min.
9. Add stop solution and read plate using plate reader at 450 nm.

4. Summary

Inflammasomes are key players in the regulation of cellular stress and immune responses. The PAMP ligands of certain key inflammasome receptors, such as NLPR3 are still poorly characterized. In contrast, AIM2 recognizes and binds to microbial DNA allowing it to function as a PRR and activate inflammasomes. Recent studies have demonstrated that AIM2 detects mislocalized self-DNA thereby allowing it to sense cellular perturbations leading to loss of cell viability and disruption of nuclear envelope integrity. AIM2 has been mostly studied in cells of the immune system in the context of infection and immune deregulation, however, its role in other tissues awaits dedicated studies. Another important question that is emerging is the contribution of AIM2 to inflammasome to autoinflammation and autoimmunity. The central role of interferons in immunity, and antiviral responses in particular, has long been understood. Perhaps unexpectedly, DNA has emerged as a key indicator of the presence of pathogens and cell damage. As the role various DNA binding proteins play in innate immunity, we have begun to understand the features of DNA that make it a suitable PAMP. Developing in parallel with studies of PRRs are methods to study immune cells response to DNA PAMPs. The next generation of such assays will allow functional assessment of several innate immune pathways in individual cells.

Acknowledgment

M.A.C. is supported by supported by a grants AI102696 NIH/NIAID and P30 AI094189-01A1 NIH/NIAID.

References

Ahmad, R., Sindhu, S. T., Toma, E., Morisset, R., & Ahmad, A. (2002). Elevated levels of circulating interleukin-18 in human immunodeficiency virus-infected individuals: Role of peripheral blood mononuclear cells and implications for AIDS pathogenesis. *Journal of Virology, 76*(24), 12448–12456.

Buchmann, K. (2014). Evolution of innate immunity: Clues from invertebrates via fish to mammals. *Frontiers in Immunology, 5*, 459. https://doi.org/10.3389/fimmu.2014.00459.

Ceballos-Olvera, I., Sahoo, M., Miller, M. A., Del Barrio, L., & Re, F. (2011). Inflammasome-dependent pyroptosis and IL-18 protect against Burkholderia pseudomallei lung infection while IL-1beta is deleterious. *PLoS Pathogens*, 7(12), e1002452. https://doi.org/10.1371/journal.ppat.1002452.

Cerullo, V., Seiler, M. P., Mane, V., Brunetti-Pierri, N., Clarke, C., Bertin, T. K., et al. (2007). Toll-like receptor 9 triggers an innate immune response to helper-dependent adenoviral vectors. *Molecular Therapy, 15*(2), 378–385. https://doi.org/10.1038/sj.mt.6300031.

Chattergoon, M. A., Latanich, R., Quinn, J., Winter, M. E., Buckheit, R. W., 3rd, Blankson, J. N., et al. (2014). HIV and HCV activate the inflammasome in monocytes and macrophages via endosomal toll-like receptors without induction of type 1 interferon. *PLoS Pathogens, 10*(5), e1004082. https://doi.org/10.1371/journal.ppat.1004082.

Chattergoon, M. A., Levine, J. S., Latanich, R., Osburn, W. O., Thomas, D. L., & Cox, A. L. (2011). High plasma interleukin-18 levels mark the acute phase of hepatitis C virus infection. *The Journal of Infectious Diseases, 204*(11), 1730–1740. https://doi.org/10.1093/infdis/jir642.

Chen, I. Y., & Ichinohe, T. (2015). Response of host inflammasomes to viral infection. *Trends in Microbiology, 23*(1), 55–63. https://doi.org/10.1016/j.tim.2014.09.007.

Chiu, Y. H., Macmillan, J. B., & Chen, Z. J. (2009). RNA polymerase III detects cytosolic DNA and induces type I interferons through the RIG-I pathway. *Cell, 138*(3), 576–591. https://doi.org/10.1016/j.cell.2009.06.015.

Coll, R. C., O'Neill, L., & Schroder, K. (2016). Questions and controversies in innate immune research: What is the physiological role of NLRP3? *Cell Death Discovery, 2*, 16019. https://doi.org/10.1038/cddiscovery.2016.19.

Collins, A. C., Cai, H., Li, T., Franco, L. H., Li, X. D., Nair, V. R., et al. (2015). Cyclic GMP-AMP synthase is an innate immune DNA sensor for Mycobacterium tuberculosis. *Cell Host & Microbe*, *17*(6), 820–828. https://doi.org/10.1016/j.chom.2015.05.005.

Costa, A., Gupta, R., Signorino, G., Malara, A., Cardile, F., Biondo, C., et al. (2012). Activation of the NLRP3 inflammasome by group B streptococci. *Journal of Immunology*, *188*(4), 1953–1960. https://doi.org/10.4049/jimmunol.1102543.

Crow, Y. J., & Manel, N. (2015). Aicardi-Goutieres syndrome and the type I interferonopathies. *Nature Reviews. Immunology*, *15*(7), 429–440. https://doi.org/10.1038/nri3850.

Dansako, H., Ueda, Y., Okumura, N., Satoh, S., Sugiyama, M., Mizokami, M., et al. (2016). The cyclic GMP-AMP synthetase-STING signaling pathway is required for both the innate immune response against HBV and the suppression of HBV assembly. *The FEBS Journal*, *283*(1), 144–156. https://doi.org/10.1111/febs.13563.

DeFilippis, V. R., Alvarado, D., Sali, T., Rothenburg, S., & Fruh, K. (2010). Human cytomegalovirus induces the interferon response via the DNA sensor ZBP1. *Journal of Virology*, *84*(1), 585–598. https://doi.org/10.1128/JVI.01748-09.

Delale, T., Paquin, A., Asselin-Paturel, C., Dalod, M., Brizard, G., Bates, E. E., et al. (2005). MyD88-dependent and -independent murine cytomegalovirus sensing for IFN-alpha release and initiation of immune responses in vivo. *Journal of Immunology*, *175*(10), 6723–6732.

Delaloye, J., Roger, T., Steiner-Tardivel, Q. G., Le Roy, D., Knaup Reymond, M., Akira, S., et al. (2009). Innate immune sensing of modified vaccinia virus Ankara (MVA) is mediated by TLR2-TLR6, MDA-5 and the NALP3 inflammasome. *PLoS Pathogens*, *5*(6), e1000480. https://doi.org/10.1371/journal.ppat.1000480.

Fernandes-Alnemri, T., Yu, J. W., Juliana, C., Solorzano, L., Kang, S., Wu, J., et al. (2010). The AIM2 inflammasome is critical for innate immunity to Francisella tularensis. *Nature Immunology*, *11*(5), 385–393. https://doi.org/10.1038/ni.1859.

Gao, P., Ascano, M., Wu, Y., Barchet, W., Gaffney, B. L., Zillinger, T., et al. (2013). Cyclic [G(2′,5′)pA(3′,5′)p] is the metazoan second messenger produced by DNA-activated cyclic GMP-AMP synthase. *Cell*, *153*(5), 1094–1107. https://doi.org/10.1016/j.cell.2013.04.046.

Guo, H., Callaway, J. B., & Ting, J. P. (2015). Inflammasomes: Mechanism of action, role in disease, and therapeutics. *Nature Medicine*, *21*(7), 677–687. https://doi.org/10.1038/nm.3893.

Hansen, K., Prabakaran, T., Laustsen, A., Jorgensen, S. E., Rahbaek, S. H., Jensen, S. B., et al. (2014). Listeria monocytogenes induces IFNbeta expression through an IFI16-, cGAS- and STING-dependent pathway. *The EMBO Journal*, *33*(15), 1654–1666. https://doi.org/10.15252/embj.201488029.

He, Y., Hara, H., & Nunez, G. (2016). Mechanism and regulation of NLRP3 inflammasome activation. *Trends in Biochemical Sciences*, *41*(12), 1012–1021. https://doi.org/10.1016/j.tibs.2016.09.002.

Horan, K. A., Hansen, K., Jakobsen, M. R., Holm, C. K., Soby, S., Unterholzner, L., et al. (2013). Proteasomal degradation of herpes simplex virus capsids in macrophages releases DNA to the cytosol for recognition by DNA sensors. *Journal of Immunology*, *190*(5), 2311–2319. https://doi.org/10.4049/jimmunol.1202749.

Hornung, V., Ablasser, A., Charrel-Dennis, M., Bauernfeind, F., Horvath, G., Caffrey, D. R., et al. (2009). AIM2 recognizes cytosolic dsDNA and forms a caspase-1-activating inflammasome with ASC. *Nature*, *458*(7237), 514–518. https://doi.org/10.1038/nature07725.

Huang, Y., Liu, L., Ma, D., Liao, Y., Lu, Y., Huang, H., et al. (2017). Human cytomegalovirus triggers the assembly of AIM2 inflammasome in THP-1-derived macrophages. *Journal of Medical Virology*, *89*(12), 2188–2195. https://doi.org/10.1002/jmv.24846.

Ishikawa, H., Ma, Z., & Barber, G. N. (2009). STING regulates intracellular DNA-mediated, type I interferon-dependent innate immunity. *Nature*, *461*(7265), 788–792. https://doi.org/10.1038/nature08476.

Kanneganti, T. D., Body-Malapel, M., Amer, A., Park, J. H., Whitfield, J., Franchi, L., et al. (2006). Critical role for Cryopyrin/Nalp3 in activation of caspase-1 in response to viral infection and double-stranded RNA. *The Journal of Biological Chemistry*, *281*(48), 36560–36568. https://doi.org/10.1074/jbc.M607594200.

Kato, H., Takahasi, K., & Fujita, T. (2011). RIG-I-like receptors: Cytoplasmic sensors for non-self RNA. *Immunological Reviews*, *243*(1), 91–98. https://doi.org/10.1111/j.1600-065X.2011.01052.x.

Kawai, T., & Akira, S. (2010). The role of pattern-recognition receptors in innate immunity: Update on toll-like receptors. *Nature Immunology*, *11*(5), 373–384. https://doi.org/10.1038/ni.1863.

Krug, A., French, A. R., Barchet, W., Fischer, J. A., Dzionek, A., Pingel, J. T., et al. (2004). TLR9-dependent recognition of MCMV by IPC and DC generates coordinated cytokine responses that activate antiviral NK cell function. *Immunity*, *21*(1), 107–119. https://doi.org/10.1016/j.immuni.2004.06.007.

Lam, E., Stein, S., & Falck-Pedersen, E. (2014). Adenovirus detection by the cGAS/STING/TBK1 DNA sensing cascade. *Journal of Virology*, *88*(2), 974–981. https://doi.org/10.1128/JVI.02702-13.

Li, X., Shu, C., Yi, G., Chaton, C. T., Shelton, C. L., Diao, J., et al. (2013). Cyclic GMP-AMP synthase is activated by double-stranded DNA-induced oligomerization. *Immunity*, *39*(6), 1019–1031. https://doi.org/10.1016/j.immuni.2013.10.019.

Liu, S., Cai, X., Wu, J., Cong, Q., Chen, X., Li, T., et al. (2015). Phosphorylation of innate immune adaptor proteins MAVS, STING, and TRIF induces IRF3 activation. *Science*, *347*(6227), aaa2630. https://doi.org/10.1126/science.aaa2630.

Lugrin, J., & Martinon, F. (2017). Detection of ASC oligomerization by western blotting. *Bio Protocol*, 7(10), e2292. https://doi.org/10.21769/BioProtoc.2292.

Man, S. M., Karki, R., Malireddi, R. K., Neale, G., Vogel, P., Yamamoto, M., et al. (2015). The transcription factor IRF1 and guanylate-binding proteins target activation of the AIM2 inflammasome by Francisella infection. *Nature Immunology*, *16*(5), 467–475. https://doi.org/10.1038/ni.3118.

Masters, S. L., Dunne, A., Subramanian, S. L., Hull, R. L., Tannahill, G. M., Sharp, F. A., et al. (2010). Activation of the NLRP3 inflammasome by islet amyloid polypeptide provides a mechanism for enhanced IL-1beta in type 2 diabetes. *Nature Immunology*, *11*(10), 897–904. https://doi.org/10.1038/ni.1935.

Muruve, D. A., Petrilli, V., Zaiss, A. K., White, L. R., Clark, S. A., Ross, P. J., et al. (2008). The inflammasome recognizes cytosolic microbial and host DNA and triggers an innate immune response. *Nature*, *452*(7183), 103–107. https://doi.org/10.1038/nature06664.

Netea, M. G., van de Veerdonk, F. L., van der Meer, J. W., Dinarello, C. A., & Joosten, L. A. (2015). Inflammasome-independent regulation of IL-1-family cytokines. *Annual Review of Immunology*, *33*, 49–77. https://doi.org/10.1146/annurev-immunol-032414-112306.

Nour, A. M., Reichelt, M., Ku, C. C., Ho, M. Y., Heineman, T. C., & Arvin, A. M. (2011). Varicella-zoster virus infection triggers formation of an interleukin-1beta (IL-1beta)-processing inflammasome complex. *The Journal of Biological Chemistry*, *286*(20), 17921–17933. https://doi.org/10.1074/jbc.M110.210575.

Rasmussen, S. B., Sorensen, L. N., Malmgaard, L., Ank, N., Baines, J. D., Chen, Z. J., et al. (2007). Type I interferon production during herpes simplex virus infection is controlled by cell-type-specific viral recognition through toll-like receptor 9, the mitochondrial antiviral signaling protein pathway, and novel recognition systems. *Journal of Virology*, *81*(24), 13315–13324. https://doi.org/10.1128/JVI.01167-07.

Rathinam, V. A., Jiang, Z., Waggoner, S. N., Sharma, S., Cole, L. E., Waggoner, L., et al. (2010). The AIM2 inflammasome is essential for host defense against cytosolic bacteria and DNA viruses. *Nature Immunology*, *11*(5), 395–402. https://doi.org/10.1038/ni.1864.
Reinholz, M., Kawakami, Y., Salzer, S., Kreuter, A., Dombrowski, Y., Koglin, S., et al. (2013). HPV16 activates the AIM2 inflammasome in keratinocytes. *Archives of Dermatological Research*, *305*(8), 723–732. https://doi.org/10.1007/s00403-013-1375-0.
Samarani, S., Allam, O., Sagala, P., Aldabah, Z., Jenabian, M. A., Mehraj, V., et al. (2016). Imbalanced production of IL-18 and its antagonist in human diseases, and its implications for HIV-1 infection. *Cytokine*, *82*, 38–51. https://doi.org/10.1016/j.cyto.2016.01.006.
Storek, K. M., Gertsvolf, N. A., Ohlson, M. B., & Monack, D. M. (2015). cGAS and Ifi204 cooperate to produce type I IFNs in response to Francisella infection. *Journal of Immunology*, *194*(7), 3236–3245. https://doi.org/10.4049/jimmunol.1402764.
Sun, L., Wu, J., Du, F., Chen, X., & Chen, Z. J. (2013). Cyclic GMP-AMP synthase is a cytosolic DNA sensor that activates the type I interferon pathway. *Science*, *339*(6121), 786–791. https://doi.org/10.1126/science.1232458.
Tabeta, K., Georgel, P., Janssen, E., Du, X., Hoebe, K., Crozat, K., et al. (2004). Toll-like receptors 9 and 3 as essential components of innate immune defense against mouse cytomegalovirus infection. *Proceedings of the National Academy of Sciences of the United States of America*, *101*(10), 3516–3521. https://doi.org/10.1073/pnas.0400525101.
Torii, Y., Kawada, J. I., Murata, T., Yoshiyama, H., Kimura, H., & Ito, Y. (2017). Epstein-Barr virus infection-induced inflammasome activation in human monocytes. *PLoS One*, *12*(4), e0175053. https://doi.org/10.1371/journal.pone.0175053.
Torre, D., Speranza, F., Martegani, R., Pugliese, A., Castelli, F., Basilico, C., et al. (2000). Circulating levels of IL-18 in adult and paediatric patients with HIV-1 infection. *AIDS*, *14*(14), 2211–2212.
Unterholzner, L., Keating, S. E., Baran, M., Horan, K. A., Jensen, S. B., Sharma, S., et al. (2010). IFI16 is an innate immune sensor for intracellular DNA. *Nature Immunology*, *11*(11), 997–1004. https://doi.org/10.1038/ni.1932.
Veenhuis, R. T., Astemborski, J., Chattergoon, M. A., Greenwood, P., Jarosinski, M., Moore, R. D., et al. (2017). Systemic elevation of proinflammatory interleukin 18 in HIV/HCV coinfection versus HIV or HCV monoinfection. *Clinical Infectious Diseases*, *64*(5), 589–596. https://doi.org/10.1093/cid/ciw771.
Watson, R. O., Bell, S. L., MacDuff, D. A., Kimmey, J. M., Diner, E. J., Olivas, J., et al. (2015). The cytosolic sensor cGAS detects Mycobacterium tuberculosis DNA to induce type I interferons and activate autophagy. *Cell Host & Microbe*, *17*(6), 811–819. https://doi.org/10.1016/j.chom.2015.05.004.
Woodward, J. J., Iavarone, A. T., & Portnoy, D. A. (2010). C-di-AMP secreted by intracellular listeria monocytogenes activates a host type I interferon response. *Science*, *328*(5986), 1703–1705. https://doi.org/10.1126/science.1189801.
Wu, J. J., Li, W., Shao, Y., Avey, D., Fu, B., Gillen, J., et al. (2015). Inhibition of cGAS DNA sensing by a herpesvirus virion protein. *Cell Host & Microbe*, *18*(3), 333–344. https://doi.org/10.1016/j.chom.2015.07.015.
Wu, J., Sun, L., Chen, X., Du, F., Shi, H., Chen, C., et al. (2013). Cyclic GMP-AMP is an endogenous second messenger in innate immune signaling by cytosolic DNA. *Science*, *339*(6121), 826–830. https://doi.org/10.1126/science.1229963.
Yao, Y., Chen, S., Cao, M., Fan, X., Yang, T., Huang, Y., et al. (2017). Antigen-specific CD8 (+) T cell feedback activates NLRP3 inflammasome in antigen-presenting cells through perforin. *Nature Communications*, *8*, 15402. https://doi.org/10.1038/ncomms15402.
Zhang, Z., Yuan, B., Bao, M., Lu, N., Kim, T., & Liu, Y. J. (2011). The helicase DDX41 senses intracellular DNA mediated by the adaptor STING in dendritic cells. *Nature Immunology*, *12*(10), 959–965. https://doi.org/10.1038/ni.2091.

CHAPTER SEVENTEEN

Biochemical and microscopic analysis of inflammasome complex formation

Patrycja Swacha[b], Nelson O. Gekara[a,b,*], Saskia F. Erttmann[c]
[a]Department of Molecular Biosciences, The Wenner-Gren Institute, Stockholm University, Stockholm, Sweden
[b]The Laboratory for Molecular Infection Medicine Sweden (MIMS), Umeå University, Umeå, Sweden
[c]Department of Molecular Biology, Umeå University, Umeå, Sweden
*Corresponding author: e-mail address: nelson.gekara@su.se

Contents

Abstract

Inflammasomes are multiprotein signaling platforms responsible for the maturation of pro-IL-1β and pro-IL-18 as well as the induction of an inflammatory cell death termed pyroptosis. Most inflammasomes consist of an upstream sensor, in most cases an adaptor protein (ASC) and inflammatory caspases such as caspase-1. Upon activation, sensor proteins oligomerize with adaptor proteins, forming large complexes called specks. These complexes can be stabilized and detected by Western blotting or fluorescence microscopy providing a direct evidence of inflammasome activation. Here we describe protocols for two complementary methods for detecting inflammasome complexes: (1) biochemical isolation and detection of ASC oligomers by Western blot analysis and (2) microscopic visualization of active caspase-1—ASC complexes. These protocols have successfully been applied in our recent study to unveil new regulatory mechanisms for different inflammasomes including the DNA sensor AIM2 (Erttmann et al., 2016).

Methods in Enzymology, Volume 625
ISSN 0076-6879
https://doi.org/10.1016/bs.mie.2019.05.014

1. Introduction

A major role of the innate immune system is to protect from invading pathogens and maintain tissue homeostasis. In order to fulfill this role, the immune system needs to recognize signals associated with pathogen entry and damage. This function is mediated by Pattern Recognition Receptors (PRRs), protein sensors that monitor the extracellular space or some subcellular compartments for the presence of foreign molecules as well as altered or misplaced self-molecules, which emerge as result of tissue damage. Recognition of these molecules by PRRs triggers signaling cascades leading to the production of pro-inflammatory cytokines and type I interferons (Takeuchi & Akira, 2010). In addition, upon activation, some of these PRRs form multiprotein signaling complexes called inflammasomes. Most inflammasomes consist of three components: (1) a sensor protein (member of a (NOD)-like family or absent in melanoma 2 (AIM2) receptor), (2) the adaptor molecule apoptosis-associated speck-like protein containing CARD (ASC) and (3) pro-caspase-1 (Lamkanfi & Dixit, 2014; Martinon et al., 2002).

Activation of inflammasomes occurs in two steps. The first step, referred to as priming step involves triggering of receptors such as Toll-like receptors (TLRs). This leads to the production of pro-inflammatory mediators including members of the interleukin (IL)-1 family of cytokines such as IL-1β and other inflammasome components including innate immune sensors, e.g., NLRP3. The second step, the actual inflammasome activation, is triggered by various stimuli and results in multimerization of the sensor protein with the adaptor protein ASC followed by the recruitment and subsequent autocatalytic cleavage of pro-caspase-1. Active caspase-1 processes pro-IL-1β and pro-IL-18 into their biologically active forms. IL-1β and IL-18 induce broad effects essential for efficient anti-microbial clearance (Garlanda et al., 2013; von Moltke et al., 2013). Active caspase-1 is also involved in the induction of a pro-inflammatory cell death called pyroptosis (Lamkanfi & Dixit, 2014).

Induction of pyroptosis is dependent on pore formation through gasdermin D oligomeric complexes in the plasma membrane allowing the release of IL-1β, IL-18 and caspase-1 into the extracellular space (Liu et al., 2016). This event is one of the hallmarks of inflammasome activation. Detection of cleaved caspase-1 and IL-1β as well as measurement of IL-1β concentration are the most commonly used methods in inflammasome studies. Western blot analyses of caspase-1 and IL-1β in cell lysates and supernatants provide information about both cleavage and secretion of these molecules while an enzyme-linked immunosorbent assay (ELISA) allows precise quantification of IL-1β. It is important to mention that most of the commercially available kits detect IL-1β as well as its immature form, which may be released upon cell death, independently of inflammasome activation.

Another hallmark of inflammasome activation is the formation of large ASC complexes termed specks (Broz et al., 2010). Most inflammasomes require oligomerization of ASC for the recruitment of pro-caspase-1 and the presence of ASC specks is a direct evidence of inflammasome activation. These micrometer-sized, insoluble aggregates can be visualized in cells using fluorescence microscopy or detected in cell lysates by Western blot analysis. The following protocols describe the detection of ASC aggregates *via* Western blotting as well as by immunofluorescence analysis. We describe the stimulatory conditions for AIM2 inflammasome activation in bone marrow-derived macrophages (BMDMs) and the detection of ASC oligomers by Western blotting. Moreover, we describe the microscopic detection of ASC specks and active caspase-1 foci in BMDMs after inflammasome activation triggered by infection with *Pseudomonas aeruginosa* an activator of the NLRC4 inflammasome.

2. Analysis of ASC inflammasome complexes by Western blotting

Absent in melanoma 2 (AIM2) is a cytosolic sensor responding to dsDNA of viruses or bacteria as well as self-DNA, which has leaked into the cytosol upon genomic instability (Hornung et al., 2009; Lugrin & Martinon, 2018). In the following protocol we use poly(deoxyadenylic-deoxythymidylic) acid (poly(dA:dT))—a synthetic dsDNA molecule—for the activation of the AIM2 inflammasome in bone marrow-derived macrophages.

BMDMs are primary cells able to react quickly to many inflammasome activators. Activation of the AIM2 inflammasome and detection of ASC oligomers can also be performed in monocytes and dendritic cells. Alternatively, immortalized cell lines may be used. However, primary cells are preferred as they are more physiologically relevant and do not possess alterations in signaling pathways, which may occur as an effect of immortalization.

Moreover, some cell lines lack the expression of inflammasome components and therefore are unresponsive toward some stimuli.

AIM2 directly binds poly(dA:dT). Following activation, the receptor forms a complex with ASC resulting in the recruitment and activation of pro-caspase-1. Based on our experience, we recommend treating cells with caspase-1 inhibitor before stimulation to prevent pyroptosis and the release of inflammasome complexes into the extracellular space. This allows ASC oligomers to be retained within the cell thereby increasing detection in cell lysates.

To preserve ASC complexes in their native state, we use non-denaturing lysis buffer supplemented with protease inhibitors. Stabilization of ASC complexes is achieved using the cross-linking agent disuccinimidyl suberate (DSS). DSS has two reactive sites linked with an 8-carbon spacer region (Leitner et al., 2010). The reactive sites of DSS bind to lysine residues thereby forming covalent bonds between two polypeptide chains, which enable stabilization and detection of ASC oligomers.

2.1 Equipment

37 °C, 5% CO_2 cell culture incubator (RS Biotech)
Tissue culture laminar flow hood (Bioair Instruments)
Manual pipettes, different volumes (Sartorius)
Pipet tips (VWR)
Serological pipets (VWR)
6-well tissue culture plates (TPP)
Cell scrapers (TPP)
Safe-Lock microcentrifuge tubes (Eppendorf)
Syringes, 1 mL (BD)
21 gauge needles (BD)
Bench top refrigerated centrifuge for 1.5 mL tubes (Eppendorf, Centrifuge 5424 R)
Rotation shaker (Rotamix RM-1, ELMI Ltd.)
Heating block (ThermoMixer C, Eppendorf)

2.2 Reagents

IMDM with Hepes (Gibco, Thermo Fisher Scientific)
Fetal bovine serum (FBS, heat inactivated; Gibco, Thermo Fisher Scientific)
1 × DPBS (Gibco, Thermo Fisher Scientific)

Pam_3CSK_4 (InvivoGen)
LPS (LPS-EB, LPS from *E. coli* O111:B4, InvivoGen)
OptiMEM (Gibco, Thermo Fisher Scientific)
Caspase-1 inhibitor II (Santa Cruz Biotechnology)
poly(dA:dT) naked (InvivoGen)
Lipofectamine 2000 (Invitrogen)
Disuccinimidyl suberate (DSS) (Sigma)
Dimethyl sulfoxide (DMSO) (Sigma)
Rat anti-mouse ASC antibody (CARD5, clone 8E4.1) (Genentech)
Monoclonal anti-β-actin antibody produced in mouse (clone AC-74, ascites fluid) (Sigma)
Anti-mouse IgG, HRP-linked antibody (CST)
Anti-rat IgG, HRP-linked antibody (CST)
HEPES $\geq$99.5% (Merck)
Potassium chloride (KCl) (Riedel-de Haën)
IGEPAL CA-630 (Nonidet P-40) (Merck)
TRIS (Ultra-Pure, VWR)
Sodium dodecyl sulfate (SDS) $\geq$98.5% (Merck)
Glycerol $\geq$99% (GC) (Merck)
Bromophenol blue (International Biotechnologies, INC (IBI))
β-mercaptoethanol (Roth)
Protease inhibitor cocktail: cOmplete™, Mini, EDTA-free Protease Inhibitor Cocktail (Roche); stock solution: One tablet dissolved in 2 ml dH_2O, stored at −20 °C
Lysis buffer
    20 mM HEPES-KOH pH 7.4
    150 mM KCl
    1% IGEPAL CA-630
    1:25 protease inhibitor cocktail
6 × Laemmli buffer
    0.5 M TRIS-HCL pH 6.8
    0.35 M SDS
    30% glycerol
    0.175 mM Bromophenol blue
    10% β-mercaptoethanol
2 × Lammli buffer
    Obtained by diluting 6 × Laemmli buffer in dH_2O (1 part 6 × laemmli buffer to 2 parts dH_2O)

2.3 Protocol

1. Seed 1.5×10^6 bone marrow-derived macrophages in 2 mL IMDM containing 10% FBS per well of a 6-well plate. Incubate cells at 37 °C and 5% CO_2.

   Note: we recommend using 3×10^6 cells (2 wells) for each stimulation condition.
2. 12–16 h later remove the medium and wash cells twice with DPBS.
3. Stimulate cells for 4.5 h with 500 ng/mL LPS or 500 ng/mL Pam_3CSK_4 in 1 mL OptiMEM at 37 °C and 5% CO_2.
4. Optional: 30 min before stimulation add 50 μM caspase-1 inhibitor II to prevent the release of inflammasome complexes into the extracellular space.
5. Transfect cells by addition of 1 μg/mL poly(dA:dT) with 2.5 μL lipofectamine 2000 in 200 μL OptiMEM and incubate cells for 2 h at 37 °C and 5% CO_2.
6. Optional: Harvest supernatants to assess the release of IL-1β and caspase-1 by Western blotting. This step cannot be performed if caspase-1 inhibitor II was added before stimulation.
   - All procedures described in step 7–13 need to be performed on ice.
7. Wash cells twice with ice-cold DPBS.
8. Lyse cells in 500 μL ice-cold lysis buffer (see recipe above).
9. Scrape cells off and transfer lysates into fresh 1.5 mL microcentrifuge tubes. Pool duplicates from each stimulation condition together at this step.
10. Shear crude cell lysates 10 times using a 21 G needle.
11. Collect 50 μL of each cell lysate as input control of the ASC isolation. Add 10 μL of 6 × Laemmli buffer and store samples at −20 °C.
12. Centrifuge samples at 3300 × *g* for 10 min at 4 °C.
13. Remove supernatants and wash pellets three times with 1 mL DPBS.
14. Remove supernatants and resuspend pellets in 500 μL DPBS.
15. Add 10 μL of freshly prepared 100 mM DSS solution (dissolved in DMSO) to the pellets resuspended in DPBS to obtain a final concentration of 2 mM.
16. Incubate cell pellets together with the cross-linker on a rotating shaker for 30 min at room temperature.
17. To separate unbound DSS from the pellets, centrifuge the samples for 10 min at 3300 × *g* and 4 °C and remove supernatants carefully to not disrupt the cross-linked pellets.

    Note: This step might be repeated several times to ensure the complete removal of liquid over the pellets.

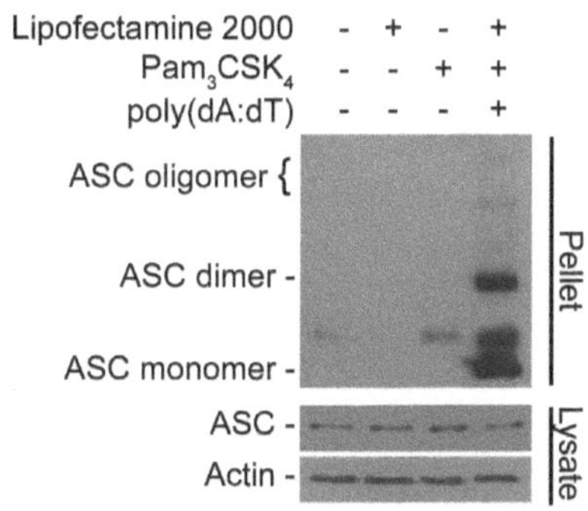

Fig. 1 Detection of ASC oligomers using Western blot analysis. Immunoblot analysis of ASC oligomers in BMDMs stimulated (or not) with 500 ng/mL Pam_3CSK_4 for 4.5 h and then transfected (or not) with 1 µg/mL poly(dA:dT) for 2 h. Probing for ASC in pellet fraction allows the detection of ASC oligomers. Detection of ASC and β-actin in lysate fraction serves as input (loading) control.

18. Resuspend each pellet in 35 µL of 2 × Laemmli buffer and store samples at −20 °C.

19. Heat samples for 5 min at 95 °C, separate ASC oligomers using a 10% SDS-PAGE gel and detect ASC by Western blotting.

20. Heat input samples (obtained in step 11) for 5 min at 95 °C, separate them on a 13.5% SDS-PAGE gel and detect ASC and β-actin.

An example of ASC oligomers in cell lysates of Pam_3CSK_4-primed BMDMs transfected with poly(dA:dT) is depicted in Fig. 1.

3. Analysis of caspase-1—ASC complexes by immunofluorescence microscopy

Besides the detection of ASC oligomers *via* Western blotting, inflammasome activation can be detected using immunofluorescence microscopic analysis of active caspase-1 and ASC specks. This methodology employs the green fluorescent probe FAM-YVAD-FMK to allow detection of active caspase-1. FAM-YVAD-FMK is a non-cytotoxic fluorescent-labeled inhibitor of caspase (FLICA) probe that binds to the active site of caspase-1, as well as caspase-4 and -5 (Amstad et al. 2001; Bedner et al., 2000). Active caspases recognize and cleave proteins, which contain a three or four amino acid sequence with an aspartic acid residue in position 1 (Thornberry et al., 1997). The caspase-1-specific FLICA probe contains the amino acid sequence tyrosine-valine-alanine-aspartic acid (YVAD), which is flanked by carboxyfluorescein (FAM) and fluoromethyl ketone

(FMK). While active caspase-1 recognizes and binds to YVAD, the fluorometyl ketone group of the probe reacts with caspase-1. The resulting covalent bond between the probe and caspase-1 inhibits further proteolytic activities of the enzyme. Using the active caspase-1-specific green fluorescent FLICA probe followed by immunofluorescence staining of ASC, allows microscopic visualization of active caspase-1—ASC complexes. This procedure together with Western blot analyses of chemically cross-linked ASC oligomers has been applied in our recent study to unravel mechanisms of inflammasome regulation (Erttmann et al., 2016).

3.1 Equipment

Eclipse C1 Plus Eclipse confocal microscope (Nikon)
37 °C, 5% CO_2 cell culture incubator (RS Biotech)
Tissue culture laminar flow hood (Bioair Instruments)
Compact Digital Rocker (Thermo Fisher Scientific)
Manual pipettes, different volumes (Sartorius)
Pipet tips (VWR)
Serological pipets (VWR)
24-well tissue culture plates (TPP)
Glass coverslips 13 μm Ø (VWR)
Forceps to lift glass coverslips
Microscope slides (MENZEL-GLÄSER, Thermo Fisher Scientific)

3.2 Reagents

IMDM with Hepes (Gibco, Thermo Fisher Scientific)
Fetal bovine serum (FBS, heat inactivated; Gibco, Thermo Fisher Scientific)
1 × DPBS (Gibco, Thermo Fisher Scientific)
OptiMEM (Gibco, Thermo Fisher Scientific)
LPS (LPS-EB, LPS from *E. coli* O111:B4, InvivoGen)
Paraformaldehyde (PFA), powder (Riedel-de Haën)
Saponin (United States Biochemical Corporation)
Bovine serum albumin (BSA, albumin fraction V) (Roth)
FAM-YVAD-FMK (ImmunoChemistry Technologies)
Rat anti-mouse ASC antibody (CARD5, clone 8E4.1) (Genentech)
Alexa Fluor 546 goat anti-rat IgG secondary antibody (life technologies, Thermo Fisher Scientific)
4′,6-Diamidino-2-phenylindole (DAPI) (5 mg/mL in dH_2O) (Invitrogen, Thermo Fisher Scientific)

SlowFade™ Diamond Antifade Mountant (Invitrogen, Thermo Fisher Scientific)
BD Bacto™ peptone (BD Biosciences)
BD Bacto™ yeast extract (BD Biosciences)
Sodium chloride (Merck)
Dimethyl sulfoxide (DMSO), cell culture reagent (Santa Cruz)
Pseudomonas aeruginosa (strain Xen41, PAO1 derivative)
LB medium
  10 g BD Bacto™ peptone
  5 g BD Bacto™ yeast extract
  5 g sodium chloride
FAM-YVAD-FMK stock solution
  Reconstitute one vial of FAM-YVAD-FMK with 50 μL DMSO.
  Stock solution can be stored at 4 °C in the dark.
  Final concentration for labeling intracellular caspase-1 is reached by the addition of the FAM-YVAD-FMK stock solution in a 1:150 ratio to the culture medium.
4% (w/v) PFA in PBS
  4 g paraformaldehyde
  Ad 100 ml with DPBS
  Storage at −80 °C
Blocking buffer
  3% (w/v) BSA
  0.1% (w/v) saponin
  Dissolved in DPBS and stored at 4 °C

3.3 Protocol

While this protocol describes active caspase-1 and ASC foci detection in response to activation of the NLRC4 inflammasome by *Pseudomonas aeruginosa*, this protocol can be used to study activation of other inflammasomes such as AIM2 (by DNA) or NLRP3 (by ATP).

1. Put a sterilized glass coverslip in each well of a 24-well plate using forceps.
2. Seed BMDMs at a density of 3×10^5 cells/well in a 24-well plate in IMDM containing 10% FBS and incubate them at 37 °C and 5% CO_2.
3. After 16 h, wash cells twice with 500 μL DPBS per well.
4. Stimulate cells with 500 ng/mL LPS in 300 μL OptiMEM per well for 4 h 15 min at 37 °C and 5% CO_2 to achieve inflammasome priming and therewith the expression of inflammasome components.

5. To efficiently induce inflammasome activation using *P. aeruginosa* infection, the bacteria need to be sub-cultured and used when in exponential growth phase. This is obtained by culturing a single colony of *P. aeruginosa* overnight in 3mL LB medium at 37°C and 150rpm. During the priming step of inflammasome activation *P. aeruginosa* is sub-cultured for 2h at 37°C and 150rpm by diluting the overnight culture 1:20 with LB medium. Afterward *P. aeruginosa* is washed with DPBS and resuspended in OptiMEM according to the required infection dose.

   Note: in our hands, an OD_{600} of 0.4 corresponds to a density of 3×10^8 bacteria per mL.
6. Add 2μL of FAM-YVAD-FMK stock solution per well (1:150 dilution) and incubate cells for 15min at 37°C and 5% CO_2.

   Note: we recommend controls, which instead of FAM-YVAD-FMK receive 2μL DMSO to control the specificity of the probe.
7. Infect cells with *P. aeruginosa* at a multiplicity of infection of 20 for 1h by adding a small volume (such as 20μL) containing the required bacteria per well. For uninfected control wells add an equivalent volume of OptiMEM.
8. After 1h of infection, discard cell culture supernatants and wash each well twice with 500μL of DPBS.
9. For cell fixation add 300μL of 4% PFA in PBS and incubate for 15min at 37°C.
10. Wash fixed cells three times with DPBS.
11. Incubate cells with anti-ASC antibody diluted 1:500 in blocking buffer for 30min at room temperature under rocking.

    Note: permeabilization of the cell membrane is reached by addition of saponin to the blocking buffer allowing entry of the primary antibody to intracellular compartments.
12. Wash cells three times with DPBS to remove unbound primary antibody.
13. Incubate cells with Alexa Fluor 546 goat anti-rat IgG secondary antibody diluted 1:250 in blocking buffer for 30min at room temperature and in the dark under rocking.

    Note: we recommend controls without secondary antibody.
14. Wash cells three times with DPBS to remove unbound secondary antibody.
15. To stain nuclei, incubate cells in 500ng/mL DAPI in DPBS for 5min at room temperature and in the dark under rocking.

    Note: this is required to be able to count the number of cells in relation to formed active caspase-1—ASC foci.

16. Wash cells five times with DPBS to remove unbound DAPI.
17. To remove salt crystals wash slides once with dH_2O before mounting.
18. Put a drop of mounting medium (SlowFade Diamond Antifade Mountant) on a microscope slide.
19. Use forceps to gently lift up a coverslip from a well and remove the remaining liquid carefully.
20. Turn the coverslip cell side facing down and place it onto the mounting medium.

    Note: avoid bubbles in the mountant by preventing shaking or inverting the bottle of mounting medium.

    - FAM-FLICA optimally excites at 488–492 nm and has a peak emission at 515–535 nm. Therefore, active-caspase-1 is detected by using a FITC filter set.
    - Detection of ASC foci *via* Alexa Fluor 456 is achieved by standard microscopy using the TRITC filter set. Alexa Fluor 456 has its excitation maximum at 556 nm and its emission maximum at 573 nm.
    - DAPI has its absorption and emission maximum at 358 and 461 nm, respectively. For fluorescence microscopy, DAPI is excited with ultraviolet light and detected through a blue/cyan filter.

    Note: due to very rapid bleaching of the stained samples, the microscopic visualization should occur within 24 h post staining.

Please find an example of results obtained by the described procedure in Fig. 2.

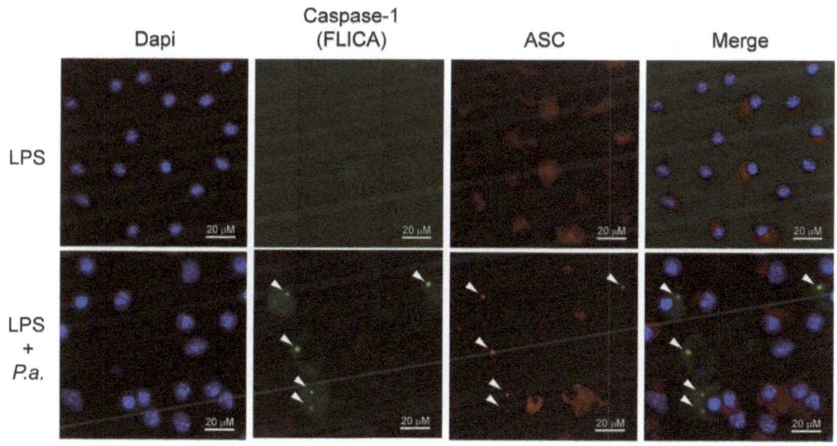

Fig. 2 Detection of active caspase-1—ASC complexes by fluorescence microscopy. Microscopic visualization of active caspase-1 (green)/ASC (red) complexes (overlay (merge) yellow) in LPS-primed BMDMs subsequently infected (or not) with *P. aeruginosa* (MOI 20, 60 min). DAPI was used to stain nuclei. Arrow heads indicate foci formation.

Acknowledgment

This work was supported by the Laboratory for Molecular Infection Medicine Sweden (MIMS), Umeå University, Stockholm University, the Swedish Research Council (grants 2015-02857 and 2016-00890 to N.O.G) and the Swedish Cancer Foundation (grant, CAN 2017/421 to N.O.G).

References

Amstad, P. A., et al. (2001). Detection of caspase activation in situ by fluorochrome-labeled caspase inhibitors. *Biotechniques*, *31*(3), 608–610. 612, 614.

Bedner, E., et al. (2000). Activation of caspases measured in situ by binding of fluorochrome-labeled inhibitors of caspases (FLICA): Correlation with DNA fragmentation. *Experimental Cell Research*, *259*(1), 308–313.

Broz, P., et al. (2010). Redundant roles for inflammasome receptors NLRP3 and NLRC4 in host defense against salmonella. *The Journal of Experimental Medicine*, *207*(8), 1745–1755.

Erttmann, S. F., et al. (2016). Loss of the DNA damage repair kinase ATM impairs inflammasome-dependent anti-bacterial innate immunity. *Immunity*, *45*(1), 106–118.

Garlanda, C., et al. (2013). The interleukin-1 family: Back to the future. *Immunity*, *39*(6), 1003–1018.

Hornung, V., et al. (2009). AIM2 recognizes cytosolic dsDNA and forms a caspase-1-activating inflammasome with ASC. *Nature*, *458*(7237), 514–518.

Lamkanfi, M., & Dixit, V. M. (2014). Mechanisms and functions of inflammasomes. *Cell*, *157*(5), 1013–1022.

Leitner, A., et al. (2010). Probing native protein structures by chemical cross-linking, mass spectrometry, and bioinformatics. *Molecular & Cellular Proteomics*, *9*(8), 1634–1649.

Liu, X., et al. (2016). Inflammasome-activated gasdermin D causes pyroptosis by forming membrane pores. *Nature*, *535*(7610), 153–158.

Lugrin, J., & Martinon, F. (2018). The AIM2 inflammasome: Sensor of pathogens and cellular perturbations. *Immunological Reviews*, *281*(1), 99–114.

Martinon, F., et al. (2002). The inflammasome: A molecular platform triggering activation of inflammatory caspases and processing of proIL-beta. *Molecular Cell*, *10*(2), 417–426.

Takeuchi, O., & Akira, S. (2010). Pattern recognition receptors and inflammation. *Cell*, *140*(6), 805–820.

Thornberry, N. A., et al. (1997). A combinatorial approach defines specificities of members of the caspase family and granzyme B. Functional relationships established for key mediators of apoptosis. *The Journal of Biological Chemistry*, *272*(29), 17907–17911.

von Moltke, J., et al. (2013). Recognition of bacteria by inflammasomes. *Annual Review of Immunology*, *31*, 73–106.

CHAPTER EIGHTEEN

Comet and micronucleus assays for analyzing DNA damage and genome integrity

Hui Jiang[a], Swarupa Panda[a], Nelson O. Gekara[a,b,*]
[a]The Laboratory for Molecular Infection Medicine Sweden (MIMS), Umeå Centre for Microbial Research (UCMR), Umeå University, Umeå, Sweden
[b]Department of Molecular Biosciences, The Wenner-Gren Institute, Stockholm University, Stockholm, Sweden
*Corresponding author: e-mail address: nelson.gekara@su.se

Contents

Abstract

Detection of DNA damage in cells is fundamental for the study of DNA repair and genome-instability associated processes including carcinogenesis. Many studies often rely on cytotoxicity assays to estimate genotoxicity. However, measurements of cytotoxicity, a delayed outcome requiring high threshold genotoxicity to induce, does not provide information about the subtle, early genotoxic effects relevant for mechanistic understanding of DNA repair processes. Here describe how to combine two simple procedures for monitoring the presence of DNA damage in individual eukaryotic cells using: (1) the Comet assay for measuring initial DNA breaks and (2) the Micronucleus assay for detecting delayed outcome DNA breaks in dividing cells. We discuss the principles, experimental design considerations and troubleshooting tips for optimizing these methods. They require standard molecular biology instruments and a fluorescent microscope.

Methods in Enzymology, Volume 625
ISSN 0076-6879
https://doi.org/10.1016/bs.mie.2019.05.015

1. Introduction

DNA damage from endogenous and exogenous genotoxic stress is persistent threat to the integrity of all organisms (Ciccia & Elledge, 2010). Cells resolve different types of DNA lesions via distinct pathways including mismatch repair (MMR), base excision repair (BER), nucleotide excision repair (NER), single-strand break repair (SSBR) and double strand breaks repair (DSBR) (Caldecott, 2008; David, O'Shea, & Kundu, 2007; Harfe & Jinks-Robertson, 2000; Sancar, Lindsey-Boltz, Unsal-Kacmaz, & Linn, 2004; Scharer, 2013; Wyman & Kanaar, 2006). Of the different types of DNA lesions, Double strand DNA breaks (DSBs) are the most deleterious and results in genome fragmentation and cell death. If repaired improperly DSBs can also cause chromosome translocation or deletions which may result in malignant transformation and cancer (Basu, 2018; Roos & Kaina, 2006, 2013; Wang, 2001).

The most direct approach to detecting DSB in cells is by assaying for the presence of genomic DNA fragments in the cell. The most commonly used technique is the Comet assay, a simple and sensitive method that allows the visualization and quantitative evaluation chromosome fragments at a single cell level (Fairbairn, Olive, & O'Neill, 1995).

The comet assay, also known as the single cell gel electrophoresis assay, was firstly developed by Ostling and Johanson (1984). Briefly, damaged DNA fragments migrate faster thereby separating from undamaged intact genomic DNA when electrophoresed in the gel. The more DNA breaks in the cell, the more chromosome fragments generated hence the longer the tails formed by electrophoresis. Therefore, the intensity and length of the comet tail relative to the head reflects the number DSB in the cell.

One of the late cellular outcomes of DSB is the generation of micronuclei (MN). How are micronuclei formed? In higher eukaryotes, cell division involves "open" mitosis, involving the complete dissolution of the nuclear membrane and subsequent reassembly as the cell partition the replicated DNA into daughter cells. During this process, whole or broken chromosome fragments can miss-segregate from the main chromatin mass and at the end of mitosis these, miss-segregated chromosome fragments can recruit nuclear membrane components to form micronuclei compartmentally distinct from the primary nucleus, see the schematic representation of micronucleus formation in Fig. 1 (Fenech et al., 2011; Gekara, 2017). Micronuclei also known as Howell–Jolly bodies and first identified in red cell precursors

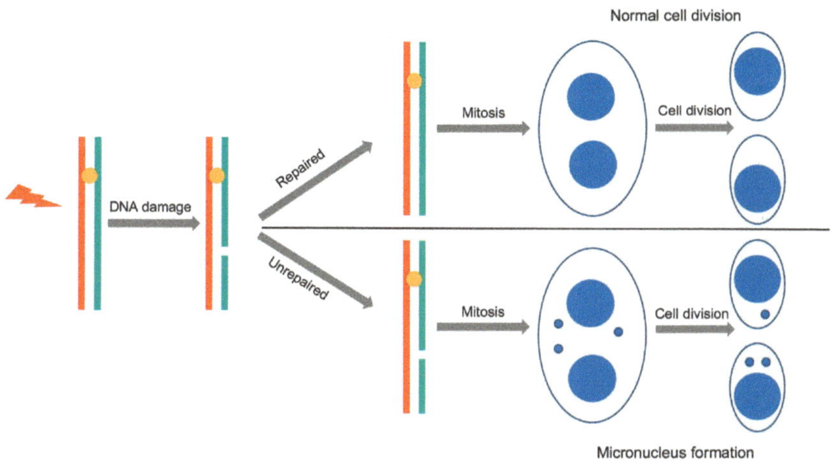

Fig. 1 Schematic representation of micronucleus formation after DNA damage.

by William Howell and Justin Jolly 100 years ago (Howell, 1890; Jolly, 1907) are now well acknowledged as key hallmarks of genome instability. More recently micronuclei have been proposed as platforms for innate immune recognition of genomic DNA by the cytosolic DNA sensor Cyclic GMP-AMP synthase (cGAS) following DNA damage (Gekara, 2017; Harding et al., 2017; Jiang et al., 2018; Mackenzie et al., 2017). Essentially, micronuclei are DNA fragments and can be easily stained by nucleic acid dye. Because micronuclei are structurally smaller, they can readily be distinguished from the main nucleus by standard fluorescence microscopy. Here, we provide detailed protocols for the comet and micronuclei assays and illustrate their application in the analysis of genome integrity.

2. The comet assay

Among several versions of the comet assays, the alkaline comet assay is widely used since it is able to detect single-strand breaks (SSBs), double-strand breaks (DSBs), and alkaline-labile sites (ALS). In this protocol, we described procedures for measuring DNA damage using an alkaline comet assay.

2.1 Materials

- 2-Hydroxyethyl agarose (Low gelling temperature-agarose, Sigma; cat. no. A4018)
- Phosphate-buffered saline (Ca^{2+} and Mg^{2+} free)

- 0.5 M EDTA (pH 8.0)
- 5 M NaOH.
- Propidium iodide (Sigma; cat. no. P4170) or DAPI may also be used,
- Sodium laurylsarcosinate (Sigma; cat. no. 61747)
- OxiSelect™ 3-well Comet Assay Slides (Cell biolabs; cat. no. STA-353)
- Horizontal gel electrophoresis chamber and power
- Fluorescence microscopy
- Image J software

2.2 Methods

2.2.1 Preparation of solutions for comet assay

- Alkaline lysis solution: 1.2 M NaCl, 100 mM EDTA, 0.1% sodium laurylsarcosinate, 0.26 M NaOH (pH >13); store at 4 °C. This solution should be prepared freshly for each experiment.
- Alkaline electrophoresis and rinse solution: 0.03 M NaOH, 2 mM EDTA (pH >13).

2.2.2 Preparation of comet agarose

Dissolve 1 g 2-Hydroxyethyl agarose in 100 mL distilled water by incubating at 100 °C in water bath for 5 min to make 1% comet agarose (Microwaving is not recommended). Maintain agarose at 40 °C in water bath.

2.2.3 Preparation of cell samples

Shortly before starting the comet assay, harvest cells (e.g., following exposure to genotoxic agents) and keep on ice to stop DNA repair.

- Using ice-cold PBS, collect and wash the cell samples by centrifugation (1500 rpm, 5 min) at 4 °C.
- Resuspend cells (1×10^5 cells/mL) in ice cold 1 × PBS (Ca^{2+} and Mg^{2+} free, to inhibit endonuclease activities).
- Include the appropriate controls, i.e., negative control (undamaged cells) and positive control (e.g., cells treated with genotoxins or γ-irradiation).
- Mix 1×10^5 cells/mL in cold PBS with 1% low-melting agarose at 40 °C at a ratio of 1:3 (v/v).
- Pipette the cell/agarose mixture onto a well of CometSlide (100 μL/well) and use the pipette tip to spread agarose/cells over sample area to ensure complete coverage of the sample area.
- Before lysing the cells, make sure the cell/agarose mixture has gelled (usually gelling for 20 min in high humidity environments).

2.2.4 Lysis of cells

- Before lysing cells, the lysis solution should be freshly prepared and chilled at 4 °C for at least 30 min.
- Submerge slide in prechilled lysis solution and leave at 4 °C for overnight in the dark.

  This step should be done carefully by slowly submerging the slides into solutions and maintaining the slides in horizontal position.

2.2.5 Gel electrophoresis

- After cell lysis, gently remove slides from the lysis solution at 4 °C. Then immerse the slides in Alkaline electrophoresis and rinse solution for 10 min in room temperature.
- Place the slides on the horizontal electrophoresis chamber already filled with Alkaline electrophoresis and rinse solution. Make sure the buffer covers the slides.
- Set the power supply to 20 V (~0.6 V/cm) and current at about 40 mA. Then allow electrophoresis to run for 20 min.

2.2.6 Staining

- Remove slides from electrophoresis chamber and carefully rinse with distilled water twice.
- Use filter paper to absorb water from the slides. Add 100 μL of a 10 μg/mL stock solution of propidium iodide directly onto one well of the slide and incubate 30 min.
- Rinse slides with distilled water twice.
- Keep the slides moist for further analysis (if in dry condition, the gel will shrink).

2.3 Analysis of the result

Use a fluorescence microscope with an excitation of 530 nm to observe the DNA damage at suitable magnification (usually 50–100 comets in one field).

The Tail Moment is the most common parameter to for evaluating extent of DNA damage. As shown in Fig. 2, use image analysis software such as Image J to quantitate tail moment length, percentage DNA in tail and tail moment (Tail Moment = Tail DNA% × Tail Moment Length). For each sample, 100–200 cells should be analyzed. In Fig. 3, we use γ-irradiation (IR) to induce DNA damage in HEK293 cells for different time to monitor the DNA repair process.

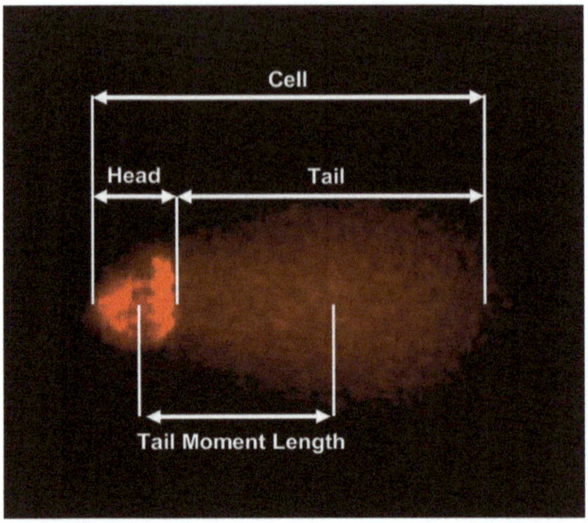

Fig. 2 An example of damaged DNA in Comet. Representative comet tails of HEK293 cells exposed to γ-irradiation (10 Gy).

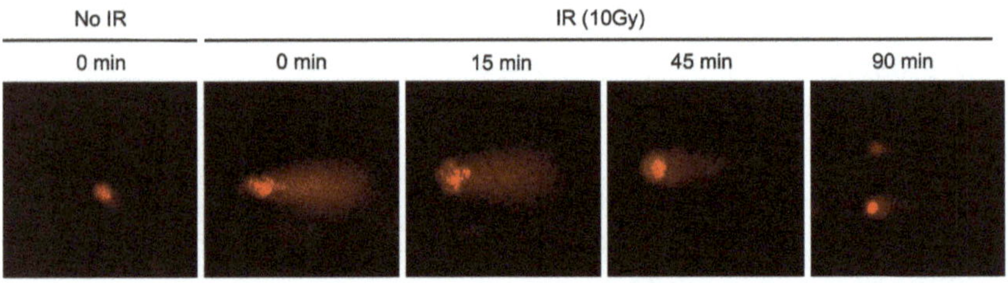

Fig. 3 Monitoring the DNA damage repair process by Comet assay. HEK293 cells were γ-irradiated in a 137Cs gamma-ray source (Gammacell 40 irradiator, MDS Nordion) with indicated dose. During irradiation all the cells are keep in ice to stop the DNA repair process. Thereafter, cells were transfer to 37 °C to allow DNA repair and then harvested at indicated time points for analysis.

3. The micronucleus assay

3.1 Materials

- Phosphate-buffered saline (Ca^{2+} and Mg^{2+} free)
- 4% paraformaldehyde
- 0.5% Triton X-100
- DNA dye 4′,6-diamidino-2-phenylindole (DAPI)
- Nocodazole (Sigma; cat. no. SML1665)

- Dako fluorescence mounting medium (Agilent)
- Coverslips
- Confocal microscope

3.2 Methods

The DNA binding innate immune sensor cGAS is abundant in micronuclei hence GFP-tagged cGAS can be used as a marker to monitor micronuclei generation (Jiang et al., 2018). For illustration here below, we describe micronuclei formation in HEK293 that stably express GFP-cGAS (HEK293-GFP-cGAS) or GFP reporter tag with a nuclear localization sequence NLS).

- Seed 1×10^5 cells and cultured on glass coverslips in 6-well plate.
- Synchronize the cells with 100 nM nocodazole for about 12 h to arrest the cell cycle at G2 phase.
- Expose cells to DNA damage (e.g., genotoxins or γ-irradiation).
- Released the synchronized G2 phase cells into mitosis (wash three times with PBS and add warm fresh medium).
- After 24 h culture, fix the cells on coverslips by incubating them in 4% paraformaldehyde for 30 min at room temperature.
- Permeabilize cells in 0.5% Triton X-100 for 5 min.
- Incubate coverslips with DAPI (1 μg/mL) for 10 min in the dark at room temperature.
- Wash the coverslips three times with PBS.
- Mount coverslips using Dako fluorescence mounting medium (Agilent).
- Image cells using the confocal microscope Fig. 4.

3.3 Analysis of the result

Micronuclei are defined as discrete DNA aggregates separate from the primary nucleus in cells where interphase primary nuclear morphology is normal. Cells with an apoptotic or necrotic appearance should be excluded. Recent studies show that innate immune DNA sensor cGAS localizes to micronuclei (Jiang et al., 2018; Mackenzie et al., 2017). Fig. 4 shows that after irradiation, the number of micronuclei in the cell is significantly increased and that GFP-cGAS foci co-localize with micronuclei (DAPI staining). For quantification, count all micronuclei as well as the total number of cells in one microscopic field, then calculate ratio: micronuclei (MN)/cell. For each sample, at least 50 different microscopic fields with over 500 cells should be analyzed.

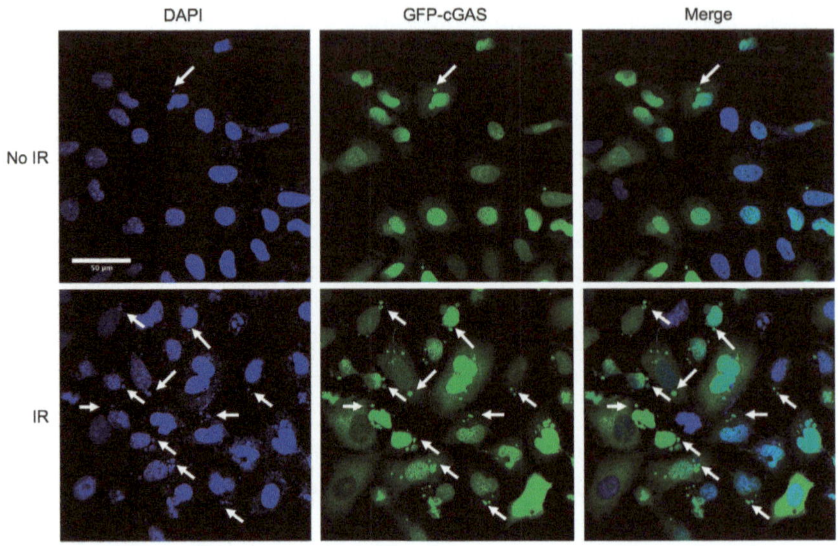

Fig. 4 DNA damage-induced micronuclei generation. The synchronized HEK293 cells stably expressing GFP-cGAS were exposed (or not) to γ-irradiation and cultured for 24 h. Cells were fixed, permeabilized in 0.5% Triton X-100, stained with the DNA dye DAPI, then analyzed by confocal microscopy for micronuclei. The white arrow shows cGAS foci (green spot) co-localized with micronuclei (blue). After DNA damage, the cGAS-micronuclei foci is dramatically increased. Scale bar: 50 μm.

4. Summary

In this protocol, we describe two simple methods for measuring DNA damage in eukaryotic cells. The Comet assay for initial DNA damage event and micronucleus assays for later DNA damage event. Both approaches are valuable and useful not only for experimental studies but also for clinical studies. For comet assay, optimization of the time course is the key to understanding of DNA damage repair process. For micronuclei assays, the appropriate cell seeding density and sample collection time point are very important for successful experiment (Micronuclei arise from cell proliferation, hence high cell density will inhibit cell proliferation). Taken together, these two technologies are among the most straightforward, sensitive methods for monitoring DNA damage/repair processes.

Acknowledgments

This work was supported by Stockholm University, the Laboratory for Molecular Infection Medicine Sweden (MIMS), Umeå University, the Swedish Research Council (grants 2015-02857 and 2016-00890 to N.O.G.) and the Swedish Cancer Foundation (grant, CAN 2017/421 to N.O.G.).

References

Basu, A. K. (2018). DNA damage, mutagenesis and cancer. *International Journal of Molecular Sciences*, *19*, 970.

Caldecott, K. W. (2008). Single-strand break repair and genetic disease. *Nature Reviews. Genetics*, *9*, 619–631.

Ciccia, A., & Elledge, S. J. (2010). The DNA damage response: Making it safe to play with knives. *Molecular Cell*, *40*, 179–204.

David, S. S., O'Shea, V. L., & Kundu, S. (2007). Base-excision repair of oxidative DNA damage. *Nature*, *447*, 941–950.

Fairbairn, D. W., Olive, P. L., & O'Neill, K. L. (1995). The comet assay: A comprehensive review. *Mutation Research*, *339*, 37–59.

Fenech, M., Kirsch-Volders, M., Natarajan, A. T., Surralles, J., Crott, J. W., Parry, J., et al. (2011). Molecular mechanisms of micronucleus, nucleoplasmic bridge and nuclear bud formation in mammalian and human cells. *Mutagenesis*, *26*, 125–132.

Gekara, N. O. (2017). DNA damage-induced immune response: Micronuclei provide key platform. *The Journal of Cell Biology*, *216*, 2999–3001.

Harding, S. M., Benci, J. L., Irianto, J., Discher, D. E., Minn, A. J., & Greenberg, R. A. (2017). Mitotic progression following DNA damage enables pattern recognition within micronuclei. *Nature*, *548*, 466–470.

Harfe, B. D., & Jinks-Robertson, S. (2000). DNA mismatch repair and genetic instability. *Annual Review of Genetics*, *34*, 359–399.

Howell, W. H. (1890). The life-history of the formed elements of the blood, especially the red blood corpuscles. *Journal of Morphology*, *4*, 57–116.

Jiang, H., Panda, S., Xue, X., Liang, F., Sung, P., & Gekara, N. O. (2018). The innate immune DNA sensor cGAS is a negative regulator of DNA repair hence promotes genome instability and cell death. *Biorxiv*. 465401.

Jolly, J. (1907). Recherches sur la formation des globules rouges des mammifères. In *9. Archives d'anatomie microscopique* (pp. 133–314). http://www.sudoc.abes.fr/xslt/CMD?DB=2.1&ACT=SRCHA&PRS=HOL&HLIB=751052312&IKT=8910&TRM=492828151&COOKIE=U10178,Klecteurweb,I250,B341720009+,SY,NLECTEUR+WEBOPC,D2.1,Eb0b33e21-ac,A,H,R94.255.148.162,FY.

Mackenzie, K. J., Carroll, P., Martin, C. A., Murina, O., Fluteau, A., Simpson, D. J., et al. (2017). cGAS surveillance of micronuclei links genome instability to innate immunity. *Nature*, *548*, 461–465.

Ostling, O., & Johanson, K. J. (1984). Microelectrophoretic study of radiation-induced DNA damages in individual mammalian cells. *Biochemical and Biophysical Research Communications*, *123*, 291–298.

Roos, W. P., & Kaina, B. (2006). DNA damage-induced cell death by apoptosis. *Trends in Molecular Medicine*, *12*, 440–450.

Roos, W. P., & Kaina, B. (2013). DNA damage-induced cell death: From specific DNA lesions to the DNA damage response and apoptosis. *Cancer Letters*, *332*, 237–248.

Sancar, A., Lindsey-Boltz, L. A., Unsal-Kacmaz, K., & Linn, S. (2004). Molecular mechanisms of mammalian DNA repair and the DNA damage checkpoints. *Annual Review of Biochemistry*, *73*, 39–85.

Scharer, O. D. (2013). Nucleotide excision repair in eukaryotes. *Cold Spring Harbor Perspectives in Biology*, *5*, a012609.

Wang, Z. (2001). DNA damage-induced mutagenesis : A novel target for cancer prevention. *Molecular Interventions*, *1*, 269–281.

Wyman, C., & Kanaar, R. (2006). DNA double-strand break repair: All's well that ends well. *Annual Review of Genetics*, *40*, 363–383.

CHAPTER NINETEEN

Workflows and considerations for investigating protein interactions of viral DNA sensors

Timothy R. Howard†, **Bokai Song**†, **Ileana M. Cristea***
Department of Molecular Biology, Princeton University, Princeton, NJ, United States
*Corresponding author: e-mail address: icristea@princeton.edu

Contents

Abstract

DNA sensors are a core component of innate immunity in mammalian cells. In response to pathogen infection, these specialized proteins sense pathogenic DNA from bacteria or viruses and initiate immune signaling cascades. These defense mechanisms rely on the rapid formation and temporal regulation of protein-protein interactions. Similarly, protein interactions underlie virus immune evasion mechanisms, as proteins from diverse viruses associate with and inhibit DNA sensors. Here, we describe experimental protocols for identifying protein interactions of DNA sensors, and discuss considerations for optimal isolation of protein complexes when targeting either endogenous or tagged proteins. Additionally, as viral infections and immune responses are known to induce prominent changes in cellular protein abundances, we provide a workflow for investigating these protein associations in the context of proteome alterations.

† Co-first authors.

Methods in Enzymology, Volume 625
ISSN 0076-6879
https://doi.org/10.1016/bs.mie.2019.04.026

1. Introduction

The ability of mammalian cells to sense pathogenic DNA is a fundamental mechanism of defense against numerous microorganisms that are part of our ecosystem. Outside of RNA viruses, all other pathogens, including bacteria and DNA viruses, carry their genetic information in the form of DNA and rely on their DNA replication for the spread of infection. In response, mammalian cells have evolved a multitude of defense mechanisms that employ specialized proteins, known as pattern recognition receptors (PRRs), to detect pathogen-associated molecular patterns (PAMPs), including pathogenic double-stranded DNA (dsDNA). Diverse PRRs survey the cellular space, providing means for pathogenic DNA sensing within different cellular compartments. Upon binding to dsDNA, PRRs initiate defense programs, which include innate immune signaling, inflammatory responses, and apoptosis. For this, PRRs rely on adaptor and effector proteins to initiate immune signaling programs, amplify signals, and communicate information between subcellular compartments. Therefore, the interactions of DNA sensors with other cellular proteins are at the core of host antiviral response. In this chapter, we describe experimental protocols and considerations for identifying protein interactions of DNA sensors, and for investigating these associations in the context of proteome alterations during immune signaling.

Fig. 1 provides examples of prominent DNA sensors and their interacting protein partners. Endosomal compartments are monitored for the presence of foreign dsDNA by the transmembrane protein Toll-like receptor 9 (TLR9). This protein binds to both bacterial (Deretic, 2012) and viral (Krug et al., 2004) unmethylated CpG DNA motifs and initiates a signaling cascade that culminates in the expression of pro-inflammatory cytokines. Like other Toll-like receptors, the cytosolic domain of TLR9 interacts with the adaptor proteins TIRAP and MyD88 to initiate the formation of the Myddosome in the cytoplasm (Bonham et al., 2014; Hemmi et al., 2000; Kawai, Adachi, Ogawa, Takeda, & Akira, 1999). The Myddosome then incorporates the serine/threonine kinases IRAK 2 and 4 (Gillen & Nita-Lazar, 2017), which act as effector proteins via TRAF6/IRF8 to activate the transcription factor NF-κB for cytokine production (Bonham et al., 2014; Tsujimura et al., 2004).

Within the cytosol, cyclic GMP-AMP synthase (cGAS) is known as a major sensor of dsDNA. Upon binding to dsDNA in a sequence-independent manner, cGAS catalyzes the formation of 2′3′-cGAMP

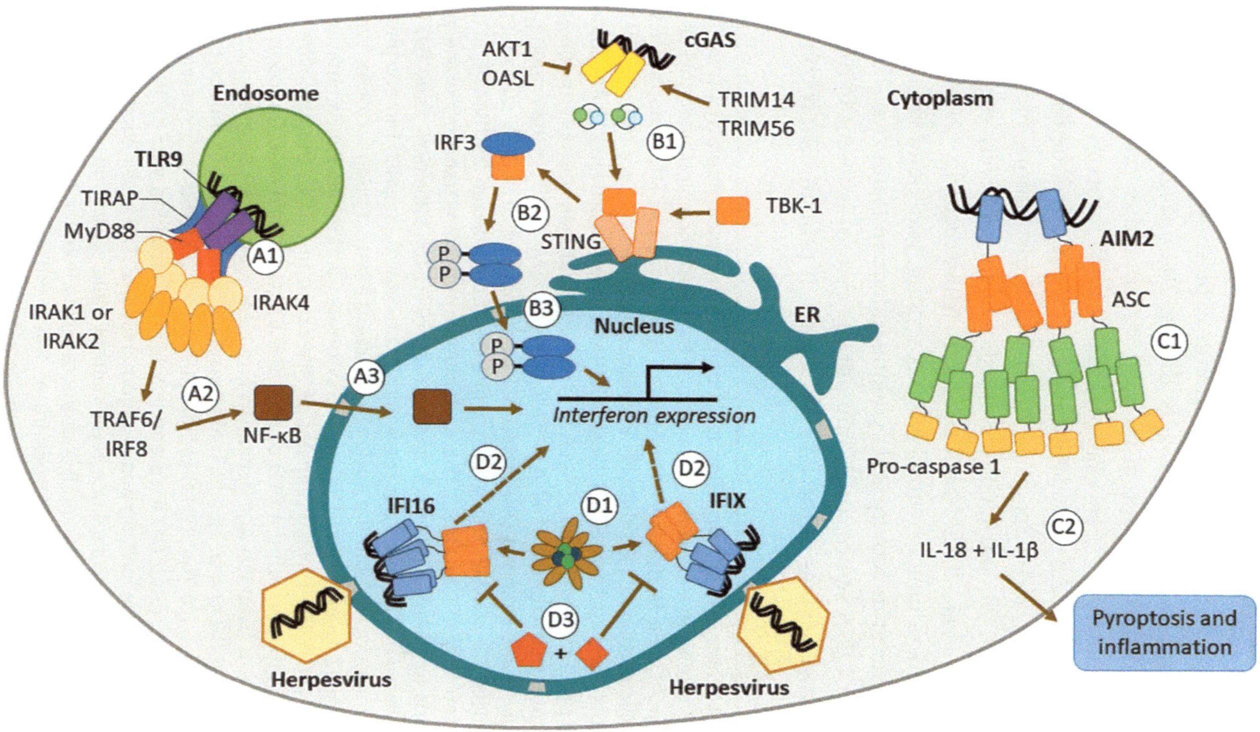

Fig. 1 See figure legend on next page.

(Sun, Wu, Du, Chen, & Chen, 2013). This small molecule acts as a secondary messenger that binds to and activates the adaptor ER-resident protein STING, causing its conformational change and dimerization (Wu et al., 2013). STING further induces interferon expression through the TBK1-IRF3 signaling axis (Ishikawa, Ma, & Barber, 2009) (Fig. 1). Recent studies have shown that cGAS protein interactions can either promote or inhibit its immune functions. For example, the TRIM proteins TRIM14 and TRIM56 reduce cGAS degradation and enhance cGAS dimerization, respectively (Chen et al., 2016; Seo et al., 2018). Other interactions, including those with AKT1, CCP5, and CCP6, were shown to negatively regulate cGAS functions (Seo et al., 2015; Xia et al., 2016). More recently, OASL, a protein known for functioning in cytosolic RNA sensing, was found to interact with cGAS and inhibit its activity and cytokine induction (Lum et al., 2018).

Another DNA sensor functioning in the cytosol is absent in melanoma 2 (AIM2) that belongs to the PYHIN family of protein, i.e., it contains an

Fig. 1 DNA sensor immune signaling relies on protein-protein interactions. (A) Toll-like receptor 9 resides in endosome membranes to detect incoming unmethylated CpG DNA. (Step A1) The Myddosome forms after TLR9 binds to DNA, dimerizes, contacts the cytosolic protein MyD88 through interaction with the adaptor protein TIRAP, and recruits IRAK4 and IRAK1 or IRAK2. (Step A2) IRAK1 or IRAK2 activates TRAF6 to remove inhibitory proteins from NF-κB. (Step A3) Activated NF-κB translocates into the nucleus and promotes transcription of *IFNA* or *IFNB*. (B) Cyclic GMP-AMP synthase monitors the cytosol for double-stranded DNA. (Step B1) Upon binding to dsDNA, cGAS dimerizes and produces 2′3′-cGAMP. This activity can either be enhanced by TRIM14 or TRIM56 or repressed by AKT1 or OASL. The endoplasmic reticulum transmembrane protein STING binds to cGAMP, dimerizes, and recruits TBK-1 for activation. (Step B2) IRF3 is phosphorylated by TBK-1 and dimerizes within the cytosol. (Step B3) Dimeric phosphorylated IRF3 translocates into the nucleus and acts as a transcription factor for Type I IFN genes. (C) AIM2 recognizes dsDNA in the cytoplasm and initiates inflammasome formation. (Step C1) The HIN200 domains of AIM2 bind to dsDNA thereby allowing the AIM2 PYD to interact with the PYD of ASC. The CARD domain of ASC recognizes that of Pro-caspase 1. (Step C2) Autoactivation and cleavage of Pro-caspase 1 leads to the maturation of IL-18 and IL-1β which are excreted from the cell to induce pyroptosis and inflammation. (D) The PYHIN proteins IFI16 and IFIX bind to herpesvirus dsDNA within the nucleus. (Step D1) IFI16 and IFIX bind to incoming viral dsDNA with their HIN200 domains while forming homo-oligomers with PYD. These oligomers then interact with ND10 bodies to facilitate repression of virus replication. (Step D2) Through unknown mechanisms, IFI16 and IFIX promote innate immune signaling through interferon expression. (Step D3) Herpesviruses have evolved distinct mechanisms for inactivating IFI16 and IFIX: HCMV uses the tegument protein pUL83 to bind PYD and prevent oligomerization, while the HSV-1 immediate-early protein ICP0 targets PYD for degradation via the proteasome.

N-terminal PYRIN domain (PYD) that mediates homotypic protein-protein interactions and a C-terminal HIN200 domain capable of binding double-stranded DNA. It is proposed that, during cellular homeostasis, AIM2 is maintained inactive via an autoinhibition mechanism that involves folding of the HIN200 domain over the PYD (Jin, Perry, Smith, Jiang, & Xiao, 2013). Upon DNA sensing, the AIM2 PYD becomes free to initiate a chain reaction of protein interactions that results in a structure known as the inflammasome. Specifically, the AIM2 PYD associates with the PYD of ASC, also known as apoptosis-associated Speck-like protein containing a caspase activation and recruitment domain (CARD), which in turn uses its CARD to recruit and activate Caspase 1 (Fernandes-Alnemri, Yu, Datta, Wu, & Alnemri, 2009; Hornung et al., 2009) (Fig. 1). The resulting inflammasome structure amplifies the danger signal from AIM2 to ultimately result in pyroptosis and inflammation caused by maturation and secretion of cytokines IL-1β and IL-18 (Fernandes-Alnemri et al., 2009).

Although it was long thought that PRRs that sense dsDNA are exclusively partitioned outside of the nucleus in order to avoid mistaking self-DNA for that of a pathogen, investigations during the past decade have demonstrated the existence of nuclear DNA sensing mechanisms. The interferon inducible proteins IFI16 and IFIX were established as the first proteins capable of sensing pathogen DNA within the nucleus (Diner, Li, et al., 2015; Li, Diner, Chen, & Cristea, 2012; Unterholzner et al., 2010). Like AIM2, these proteins are members of the PYHIN family, but they do not seem to predominantly use their PYRIN domains to recruit other PYD-containing proteins. Rather, they form homo-oligomers upon binding DNA. In this activated state, IFI16 and IFIX promote cytokine expression (Diner, Li, et al., 2015; Unterholzner et al., 2010) and bind to other nuclear proteins with known antiviral roles, including ND10 components PML, hDAXX, and ATRX (Diner, Li, et al., 2015; Diner, Lum, Javitt, & Cristea, 2015; Diner, Lum, Toettcher, & Cristea, 2016; Merkl & Knipe, 2019), as well as the previously discussed cGAS (Diner et al., 2016; Orzalli et al., 2015). Several studies have demonstrated that ND10 factors epigenetically repress herpes simplex virus type 1 (HSV-1) replication (Cabral, Oh, & Knipe, 2018; Lukashchuk & Everett, 2010), and a leading hypothesis suggests that association with IFI16 and IFIX can help to direct ND10 bodies to incoming viral genomes (Cabral et al., 2018; Diner et al., 2016; Everett, 2016; Merkl & Knipe, 2019). Meanwhile, nuclear cGAS was shown to stabilize IFI16 and promote innate immune signaling during HSV-1 infection (Orzalli et al., 2015).

The importance of protein-protein interactions in regulating DNA sensing is further highlighted by virus-host protein interactions through which pathogens inhibit PRRs. For example, the HSV-1 infection induces the proteasome-dependent degradation of IFI16 and, although this is still debated (Cuchet-Lourenco, Anderson, Sloan, Orr, & Everett, 2013), some studies indicate that the immediate early protein ICP0 contributes to this process (Orzalli, DeLuca, & Knipe, 2012), while human cytomegalovirus uses the major tegument protein pUL83 to bind IFI16 PYD and prevent its homo-oligomerization (Cristea et al., 2010; Li, Chen, & Cristea, 2013). The DNA sensor cGAS was also found to be inhibited by viral proteins, including by the Kaposi's sarcoma-associated herpesvirus (KSHV) proteins LANA and ORF52 (Wu et al., 2015; Zhang et al., 2016), the HSV-1 protein pUL37 (Zhang et al., 2018), and the dengue virus protein NS2B (Aguirre et al., 2017). Thus, studies of protein-protein interactions can reveal critical insights into both positive and negative regulation of DNA sensors during infection.

2. Experimental considerations when studying DNA sensor interactions

A number of experimental approaches are available for the study of protein interactions. These range from methods focused on the discovery of direct interactions, such as binary and crosslinking assays, to methods that retain both direct and indirect interactions to characterize functional protein complexes, protein interaction networks, or proximity associations within a certain subcellular location. Although many of these methods have yet to be implemented in studies of DNA sensors, these are valuable approaches that can be considered depending on the biological question that is being addressed, and we point the readers to several research papers and review manuscripts that describe some of these workflows (Chojnowski et al., 2018; Greco, Diner, & Cristea, 2014; Leitner et al., 2010; Lobingier et al., 2017; Miteva, Budayeva, & Cristea, 2013; Roux, Kim, & Burke, 2013). The method that has been so far the most widely used for studying DNA sensor protein interactions has been immunoaffinity purification (IP). IP has been either coupled with western blotting, for targeted analysis of one or several predicted interactions, or with mass spectrometry (IP-MS), for the discovery of protein interactions and functional complexes. For example, IP-MS has been applied to investigations of cGAS (Liu et al., 2018; Lum et al., 2018; Seo et al., 2018), IFI16 (Diner et al., 2016; Diner, Li, et al., 2015;

Diner, Lum, et al., 2015), as well as IFIX (Crow & Cristea, 2017; Diner, Li, et al., 2015) and AIM2 (Diner, Li, et al., 2015) protein interactions. In this chapter, we provide a robust IP-MS protocol (see Sections 4.3 and 5.3) that was proven effective for studying DNA sensors. However, prior to starting the experimental workflow, several choices and experimental considerations relevant to DNA sensors must be taken into account (Fig. 2).

IP-MS can be used to investigate a protein in different *in vitro* and *in vivo* biological contexts, with the most frequent application for DNA sensors so far being their isolation from cell culture. One important aspect that should be considered for such studies is the choice of cell type. The endogenous abundance of the protein of interest can vary among different cell types. For example, cGAS and IFI16 are present at higher levels in macrophages (as shown for differentiated THP-1 cells) than in fibroblasts (HFFs), and are not expressed in HEK293T cells (Li et al., 2012; Lum et al., 2018). This has to be taken into account if the goal is to study the interactions of an endogenous protein. However, the use of different cell types can provide the tools needed for addressing different biological questions. For example, the lack of cGAS and IFI16 expression in HEK293T cells has been instrumental in determining the minimal components needed for innate immune signaling by reconstituting the signaling axis with the co-expression of the DNA sensor and the adaptor protein STING (Orzalli et al., 2015; Sun et al., 2013; Unterholzner et al., 2010). Human or mouse fibroblasts have often been chosen to study protein interactions during infection with herpesviruses, including HSV-1 and HCMV, as it is a relevant cell type that can mimic an *in vivo* infection (Crow & Cristea, 2017; Diner et al., 2016; Lum et al., 2018; Orzalli et al., 2015). Additionally, knowledge of the high abundance of a DNA sensor in a certain cell type has provided a platform for initial studies that could be then followed up in other relevant cell types. For example, the finding that IFI16 is abundant in CEM T cells has provided the ability to perform IP-MS studies of endogenous IFI16 (Li et al., 2012). This has led to the discovery of acetylations within its nuclear localization signal, which were then confirmed via targeted MS analyses in cell types where IFI16 is present at lower levels than in CEM T cells, including macrophages and fibroblasts (Li et al., 2012).

Another important factor when choosing a cell system is that proteins can have distinct subcellular localization patterns in different cell types. For example, while IFI16 is predominantly nuclear in fibroblasts and CEM T cells, it has both nuclear and cytoplasmic localizations in THP-1 cells (Hornung et al., 2009; Unterholzner et al., 2010). Similarly, although cGAS

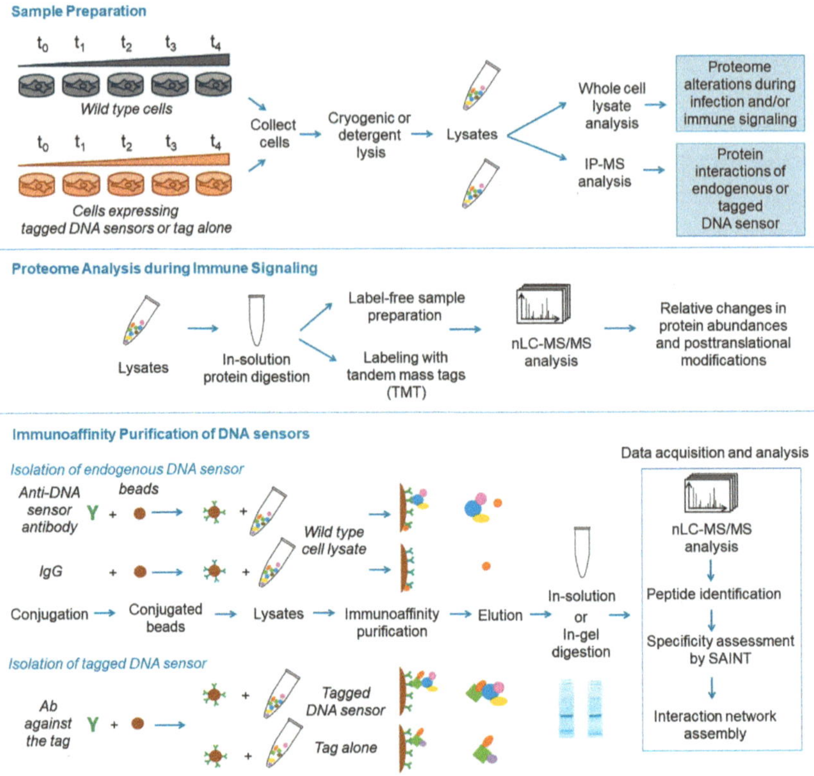

Fig. 2 Workflows for analyzing proteome changes and DNA sensor protein interactions during immune signaling. (Top) Preparation and lysis of cells prior to immunoaffinity purification and proteome analysis. Cells subjected to different experimental conditions, i.e., different time points of viral infection, are collected and lysed using either cryogenic or detergent-based techniques. Lysates are then used for either mass spectrometry proteome analysis or immunoaffinity purification of DNA sensor. (Middle) nLC-MS/MS analysis and relative quantification of protein abundances. Peptides from whole-cell lysates are produced using in-solution digestion then analyzed using either label-free or labeling quantification, such as by labeling with tandem mass tags. (Bottom) Investigation of protein-protein interactions using immunoaffinity purification coupled to nLC-MS/MS analysis. Purification of protein complexes is performed by incubating antibody-conjugated beads with cell lysates. If targeting the endogenous protein, the control IP can be performed with IgG; if the DNA sensor bait is tagged, cells expressing the tag alone can serve as a control. Eluted protein complexes are digested either in-solution or in-gel, and peptides are analyzed by nLC-MS/MS. The specificity of protein interactions can be assessed using the Significance Analysis of INTeractome (SAINT), and associations that pass the specificity filtering can be assembled into functional protein interaction networks.

is known to be predominantly cytoplasmic, several studies have now indicated its presence in the nuclei of different cell types (Liu et al., 2018; Orzalli et al., 2015). Therefore, an IP-MS study can provide a mixed pool of interactions from both subcellular compartments. This has to be accounted for when building protein interaction networks and, even prior to that, when optimizing the lysis buffers needed for the extraction of the protein, which may have different physical characteristics in the different subcellular compartments (e.g., localization-specific aggregation or oligomerization). The use of fractionations, such as nuclear-cytoplasmic fractionation or density gradient separations, followed by IP-MS studies in distinct fractions can help to hone in on certain localization-dependent functions. For example, cytoplasmic fractionation was used in a recent IP-MS study of cGAS cytoplasmic interactions and regulation (Lum et al., 2018).

Another consideration is if the protein of interest needs to be tagged for IP-MS studies. Several circumstances would require or benefit from using a tagging approach. For example, one obstacle is that antibodies raised against an endogenous protein, even if successfully used in western blotting or microscopy assays, frequently lack the affinity or specificity required for IP studies. Tagging might also be preferred when one pursues the function of a specific domain or residue. In such cases, tagged domain constructs or single-point mutants can be transfected or stably expressed in cells prior to IP-MS studies. The use of a tag would afford a similar isolation strategy and the ability to compare the interactions of different mutants and domain constructs to the tagged wild type protein. For example, generation of tagged IFI16 PYRIN domain and HIN200 domain constructs provided the ability to identify functional domain-specific interactions, including the association of the IFI16 PYRIN domain with cGAS and ND10 bodies (Diner et al., 2016).

Commonly used tags include green fluorescent protein (GFP), FLAG, hemagglutinin (HA), and Protein A. The use of a fluorescent tag, such as GFP, provides the means to consecutively investigate both protein localization (by microscopy) and interactions (by IP) (Cristea, Williams, Chait, & Rout, 2005). The isolation of a tagged protein is usually more convenient due to the high specificity of the available antibodies. However, a potential drawback is that the introduction of a tag may affect protein function, or mask a certain region on the bait protein and inhibit protein interactions. Therefore, an assessment that indicates that the tagged protein retains its normal function is necessary prior to characterization of interactions. Specifically, tagged DNA sensors should retain the ability to induce innate immune signaling and elevated antiviral cytokine expression. Different locations for the tag can be considered when taking into account the position of functional domains. In a previous study that investigated the interactions of all the four members of the PYHIN family, the authors compared N- and C-terminally tagged constructs for AIM2, IFI16, IFIX and MNDA (Diner, Li, et al., 2015). Similarly, a recent cGAS interactome study started by confirming the function of N- and C-terminally tagged cGAS constructs, which led to the discovery of its association with OASL (Lum et al., 2018).

The use of tagging approaches brings another challenge relevant for DNA sensors. Overexpressed DNA sensors may induce immune signaling and proteome changes that can impact protein interactions. This can be particularly problematic when generating stably expressing cell lines. One strategy that partly ameliorates these issues and has proven useful for DNA sensors is to use an inducible expression system (e.g., as shown for doxycycline inducible-GFP-cGAS, Lum et al. (2018) and tetracycline-inducible GFP-IFI16, Diner et al. (2016)) to express the target protein shortly before the experiments and avoid consistent cytokine induction.

The selection of the IP workflow influences the selection of the appropriate control experiments that can help determine the specificity of the identified interactions (as reviewed in Miteva et al. (2013). If a tagged protein is used, the cells in the control group should express the tag only, and be purified using the same antibody. If an endogenous protein is purified, the control IP is frequently an IgG isolation performed in the same cell type. Alternatively, the control IP can be performed using the antibody against the endogenous protein in cells following knockdown or knockout of the bait DNA sensor.

3. Protein interactions in the context of proteome changes

The dynamic regulation of the cellular proteome underlies all biological processes, including innate immune responses. Changes in protein abundances, subcellular localizations, and posttranslational modifications (PTMs) drive the formation or disruption of protein interactions and are at the core of immune signaling events. Therefore, protein interaction studies must be investigated and understood in the broader context of proteome changes (Fig. 2). For example, the host activation of DNA sensors for recognition of viral DNA and the ability of viruses to inhibit these immune mechanisms rely on finely-tuned changes in the proteome during the progression of the virus replication cycle. Several studies have started to build a broader view of temporal proteome alterations during the progression of viral infections (Beys-da-Silva et al., 2018; Drayman et al., 2017; Forrest, Hislop, Rickinson, & Zuo, 2018; Kulej et al., 2017; Weekes et al., 2014). Relevant to immune signaling, proteome changes were also investigated upon infection with HSV-1 strains that have different abilities to induce immune responses, including elevated cytokines or apoptosis (Lum et al., 2018).

In addition to changes in protein abundance, the localization of proteins can also change during infection. For example, during active immune signaling, IRF3 dimerizes after its activation by phosphorylation and translocates into nucleus, similar to NF-κB (Baldwin, 1996; Hiscott et al., 1999). Therefore, protein interactions can be temporally and spatially regulated. As indicated above, PTMs are also crucial to regulating protein function during immune signaling. For example, during HSV-1 infection, the viral tegument kinase US3 hyperphosphorylates IRF3 at an atypical site, thereby inhibiting IRF3 dimerization and blocking its function (Wang, Wang, Lin, & Zheng, 2013). Another example is the acetylation on the NLS of IFI16 by the acetyltransferase p300, which modulates the localization of IFI16 and therefore its interaction with viral DNA and other defense proteins (Li et al., 2012).

Here, in addition to the protocol for immunoaffinity purification of protein complexes, we provide a protocol of proteome analysis, thereby providing a workflow to systematically study protein interactions in the context of temporal proteome alterations. The protocols are provided in the order that a user would carry these out in the laboratory, starting with the cell disruption for bait solubilization, followed by the use of the resulting cell lysate for proteome analyses and IP-MS studies.

4. Materials

4.1 Cell collection and lysis

4.1.1 Cell preparation before lysis

1. Infected/DNA-stimulated mammalian cells in tissue culture dishes or flasks.
2. 50 mL conical tubes.
3. Dulbecco's phosphate-buffered saline (DPBS), pH 7.0–7.3, 1 × concentration (ice cold).
4. Plastic cell scraper.

5. Swinging bucket centrifuge set at 4 °C.
6. Protease and phosphatase inhibitor cocktail (P/PhIC) (Thermo Scientific).
7. Benzonase (Pierce).
8. Cell freezing buffer: 20 mM Na-HEPES, 1.2% (w/v) polyvinylpyrrolidone, pH 7.4, 1/100 (v/v), P/PhIC added to 1 × concentration immediately before use.

4.1.2 Freezing cell pellet

1. Styrofoam container with 50-mL conical tube rack insert.
2. 50-mL conical tubes.
3. Liquid nitrogen.
4. 18-gauge needle.
5. 10 mL syringe.

4.1.3 Cryogenic cell lysis

1. Retsch MM 301 Mixer Mill.
2. Two 10 mL jars and two 20 mm (tungsten carbide or stainless steel) grinding balls (Retsch, Newtown, PA).
3. 50-mL conical tubes.
4. Liquid nitrogen.
5. Foam ice bucket.
6. Long forceps.
7. Metal spatula chilled by liquid nitrogen.
8. Windex.
9. 100% Methanol.
10. 10% bleach.
11. Ultrapure water.
12. Dry ice.

4.1.4 Detergent cell lysis

1. Foam ice bucket.
2. Microcentrifuge set to 4 °C.
3. Microcentrifuge tubes
4. Wash buffer: 20 mM K-HEPES, pH 7.4, 0.11 M KOAc, 0.1% Tween-20 (v/v), 200 mM NaCl, 0.6% Triton X-100.
5. Lysis Buffer: 20 mM K-HEPES, pH 7.4, 0.11 M KOAc, 0.1% Tween-20 (v/v), 200 mM NaCl, 0.6% Triton X-100, 1 × P/PhIC, 100 U/mL benzonase.
6. Buffers might require optimization depending on cell type and protein of interest. See Section 5.3.3 for more information.

4.2 Proteome analysis

4.2.1 In-solution protein digestion

Store all solutions in glass containers that have been thoroughly washed with ultrapure water. Avoid using glassware that has been washed with detergents.

1. Whole cell lysate (see Section 5.1) or eluate from immunoaffinity isolation (see Section 5.3.2).
2. Refrigerated microcentrifuge capable of 14,000 x g (set at 20 °C).
3. LoBind pipet tips, 20–200 μL (Eppendorf).
4. Amicon Ultra-0.5 centrifugal filters, 30 kDa NMWL (Millipore).
5. MS grade water (Fisher).
6. 10% sodium deoxycholate (DOC): Prepare in MS grade water and protect from light.
7. Tris-HCl buffer: 0.2 M Tris-HCl, pH 8.0, in MS grade water. Store at 4 °C.
8. TUD wash buffer: Prepare buffer fresh before use by mixing 1.5 mL of 0.2 M Tris-HCl, 0.6 mL of 10% DOC, and 1.44 g urea. Makes enough solution for two samples, about 1.5 mL each.
9. ABC-DOC wash buffer: 0.05 M ABC, 2% DOC.
10. Trypsin, lyophilized MS-grade (Pierce). Suspended to 0.5 μg/μL, store at −80 °C.
11. Digestion Buffer: Mix 1 μL of 0.5 μg/μL trypsin stock and 99 μL of 0.05 M ABC. Prepare immediately before using.
12. 10% trifluoroacetic acid (TFA). Prepare in MS grade water, store at 4 °C.
13. Ethyl acetate. Handle carefully in a chemical hood and dispose of appropriately.
14. SpeedVac concentrator.
15. TOMY Shaker.

4.2.2 Peptide fractionation with SDB-RPS StageTips

1. Microcentrifuge.
2. LoBind pipet tips, 20–200 μL (Eppendorf).
3. 14-gauge needle (Hamilton #90514).
4. Syringe plunger, 100 μL (Hamilton #1162–02).
5. Empore SDB-RPS disks (3 M #2241).
6. 50% ethyl acetate/0.5% TFA in MS grade water.
7. 0.5% TFA in MS grade water.
8. Buffer 1: 0.10 M ammonium formate, 0.5% formic acid, 40% acetonitrile (ACN) in water.

9. Buffer 2: 0.15 M ammonium formate, 0.5% formic acid, 60% acetonitrile in water.
10. Buffer 3: 5% ammonium hydroxide and 80% acetonitrile in water.
11. FA solution: 1% formic acid and 1% acetonitrile in water.
12. Autosampler vials (Thermo Scientific).

4.2.3 Nanoliquid chromatography tandem mass spectrometry analysis

1. Nanoflow HPLC system such as Dionex Ultimate, Waters nano Acuity, or Agilent 1200 series.
2. Mobile phase A (MPA): 0.1% FA/99.9% water. Store in amber bottle for up to 6 months.
3. Mobile phase B (MPB): 0.1% FA/97% ACN/2.9% water. Store in amber bottle for up to 6 months.
4. Analytical column such as Acclaim PepMap RSLC 75 μm ID × 25 cm (Dionex).
5. Mass spectrometer such as LTQ-Orbitrap Velos or Q Exactive HF (Thermo Fisher Scientific).
6. Nanospray ESI source (Thermo Fisher Scientific).
7. SilicaTip Emitter, Tubing (OD × ID) 360 μm × 20 μm; Tip (ID) 10 μm (New Objective).

4.3 Immunoaffinity purification analysis

4.3.1 Conjugation of antibody to magnetic beads

4.3.1.1 Dynabeads conjugation

1. Dynabeads M-270 Epoxy (Invitrogen).
2. Round bottomed 2 mL microcentrifuge tubes (Eppendorf).
3. Magnetic tube rack.
4. Tube shaker (e.g., TOMY shaker).
5. Rotating tube rack in 30 °C incubator.
6. Appropriate antibody.
7. 0.1 M Sodium Phosphate Buffer (pH 7.4): 19 mM NaH_2PO_4, 81 mM Na_2HPO_4 in water.
8. 3 M Ammonium sulfate: Prepared in 0.1 M Sodium Phosphate Buffer.
9. 100 mM Glycine-HCl (pH 2.5): Prepare in water and adjust pH with HCl, sterilize with 0.2 μm filter.
10. Phosphate Buffered Saline (PBS).
11. 0.02% sodium azide (NaN_3) in PBS.
12. 0.5% Triton X-100 in PBS.
13. 10 mM Tris-HCl (pH 8.8): Prepare in water and adjust pH with HCl.
14. 100 mM Triethylamine: Prepare immediately before use in water.

4.3.1.2 Protein A/G magnetic beads conjugation

1. Pierce™ Protein A/G Magnetic Beads (ThermoFisher).
2. Round bottomed 2 mL microcentrifuge tubes (Eppendorf).
3. Magnetic tube rack.
4. Rotating tube rack at 4 °C.
5. Appropriate antibody.
6. TBST buffer: Tris-buffered saline (TBS, ThermoFisher) containing 0.05% Tween-20 detergent.

4.3.2 Immunoaffinity purification

1. Polytron homogenizer (e.g., PT 10-35 Polytron from Kinematica).
2. Centrifuge and rotor rated to spin 8000 × g at 4 °C (rotor must fit a 50 mL conical tube).
3. Rotating tube rack at 4 °C.
4. Heat block at 70 °C.
5. Tube shaker (e.g., TOMY shaker).
6. Vacuum evaporator (e.g., SpeedVac).
7. Magnetic tube rack for tubes under 2 mL and bar magnet for larger volume tubes.
8. Round bottomed microcentrifuge 2 mL tubes (for small scale IP), Protein LoBind tubes 5 mL (for large scale IP) (Eppendorf).
9. Frozen cell powder or cell lysate (see Sections 5.1.2 and 5.1.3).
10. Optimized lysis buffer and wash buffer (see Section 5.3.3).
11. Antibody conjugated magnetic beads (see Section 5.3.1).
12. Elution buffer: 1 × LDS Elution Buffer (4 × LDS elution buffer stock: 0.666 g of Tris-HCl, 0.682 g of Tris-Base, 0.8 g of LDS, and 0.006 g of EDTA (free acid) dissolved in ultrapure dH_2O to a final volume of 10 mL). Store at −20 °C.
13. 500 mM Tris(2-carboxyethyl)phosphine (TCEP).
14. 1 M chloroacetamide (CAM): dissolve in water and store at −20 °C.

5. Methods

5.1 Cell collection and lysis

Here we describe two methods of mammalian cell lysis that we have previously used for both proteome and protein-protein interaction studies: cryogenic lysis (Crow & Cristea, 2017; Diner et al., 2016; Diner, Li, et al., 2015; Diner, Lum, et al., 2015) and chemical lysis with detergent-containing buffers (Lum et al., 2018). Mechanical cell lysis may also be used, but we have found these methods to be effective at preserving the integrity of protein complexes while reducing nonspecific interactions.

5.1.1 Cell preparation before lysis

1. Aspirate media from tissue culture dish and add a small volume (~3 mL) of ice-cold DPBS. Quickly scrape cells and transfer to 50-mL conical tube. Repeat to collect any remaining cells.
2. Pellet cells at 250 × *g* at 4 °C.
3. Wash cell pellets with 20 mL DPBS and transfer to a pre-weighed 50-mL conical tube.
4. Pellet cells as before and aspirate media. If immediately lysing cells with a detergent-containing lysis buffer, continue directly to Section 5.1.3.
5. Weigh tube to determine wet cell pellet weight.
6. Prepare liquid nitrogen bath. Keep cells on ice.
7. Place a fresh 50-mL conical tube in the Styrofoam rack. Fill tube halfway with liquid nitrogen and leave uncovered. Fill the bottom of the rack with liquid nitrogen.
8. Add 100 μL of freezing buffer per gram of cells (see Note 1). Pipet dropwise into tube filled with liquid nitrogen.
9. Using the 18-gauge needle, punch holes in the conical tube cap before re-capping the tube with liquid nitrogen and cell material. Secure cap and gently agitate the rack in a fume hood to allow liquid nitrogen to slowly evaporate.

5.1.2 Cryogenic cell lysis

1. Clean the spatula, Retsch Mixer Mill jars, and grinding balls using the solutions in the following order: Windex, ultrapure water, 10% bleach, ultrapure water, and 100% methanol. Allow to air dry.
2. Use a liquid nitrogen bath in a foam ice bucket to cool both the mixing jars and the grinding balls. The jars are sufficiently cool when liquid nitrogen appears to stop bubbling. Remove materials from liquid nitrogen using long forceps.
3. Quickly transfer frozen cell pellets into mixing jar. Cell pellets can fill up to a maximum of one-third of the volume of the jar for optimal grinding—about 2.5 g frozen pellets per 10 mL jar. Place a single frozen grinding ball on the cell pellets, close the jar, and cool in liquid nitrogen once more.

4. Place jars in Retsch Mixer Mill. Remember to use an empty jar without a grinder ball as a balance, if only lysing one sample. Perform lysis with 10 cycles of 2 min and 30 s each at a frequency of 30 Hz. Re-cool jars in liquid nitrogen between each cycle, and ensure jars are securely closed.
5. Open the jar and use a chilled spatula to transfer frozen cell powder to a 50 mL conical tube chilled on dry ice. Work quickly to avoid thawing sample. Store at −80 °C.

5.1.3 Detergent cell lysis

1. If using frozen samples, thaw cell pellets on ice.
2. Add 500 μL of lysis buffer to each sample and vortex for 20 s.
3. Rock samples at room temperature for 10 min.
4. Vortex samples for 20 s.
5. Incubate samples on ice for 10 min.
6. Repeat steps 4 and 5.
7. Transfer lysates to microcentrifuge tubes and centrifuge at 8000 × *g* for 10 min at 4 °C.
8. Transfer clarified lysate to fresh microcentrifuge tubes.
9. Proceed directly to mass spectrometry preparation and analysis in the next section if performing proteome abundance quantifications. If performing protein interaction studies, please see Section 5.3 before proteomic analysis.

5.2 Mass spectrometry analysis

5.2.1 In-solution protein digestion

Day 1

1. Set microcentrifuge to 20 °C.
2. Add 400 μL TUD buffer to each eluate sample.
3. Transfer each sample to separate Amicon Ultra-0.5 filters and centrifuge at 14,000 × *g* for 10 min or until liquid level reaches minimum line, about 25 μL.
4. Discard flow-through. Add 400 μL TUD buffer to filter and repeat centrifugation (see Note 2).
5. Discard flow-through. Add 300 μL TUD buffer to filter and repeat as above.
6. Repeat with another 300 μL TUD buffer.
7. Add 200 μL of ABC-DOC buffer to filter and spin at 14,000 × *g* for 10 min. Check that the retained volume is at the minimum.

8. Transfer filters to fresh collection tubes and add 100 μL of digestion buffer. Mix on TOMY shaker for 1 min.
9. Seal tube caps with parafilm and incubate overnight in 37 °C water bath.

Day 2

1. Centrifuge filters at 14,000 × *g* for 5 min to recover digested peptides.
2. Add 25 μL MS-grade water and centrifuge at 14,000 × *g* for 5 min.
3. Repeat above step.
4. Keep flow-through with the digested peptides and discard the filter unit.
5. Add an equal volume of ethyl acetate to the sample. This should be between 100 and 150 μL, measured with LoBind pipet tip.
6. Adjust samples to 0.5% TFA with 10% TFA solution.
7. Briefly vortex samples then mix on TOMY shaker for 2 min.
8. Centrifuge at 14,000 × *g* for 5 min.
9. Recover denser aqueous phase by pipetting from the bottom of the tube (see Note 3).
10. Proceed to "Peptide fractionation with SDB-RPS StageTips" with recovered aqueous phase.

5.2.2 Peptide fractionation with SDB-RPS StageTips

1. Cut Empore SDB-RPS disks using a 14-gauge needle. Prepare one StageTip for each sample by depositing a single Empore SDB-RPS disk into the bottom of a 200 μL pipette tip and gently tamping down with a syringe plunger (see Note 4).
2. Add half of the sample to the StageTip. Centrifuge at 2000 × *g* until all solution passes through the Empore disk (see Note 5).
3. Add remaining sample to the StageTip and repeat centrifugation.
4. Wash disk with 100 μL of 50% ethyl acetate/0.5% TFA.
5. Wash disk with 100 μL of 0.5% TFA.
6. Add 50 μL Elution Buffer 1 to disk and collect eluate in an autosampler vial.
7. Repeat previous step with Elution Buffer 2.
8. Repeat once more with Elution Buffer 3.
9. Concentrate samples by vacuum centrifugation until the vial is almost completely dry.
10. Add FA solution to achieve final volume of 9 μL. Quickly vortex to mix.
11. Immediately perform nLC-MS/MS analysis or store at −80 °C.

5.2.3 Nanoliquid chromatography tandem mass spectrometry analysis

For protein identification and abundance quantification, use an nLC-MS instrument with high resolution and high mass accuracy, such as those listed in Section 4.2.3. One consideration in designing a data-dependent acquisition method is if protein abundances should be measured using label-free quantification or a labeling method, such as with tandem mass tags (TMT). If the goal of the study is to compare many samples from various treatments, infection time points, or cellular fractions, TMT can be an excellent method to multiplex samples and increase instrument efficiency and accuracy of relative quantification.

5.3 Immunoaffinity purification analysis

5.3.1 Conjugation of antibody to magnetic beads

We provide protocols for two different types of magnetic beads that we previously used when studying DNA sensors, epoxy Dynabeads and Protein A/G magnetic beads. Most of our previous experiments have been performed using Epoxy Dynabeads. However, given a recent decrease in the performance of these beads, we reevaluated different types of resins. Our comparison experiments showed that Protein A/G magnetic beads had higher antibody conjugation efficiency and better purification of bait protein. As an alternative to the above mentioned beads, if the protein of interest is tagged with GFP, immunoaffinity purification of the bait via the GFP tag can also be performed using GFP-trap beads (ChromoTek), in which case the conjugation step is not needed.

The conjugation protocol for Dynabeads and Protein A/G magnetic beads can be used for conjugating antibodies directly targeting proteins of interest, antibodies against tags (e.g., GFP, FLAG), or IgG. The antibodies should be affinity purified, and, when dealing with Dynabeads, the solution in which the antibody is stored is critical. As the binding to epoxy occurs through primary amines, ensure that the antibodies used are not stored in Tris Buffer. Also, if the antibody storing solution contains other proteins, such as BSA, it would compete with the antibody for binding to the beads and would affect the efficiency of the immunoaffinity purification.

When determining the amount of beads needed, one must consider the amount of cellular material being used and the abundance of the protein of interest. For Dynabeads, a small scale pilot immunoaffinity purification experiment using less than 1×10^7 cells may require only 1–2 mg of beads, while a larger scale IP would require 5–10 mg of beads. If the protein of interest is not highly abundant, then using 10–20 mg of beads could be appropriate. For Protein A/G magnetic beads, the beads are suspended as slurry in the vial. A small scale pilot experiment may require 20–25 μL (0.20–0.25 mg) bead slurry, while a larger scale experiment may need 50 μL (0.50 mg) or higher amount of bead slurry.

The conjugation steps can be done at room temperature. The conjugation of Dynabeads is performed overnight, and the conjugated beads can be stored at 4 °C. The Protein A/G magnetic beads should be used immediately after the 1 h conjugation step.

5.3.1.1 Dynabeads conjugation: Day 1

1. Weigh out the appropriate amount of Dynabeads into a 2 mL round-bottom tube (see Note 6).
2. Wash the beads with 1 mL of Sodium Phosphate Buffer (pH 7.4) and vortex for 30 s.
3. Mix vigorously with TOMY shaker for 15 min.
4. Magnetically separate beads from solution. Once the solution becomes clear and the beads are collected on the magnet side, remove and discard the buffer using a pipette or vacuum (see Note 7).
5. Add 1 mL of fresh Sodium Phosphate Buffer and vortex for 30 s.
6. Magnetically separate beads from solution and remove the buffer as above.
7. Conjugate beads at a ratio of ~20 μL solution per mg beads. The reaction consists of three parts of solutions: antibody solution (Note 8), Sodium Phosphate Buffer, and Ammonium Sulfate, added in that order (see Note 9).
8. Seal tube caps with parafilm and rotate the slurry overnight (16 h) at 30 ° C. *Note*: Use a rotating rack that moves the tube end-over-end to help keep the beads suspended in the antibody solution during the conjugation period.

Day 2

9. Place the tube on the magnetic rack and remove the supernatant as above. It is an option to keep the supernatant flow-through to determine the efficiency of bead conjugation via SDS-PAGE.
10. Wash the beads in 1 mL of Sodium Phosphate Buffer by gentle pipetting then magnetically separate beads (see Note 10). Once the solution is clear, remove and discard the buffer.
11. Wash the beads with 1 mL 100 mM Glycine-HCl, pH 2.5. This is a *fast* wash: remove the solution as soon as the beads are attached to the magnet and immediately proceed with the next wash step.

12. Wash the beads with 1 mL of 10 mM Tris-HCl, pH 8.8 using the same procedure as described in step 10.
13. Wash the beads in 100 mM Triethylamine. This is also a *fast* wash: proceed as is described in step 11.
14. Wash the beads four times with PBS using 1 mL for each wash. Remove the buffer following each wash.
15. Add PBS containing 0.5% Triton X-100 to the beads and mix with TOMY shaker for 15 min.
16. Wash the beads twice with 1 mL PBS. Remove the buffer following each wash.
17. Suspend the beads in PBS with 0.02% NaN_3 and store at 4 °C until their use (see Note 11).

5.3.1.2 Protein A/G magnetic beads conjugation

1. Resuspend the bead slurry completely by pipetting up and down. Aliquot out the appropriate amount of beads into a 2 mL round-bottom tube. DO NOT vortex the beads at any point until the protein complex elution step.
2. Wash the beads three times by adding 500 μL of TBST. Magnetically separate beads from solution and remove the buffer following each wash.
3. After the final wash, remove the tube from the magnet and begin conjugating the antibody.
4. The total volume of the conjugation reaction will be 500 μL. The solution consists of antibody and TBST. Add 10 μg of antibody per 25 μL (0.25 mg) beads slurry. The volume of TBST is 500 μL—volume of antibody (see Note 12).
5. Incubate the beads for 1 h on a rotating tube rack at 4 °C, and the beads should be immediately used after the conjugation (see Note 13).

5.3.2 Immunoaffinity purification

Once the cells are collected and the beads are conjugated, the immunoaffinity purification step can proceed. A small scale IP can be done in a 1 mL volume using a 2 mL round bottomed microcentrifuge tube, while a large scale IP can be done in a 3–5 mL volume using a 5 mL Protein LoBind tube. The following protocol is for cryogenically lysed samples (see Section 5.1.2). If the cells were lysed using a detergent (see Section 5.1.3) and the samples are already in a lysis buffer solution, prepare the Wash Buffer in step 1, and proceed the protocol from step 8.

1. Prepare (i) IP Buffer for suspending the cell powder and (ii) Wash Buffer for washing the magnetic beads prior to and after the immunoaffinity purification. Chill buffers on ice. The composition of the IP Buffer is optimized as described in Section 5.3.3.
2. Bring the frozen cell powder from the −80 °C storage, and keep the cells on ice.
3. Add the IP Buffer to the frozen cell powder and periodically swirl the tube until the powder is solubilized.
4. Prepare a Polytron for the homogenization of the cell lysate. Rinse the Polytron sequentially with methanol and ultrapure dH_2O, then fill a beaker with ultrapure dH_2O and allow the Polytron to run inside the beaker for 10 s.
5. To homogenize the cell lysate, run the Polytron twice for 15 s at a speed of 22.5 k RPM with cooling on ice for 30 s between the two steps (see Note 14).
6. If processing more than one sample, perform a rinse and a wash step in ultrapure dH_2O (see step 4) to avoid cross-contamination of samples.
7. Slowly spin the homogenized cell lysate at 4 °C for 5–10 min to settle the foam.
8. To separate the insoluble fraction, centrifuge the cell lysate at 8000 × *g* at 4 °C for 10 min.
9. While waiting for the completion of the centrifugation step, equilibrate beads by wash three times with 1 mL cold Wash Buffer by gentle pipetting. To avoid drying of the beads, suspend the washed beads in 100–200 μL of Wash buffer and put them on ice.
10. Once the centrifugation (step 8) is completed, move the clarified cell supernatant to a new tube by pouring or pipetting (see Note 15). Retain the pellet and 1–5% of the supernatant for evaluating IP efficiency.
11. Add the appropriate amounts of washed beads to the cell supernatant by pipetting.
12. Place the tube containing the cell lysate and the beads on a slowly rotating tube rack at 4 °C and allow for 1 h of incubation.
13. Once the 1 h incubation is completed, magnetically separate the beads and remove the flow through supernatant by pipetting, saving it in a fresh microcentrifuge tube to later evaluate IP efficiency.
14. Wash the beads three times with 1 mL Wash Buffer. After the third wash, move the bead slurry to a new 2 mL tube.
15. Once the beads are transferred to the new tube, wash the beads two more times with 1 mL Wash Buffer, and remove the Wash solution by pipetting.

16. Add 1 mL of PBS and transfer the beads to a new 1.5 mL round bottom tube. Wash once more with 1 mL PBS to remove any residual detergent and improve the efficiency of elution.
17. Add 40 μL Elution Buffer to the beads and incubate the tube at 70 °C for 10 min (see Note 16).
18. Let the tubes cool down at room temperature for 30 s and put the tubes on a TOMY shaker and mix for 10 min.
19. Magnetically separate the beads from the eluate and transfer the eluate to a new microcentrifuge tube. Retain 5% of the elution for evaluating IP efficiency by SDS-PAGE. The rest of eluate can be prepared for mass spectrometry analysis.
20. Before the mass spectrometry sample preparation, the protein samples need to be reduced and alkylated. Add TCEP (1/20 of the total volume if using 500 mM stock) and CAM (1/10 of the total volume if using 500 mM stock) to reach a final concentration of 25 mM and 50 mM, respectively. Incubate at 70 °C for 20 min (see Note 17).
21. Proceed to preparation for mass spectrometry analysis (Section 5.2) or store samples at −20 °C.

5.3.3 Optimization of lysis buffer

The composition of the lysis buffer, i.e., detergents and salt concentration, determine the stringency of the buffer. Overall, the buffer must be stringent enough to allow the solubilization of the bait protein, but it also must be mild enough to prevent the disruption of protein interactions. It is recommended to conduct small scale IPs using lysis buffers with varying stringency prior to performing larger scale experiments. A start (mild) lysis buffer can be 1 × TBT with salt, 0.6% Triton X-100, 200 mM NaCl, 1 × PIC and PhIC, and 100 U/mL Benzonase. To increase the stringency, NaCl concentration can be increase, usually up to 500 mM, and sodium deoxycholate can be added up to 0.5%. Table 1 lists examples of lysis buffers previously used to isolate DNA sensors and to study their protein interactions using various tags and cell types.

The following protocol is for cryogenically lysed samples (see Section 5.1.2). The cells can also be lysed using lysis buffers after being harvested (see Section 5.1.3) and taken straight through immunoaffinity purification.

1. Prepare cryogenically ground cell powder as described (see Section 5.1.2). Split the frozen cell powder equally into pre-chilled tubes.
2. Add different lysis buffers to each sample using 5 mL buffer per 1 g cryo-lysed cells (*see* Note 18).

Table 1 Examples of lysis buffers for isolating DNA sensors.

DNA sensor	Tag	Cell type	Lysis buffer	Reference
TLR9	Myc-tagged	HEK293T	50 mM Tris–HCl, pH 8.0, 150 mM NaCl, 5 mM EDTA, 1% digitonin, and 1 × PIC	Lee, Kang, and Kim (2016)
IFI16	**(A,B,D)** Endogenous **(C,E)**GFP-tagged	**(A)**CEM-T **(B,C,D,E)**HFF **(C)**THP-1 **(C)**HEK293	20 mM K-Hepes, pH 7.4, 0.11 M KOAc, 2 mM $MgCl_2$, 0.1% Tween-20, 1 μM $ZnCl_2$, 1 μM $CaCl_2$, 0.6% Triton X-100, 200 mM NaCl, (Pierce), 1 × PIC, 1 × PHIC, 100 U/mL Benzonase.	A. Li, et al. (2012) B. Diner, Lum, et al. (2015) C. Diner, Li, et al. (2015) D. Orzalli et al. (2015) E. Diner, et al. (2016)
IFIX	GFP-tagged	HEK293	20 mM K-Hepes, pH 7.4, 0.11 M KOAc, 2 mM $MgCl_2$, 0.1% Tween-20, 1 μM $ZnCl_2$, 1 μM $CaCl_2$, 0.6% Triton X-100, 200 mM NaCl, (Pierce), 1 × PIC, 1 × PHIC, 100 U/mL Benzonase.	A. Diner, Li, et al. (2015) B. Crow and Cristea (2017)
AIM2	GFP-tagged	HEK293	20 mM K-Hepes, pH 7.4, 0.11 M KOAc, 2 mM $MgCl_2$, 0.1% Tween-20, 1 μM $ZnCl_2$, 1 μM $CaCl_2$, 0.6% Triton X-100, 200 mM NaCl, (Pierce), 1 × PIC, 1 × PHIC, 100 U/mL Benzonase.	Diner, Li, et al. (2015)
cGAS	**(A,B)**FLAG-tagged **(C)**HA-tagged **(D)**GFP-tagged	**(A,B,C)**HEK-293T **(D)**HFF	**(A)** 20 mM K-Hepes, pH 7.4, 0.11 M KOAc, 2 mM $MgCl_2$, 0.1% Tween-20, 1 μM $ZnCl_2$, 1 μM $CaCl_2$, 0.6% Triton X-100, 200 mM NaCl, (Pierce), 1 × PIC, 1 × PHIC, 100 U/mL Benzonase. **(B)** 1% NP-40 buffer supplemented with a complete protease inhibitor cocktail (Roche). **(C)** 50 mM Tris, pH 7.4, 150 mM NaCl, 1% Triton X-100, 1 mM EDTA (pH 8.0), 1 mM PMSF, 1 mM Na_3VO_4, 1 mM of NaF. **(D)** *Cytoplasmic fraction*: (20 mM HEPES, pH 7.4, 10 mM KCl, 2 mM $MgCl_2$, 1 mM DTT, 0.5% NP-40, 1 × PIC and 1 × PhIC), 100 U/mL Benzonase. *Nuclear fraction*: cytoplasmic lysis buffer supplemented with 200 mM NaCl, 1% Triton X-100, 0.11 M KOAc, 1% Tween-20, 100 U/mL Benzonase.	A. Diner, et al. (2016) B. Seo et al. (2018) C. Liu et al. (2018) D. Lum, et al. (2018)

3. Homogenize the cell powder in different lysis buffers by intermittent 30 s vertexing for three times with a 5 min cooling on ice in between the two vortex steps.
4. Separate the insoluble fraction by centrifugation at 8000 × *g* for 10 min at 4 °C.
5. Collect and save the supernatant as the lysate fraction.
6. Resolve the pellet with 1 × NUPAGE sample buffer. Sonicate the suspension, and then boil at 95 °C for 5 min. Save the solution as the pellet fraction.
7. Load equivalent amounts (5–10%) of lysate and pellet on PAGE gel and conduct western blotting, probing with antibodies against the tag or the endogenous protein.
8. Compare the relative amounts of bait protein in the lysate and pellet fractions. The lysis buffer which gives the highest amount of bait protein in the lysate fraction should be used as the lysis buffer for the formal IP experiment.

Notes

1. Add 100 μL of freezing buffer for cell pellets that weigh less than 1 g.
2. Spray the buffer down the walls of the filter in an effort to wash all peptides down into the TUD buffer.
3. Avoid the top layer organic phase (containing ethyl acetate) and the interphase.
4. Be sure to not push the disk all the way to the end of the pipet tip—there should be a few mm of space left. Each disk is capable of binding about 25 μg of peptides. If needed, more disks can be layered on top.
5. Loading samples into and washing the Empore disks can be done by either applying pressure with a small plastic syringe or centrifuging the StageTip in a collection tube with an adapter.
6. 10 mg is the minimal amount to conjugate, as 200 μL is the least volume for the beads to be suspended in antibody solution during the conjugation (refer to step 7).
7. Removing buffer by vacuum is fast and convenient but risky because the beads may be lost. To be safe, one can always use pipette.
8. To calculate the necessary volume of antibody solution to use, for purified high affinity antibodies (e.g., in-house generated anti-GFP antibodies) use 3–5 μg Ab/mg beads. For commercially available antibodies that may not have as high affinity for binding, use 5–8 μg Ab/mg beads, and for conjugation of IgG use 8–10 μg IgG/mg beads.

9. The volume of 0.1 M Sodium Phosphate Buffer depends on that of the antibody solution. For example, when conjugating 10 mg of beads (200 μL total solution required), subtract the volume of antibody solution and 66.66 μL 3 M Ammonium Sulfate from 200 μL. The Ammonium Sulfate solution is added last so as to reach a final concentration of 1 M.
10. To avoid losing beads inside the pipette tip, after the beads attach to the magnet wall, wash the tip using the buffer inside the tube, and wait again for the beads to settle on the magnet.
11. It is recommended to use the beads within 2 weeks of conjugation, as their isolation efficiency will decrease if the beads are stored for extensive periods of time (e.g., ~40% decrease after 1 month of storage). If the beads will be used immediately, NaN_3 is not necessary.
12. 25 μL (0.25 mg) is the least amount of beads to operate on, as a lower amount of beads would be difficult to be collected on the magnet.
13. Protein A/G magnetic beads can also be used in the following way: incubate antibody and cell lysate first, and then add the beads and incubate again. Our results have shown that there is no significant difference in IP efficiency, and we only provide the protocol of pre-conjugation here in this chapter.
14. As foam is formed during this step, ensure that the volume of the cell lysate does not occupy more than a third of the tube to avoid over flow. Small scale IP samples can be homogenized in 5 or 15 mL tubes, while larger scale IP sample can be in 50 mL conical tubes.
15. Carefully analyze the appearance of the cell supernatant and ensure that there are no floating or mixed insoluble particles in the supernatant. If such particles are observed, remove them by carefully pipetting before the IP. Particles may clog the beads and interfere with the affinity purification.
16. The elution buffer is 1 × NUPAGE buffer without DTT (in-gel digestion) or 1 × LDS Elution Buffer (in-solution digestion).
17. This step does not belong to the IP process, but it is convenient to conduct during the same day.
18. Usually 0.5–1 mL is good to resolve one 15 cm dish of adherent cells.

Acknowledgments

We are grateful for funding provided by the NIH, NIGMS grant GM114141 and the Edward Mallinckrodt Foundation to I.M.C., China Scholarship Council (CSC) scholarship 201506210052 to B.S., and NIH training grant from NIGMS (T32GM007388) to T.R.H.

References

Aguirre, S., Luthra, P., Sanchez-Aparicio, M. T., Maestre, A. M., Patel, J., Lamothe, F., et al. (2017). Dengue virus NS2B protein targets cGAS for degradation and prevents mitochondrial DNA sensing during infection. *Nature Microbiology*, *2*, 17037. https://doi.org/10.1038/nmicrobiol.2017.37.

Baldwin, A. S., Jr. (1996). The NF-kappa B and I kappa B proteins: New discoveries and insights. *Annual Review of Immunology*, *14*, 649–683. https://doi.org/10.1146/annurev.immunol.14.1.649.

Beys-da-Silva, W. O., Rosa, R. L., Santi, L., Berger, M., Park, S. K., Campos, A. R., et al. (2018). Zika virus infection of human mesenchymal stem cells promotes differential expression of proteins linked to several neurological diseases. *Molecular Neurobiology*, 1–10. https://doi.org/10.1007/s12035-018-1417-x.

Bonham, K. S., Orzalli, M. H., Hayashi, K., Wolf, A. I., Glanemann, C., Weninger, W., et al. (2014). A promiscuous lipid-binding protein diversifies the subcellular sites of toll-like receptor signal transduction. *Cell*, *156*(4), 705–716. https://doi.org/10.1016/j.cell.2014.01.019.

Cabral, J. M., Oh, H. S., & Knipe, D. M. (2018). ATRX promotes maintenance of herpes simplex virus heterochromatin during chromatin stress. *eLife*, 7, e40228. https://doi.org/10.7554/eLife.40228.

Chen, M., Meng, Q., Qin, Y., Liang, P., Tan, P., He, L., et al. (2016). TRIM14 inhibits cGAS degradation mediated by selective autophagy receptor p62 to promote innate immune responses. *Molecular Cell*, *64*(1), 105–119. https://doi.org/10.1016/j.molcel.2016.08.025.

Chojnowski, A., Sobota, R. M., Ong, P. F., Xie, W., Wong, X., Dreesen, O., et al. (2018). 2C-BioID: An advanced two component BioID system for precision mapping of protein Interactomes. *iScience*, *10*, 40–52. https://doi.org/10.1016/j.isci.2018.11.023.

Cristea, I. M., Moorman, N. J., Terhune, S. S., Cuevas, C. D., O'Keefe, E. S., Rout, M. P., et al. (2010). Human cytomegalovirus pUL83 stimulates activity of the viral immediate-early promoter through its interaction with the cellular IFI16 protein. *Journal of Virology*, *84*(15), 7803–7814. https://doi.org/10.1128/JVI.00139-10.

Cristea, I. M., Williams, R., Chait, B. T., & Rout, M. P. (2005). Fluorescent proteins as proteomic probes. *Molecular & Cellular Proteomics*, *4*(12), 1933–1941. https://doi.org/10.1074/mcp.M500227-MCP200.

Crow, M. S., & Cristea, I. M. (2017). Human antiviral protein IFIX suppresses viral gene expression during herpes simplex virus 1 (HSV-1) infection and is counteracted by virus-induced proteasomal degradation. *Molecular & Cellular Proteomics*, *16*(4 suppl 1), S200–S214. https://doi.org/10.1074/mcp.M116.064741.

Cuchet-Lourenco, D., Anderson, G., Sloan, E., Orr, A., & Everett, R. D. (2013). The viral ubiquitin ligase ICP0 is neither sufficient nor necessary for degradation of the cellular DNA sensor IFI16 during herpes simplex virus 1 infection. *Journal of Virology*, *87*(24), 13422–13432. https://doi.org/10.1128/JVI.02474-13.

Deretic, V. (2012). Autophagy as an innate immunity paradigm: Expanding the scope and repertoire of pattern recognition receptors. *Current Opinion in Immunology*, *24*(1), 21–31. https://doi.org/10.1016/j.coi.2011.10.006.

Diner, B. A., Li, T., Greco, T. M., Crow, M. S., Fuesler, J. A., Wang, J., et al. (2015). The functional interactome of PYHIN immune regulators reveals IFIX is a sensor of viral DNA. *Molecular Systems Biology*, *11*(1), 787. https://doi.org/10.15252/msb.20145808.

Diner, B. A., Lum, K. K., Javitt, A., & Cristea, I. M. (2015). Interactions of the antiviral factor interferon gamma-inducible protein 16 (IFI16) mediate immune signaling and herpes simplex virus-1 immunosuppression. *Molecular & Cellular Proteomics*, *14*(9), 2341–2356. https://doi.org/10.1074/mcp.M114.047068.

Diner, B. A., Lum, K. K., Toettcher, J. E., & Cristea, I. M. (2016). Viral DNA sensors IFI16 and cyclic GMP-AMP synthase possess distinct functions in regulating viral gene expression, immune defenses, and apoptotic responses during herpesvirus infection. *mBio*, 7(6), e01553-16. https://doi.org/10.1128/mBio.01553-16.

Drayman, N., Karin, O., Mayo, A., Danon, T., Shapira, L., Rafael, D., et al. (2017). Dynamic proteomics of herpes simplex virus infection. *mBio*, *8*(6), e01612-17. https://doi.org/10.1128/mBio.01612-17.

Everett, R. D. (2016). Dynamic response of IFI16 and promyelocytic leukemia nuclear body components to herpes simplex virus 1 infection. *Journal of Virology*, *90*(1), 167–179. https://doi.org/10.1128/JVI.02249-15.

Fernandes-Alnemri, T., Yu, J. W., Datta, P., Wu, J., & Alnemri, E. S. (2009). AIM2 activates the inflammasome and cell death in response to cytoplasmic DNA. *Nature*, *458*(7237), 509–513. https://doi.org/10.1038/nature07710.

Forrest, C., Hislop, A. D., Rickinson, A. B., & Zuo, J. (2018). Proteome-wide analysis of CD8 + T cell responses to EBV reveals differences between primary and persistent infection. *PLoS Pathogens*, *14*(9). e1007110. https://doi.org/10.1371/journal.ppat.1007110.

Gillen, J. G., & Nita-Lazar, A. (2017). Composition of the myddosome during the innate immune response. *Journal of Immunology*, *198*(1 Suppl.), 75.15.

Greco, T. M., Diner, B. A., & Cristea, I. M. (2014). The impact of mass spectrometry-based proteomics on fundamental discoveries in virology. *Annual Review of Virology*, *1*(1), 581–604. https://doi.org/10.1146/annurev-virology-031413-085527.

Hemmi, H., Takeuchi, O., Kawai, T., Kaisho, T., Sato, S., Sanjo, H., et al. (2000). A toll-like receptor recognizes bacterial DNA. *Nature*, *408*(6813), 740–745. https://doi.org/10.1038/35047123.

Hiscott, J., Pitha, P., Genin, P., Nguyen, H., Heylbroeck, C., Mamane, Y., et al. (1999). Triggering the interferon response: The role of IRF-3 transcription factor. *Journal of Interferon & Cytokine Research*, *19*(1), 1–13. https://doi.org/10.1089/107999099314360.

Hornung, V., Ablasser, A., Charrel-Dennis, M., Bauernfeind, F., Horvath, G., Caffrey, D. R., et al. (2009). AIM2 recognizes cytosolic dsDNA and forms a caspase-1-activating inflammasome with ASC. *Nature*, *458*(7237), 514–518. https://doi.org/10.1038/nature07725.

Ishikawa, H., Ma, Z., & Barber, G. N. (2009). STING regulates intracellular DNA-mediated, type I interferon-dependent innate immunity. *Nature*, *461*(7265), 788–792. https://doi.org/10.1038/nature08476.

Jin, T., Perry, A., Smith, P., Jiang, J., & Xiao, T. S. (2013). Structure of the absent in melanoma 2 (AIM2) pyrin domain provides insights into the mechanisms of AIM2 autoinhibition and inflammasome assembly. *The Journal of Biological Chemistry*, *288*(19), 13225–13235. https://doi.org/10.1074/jbc.M113.468033.

Kawai, T., Adachi, O., Ogawa, T., Takeda, K., & Akira, S. (1999). Unresponsiveness of MyD88-deficient mice to endotoxin. *Immunity*, *11*(1), 115–122.

Krug, A., Luker, G. D., Barchet, W., Leib, D. A., Akira, S., & Colonna, M. (2004). Herpes simplex virus type 1 activates murine natural interferon-producing cells through toll-like receptor 9. *Blood*, *103*(4), 1433–1437. https://doi.org/10.1182/blood-2003-08-2674.

Kulej, K., Avgousti, D. C., Sidoli, S., Herrmann, C., Della Fera, A. N., Kim, E. T., et al. (2017). Time-resolved global and chromatin proteomics during herpes simplex virus type 1 (HSV-1) infection. *Molecular & Cellular Proteomics*, *16*(4 suppl 1), S92–S107. https://doi.org/10.1074/mcp.M116.065987.

Lee, Y. R., Kang, W., & Kim, Y. M. (2016). Detection of interaction between toll-like receptors and other transmembrane proteins by co-immunoprecipitation assay. In C. McCoy (Ed.), *Toll-like receptors. Methods in molecular biology: Vol. 1390.* (pp. 107–120). New York, NY: Humana Press. https://doi.org/10.1007/978-1-4939-3335-8_7.

Leitner, A., Walzthoeni, T., Kahraman, A., Herzog, F., Rinner, O., Beck, M., et al. (2010). Probing native protein structures by chemical cross-linking, mass spectrometry, and bioinformatics. *Molecular & Cellular Proteomics*, *9*(8), 1634–1649. https://doi.org/10.1074/mcp.R000001-MCP201.

Li, T., Chen, J., & Cristea, I. M. (2013). Human cytomegalovirus tegument protein pUL83 inhibits IFI16-mediated DNA sensing for immune evasion. *Cell Host & Microbe*, *14*(5), 591–599. https://doi.org/10.1016/j.chom.2013.10.007.

Li, T., Diner, B. A., Chen, J., & Cristea, I. M. (2012). Acetylation modulates cellular distribution and DNA sensing ability of interferon-inducible protein IFI16. *Proceedings of the National Academy of Sciences of the United States of America*, *109*(26), 10558–10563. https://doi.org/10.1073/pnas.1203447109.

Liu, H., Zhang, H., Wu, X., Ma, D., Wu, J., Wang, L., et al. (2018). Nuclear cGAS suppresses DNA repair and promotes tumorigenesis. *Nature*, *563*(7729), 131–136. https://doi.org/10.1038/s41586-018-0629-6.

Lobingier, B. T., Huttenhain, R., Eichel, K., Miller, K. B., Ting, A. Y., von Zastrow, M., et al. (2017). An approach to spatiotemporally resolve protein interaction networks in living cells. *Cell*, *169*(2), 350–360 e312. https://doi.org/10.1016/j.cell.2017.03.022.

Lukashchuk, V., & Everett, R. D. (2010). Regulation of ICP0-null mutant herpes simplex virus type 1 infection by ND10 components ATRX and hDaxx. *Journal of Virology*, *84*(8), 4026–4040. https://doi.org/10.1128/JVI.02597-09.

Lum, K. K., Song, B., Federspiel, J. D., Diner, B. A., Howard, T., & Cristea, I. M. (2018). Interactome and proteome dynamics uncover immune modulatory associations of the pathogen sensing factor cGAS. *Cell Systems*, 7(6), 627–642 e626. https://doi.org/10.1016/j.cels.2018.10.010.

Merkl, P. E., & Knipe, D. M. (2019). Role for a filamentous nuclear assembly of IFI16, DNA, and host factors in restriction of herpesviral infection. *mBio*, *10*(1), e02621-18. https://doi.org/10.1128/mBio.02621-18.

Miteva, Y. V., Budayeva, H. G., & Cristea, I. M. (2013). Proteomics-based methods for discovery, quantification, and validation of protein-protein interactions. *Analytical Chemistry*, *85*(2), 749–768. https://doi.org/10.1021/ac3033257.

Orzalli, M. H., Broekema, N. M., Diner, B. A., Hancks, D. C., Elde, N. C., Cristea, I. M., et al. (2015). cGAS-mediated stabilization of IFI16 promotes innate signaling during herpes simplex virus infection. *Proceedings of the National Academy of Sciences of the United States of America*, *112*(14), E1773–E1781. https://doi.org/10.1073/pnas.1424637112.

Orzalli, M. H., DeLuca, N. A., & Knipe, D. M. (2012). Nuclear IFI16 induction of IRF-3 signaling during herpesviral infection and degradation of IFI16 by the viral ICP0 protein. *Proceedings of the National Academy of Sciences of the United States of America*, *109*(44), E3008–E3017. https://doi.org/10.1073/pnas.1211302109.

Roux, K. J., Kim, D. I., & Burke, B. (2013). BioID: A screen for protein-protein interactions. *Current Protocols in Protein Science*, *74*, 1–14. https://doi.org/10.1002/0471140864.ps1923s74, Unit 19 23.

Seo, G. J., Kim, C., Shin, W. J., Sklan, E. H., Eoh, H., & Jung, J. U. (2018). TRIM56-mediated monoubiquitination of cGAS for cytosolic DNA sensing. *Nature Communications*, *9*(1), 613. https://doi.org/10.1038/s41467-018-02936-3.

Seo, G. J., Yang, A., Tan, B., Kim, S., Liang, Q., Choi, Y., et al. (2015). Akt kinase-mediated checkpoint of cGAS DNA sensing pathway. *Cell Reports*, *13*(2), 440–449. https://doi.org/10.1016/j.celrep.2015.09.007.

Sun, L., Wu, J., Du, F., Chen, X., & Chen, Z. J. (2013). Cyclic GMP-AMP synthase is a cytosolic DNA sensor that activates the type I interferon pathway. *Science*, *339*(6121), 786–791. https://doi.org/10.1126/science.1232458.

Tsujimura, H., Tamura, T., Kong, H. J., Nishiyama, A., Ishii, K. J., Klinman, D. M., et al. (2004). Toll-like receptor 9 signaling activates NF-kappaB through IFN regulatory factor-8/IFN consensus sequence binding protein in dendritic cells. *Journal of Immunology, 172*(11), 6820–6827.

Unterholzner, L., Keating, S. E., Baran, M., Horan, K. A., Jensen, S. B., Sharma, S., et al. (2010). IFI16 is an innate immune sensor for intracellular DNA. *Nature Immunology, 11*(11), 997–1004. https://doi.org/10.1038/ni.1932.

Wang, S., Wang, K., Lin, R., & Zheng, C. (2013). Herpes simplex virus 1 serine/threonine kinase US3 hyperphosphorylates IRF3 and inhibits beta interferon production. *Journal of Virology, 87*(23), 12814–12827. https://doi.org/10.1128/JVI.02355-13.

Weekes, M. P., Tomasec, P., Huttlin, E. L., Fielding, C. A., Nusinow, D., Stanton, R. J., et al. (2014). Quantitative temporal viromics: An approach to investigate host-pathogen interaction. *Cell, 157*(6), 1460–1472. https://doi.org/10.1016/j.cell.2014.04.028.

Wu, J. J., Li, W., Shao, Y., Avey, D., Fu, B., Gillen, J., et al. (2015). Inhibition of cGAS DNA sensing by a herpesvirus Virion protein. *Cell Host & Microbe, 18*(3), 333–344. https://doi.org/10.1016/j.chom.2015.07.015.

Wu, J., Sun, L., Chen, X., Du, F., Shi, H., Chen, C., et al. (2013). Cyclic GMP-AMP is an endogenous second messenger in innate immune signaling by cytosolic DNA. *Science, 339*(6121), 826–830. https://doi.org/10.1126/science.1229963.

Xia, P., Ye, B., Wang, S., Zhu, X., Du, Y., Xiong, Z., et al. (2016). Glutamylation of the DNA sensor cGAS regulates its binding and synthase activity in antiviral immunity. *Nature Immunology, 17*(4), 369–378. https://doi.org/10.1038/ni.3356.

Zhang, G., Chan, B., Samarina, N., Abere, B., Weidner-Glunde, M., Buch, A., et al. (2016). Cytoplasmic isoforms of Kaposi sarcoma herpesvirus LANA recruit and antagonize the innate immune DNA sensor cGAS. *Proceedings of the National Academy of Sciences of the United States of America, 113*(8), E1034–E1043. https://doi.org/10.1073/pnas.1516812113.

Zhang, J., Zhao, J., Xu, S., Li, J., He, S., Zeng, Y., et al. (2018). Species-specific deamidation of cGAS by herpes simplex virus UL37 protein facilitates viral replication. *Cell Host & Microbe, 24*(2), 234–248 e235. https://doi.org/10.1016/j.chom.2018.07.004.

CHAPTER TWENTY

TUBE and UbiCRest assays for elucidating polyubiquitin modifications in protein complexes

Swarupa Panda[a], Hui Jiang[a], Nelson O. Gekara[a,b,*]
[a]The Laboratory for Molecular Infection Medicine Sweden (MIMS), Umeå University, Umeå, Sweden
[b]Department of Molecular Biosciences, The Wenner-Gren Institute, Stockholm University, Stockholm, Sweden
*Corresponding author: e-mail address: nelson.gekara@su.se

Contents

Abstract

Ubiquitination is a reversible posttranslational modification that regulates nearly all cellular processes. The ubiquitin polypeptide is conjugated via its C-terminus to amine groups of lysine residues on target protein. Additionally, ubiquitins moieties can be conjugated in tandem to the initial ubiquitin via any of its internal lysine residues or N terminal methionine residue, resulting in the formation of polyubiquitin chains with distinct biophysical properties and biological functions. Elucidating the types of polyubiquitin chains present in proteins is essential for understanding their function and mechanism of regulation. Traditionally, ubiqutin modifications have been elucidated by exogenously co-expressing proteins of interest with epitope-tagged ubiquitins mutated in specific lysine residues. However, this strategy is prone experimental artifacts. In this protocol, we describe how to elucidate endogenous ubiquitin modifications. This

Methods in Enzymology, Volume 625
ISSN 0076-6879
https://doi.org/10.1016/bs.mie.2019.05.006

procedure combines TUBE (Tandem Ubiquitin Binding Entity)-based isolation of ubiquitin conjugates, digestion with linkage specific deubiquitinases and immunoblotting. This procedure is very robust can be applied to profile types and architectural organization polyubiquitin chains present on the any proteins of interest and has been instrumental in elucidating ubiquitin modifications in NOD2 signaling in our recent study (Panda & Gekara, 2018).

1. Introduction

Ubiquitin is a 8 kDa polypeptide highly conserved in eukaryotes. Ubiquitins can covalently be attached via its C-terminus to amine groups of lysine residues on target proteins. The ubiquitin molecule contains eight internal lysisne residues (K6, K11, K27, K29, K33, K48 and K63). During polyubiquitination, ubiquitins are tandemly conjugated to the initial ubiquitin via either of its internal lysines or the N-terminal methionine (M1). This culminates in the generation of polymeric ubiquitin chains with distinct conformations and functions. For example conjugation of ubiquitins via the K48 lysine results in the formation of K48-linked polymers that target proteins for proteasomal degradation (Hershko & Ciechanover, 1998). On the other hand, conjugation via the M1 or K63 leads to the formation of M1- (linear) or K63- polymers that act as scaffolding platforms for the assembly of signaling complexes (Komander & Rape, 2012; Swatek & Komander, 2016).

The current knowledge of the ubiquitin system mostly relates to homotypic polyubiquitins (contain single linkage type). However, on top of this complexity it has become clear that during polyubiquitination, ubiquitins can be attached to each other via two or more lysine residues resulting in the formation of heterotypic polyubiquitin chains with multiple linkages that adopt either mixed or branched topology (Hrdinka et al., 2016; Panda & Gekara, 2018). These layers of diversification in ubiquitin modifications exemplifies the versatility of ubiquitin signals in fine-tuning the regulation of various biological processes.

So far, studies aimed at understanding the types of polyubiquitin chains in proteins has mostly relied on exogenous overexpression of proteins of interest together with epitope-tagged ubiquitins, where one or several lysine residues are mutated to arginine. This can then be combined with pulldowns to identify which polyubiquitin chains are attached on specific proteins. Although valuable, observations made from this overexpression systems may not necessarily be reflective of the scenario under physiological settings in the specific cell that endogenously express the proteins under study. More recently, new reagents for the characterization of polyubiquitin modifications have been developed. Notable among them are the ubiquitin-linkage specific antibodies. While greatly contributing to our understanding of the roles and dynamics of polyubiquitin signals, thus far, linkage-specific antibodies are available for only five (M1, K11, K27, K48 and K63) of the eight linkage types (Matsumoto et al., 2010; Newton et al., 2008).

More recently, Tandem Ubiquitin Binding Entities (TUBEs) with a very high affinity for polyubiquitylated proteins have been developed (Aillet et al., 2012). Four types of TUBES are currently available in the market; the Non-Selective TUBEs (Pan-TUBEs), the M1 (linear), K48- and K63- linkage selective TUBEs. These reagents are very powerful tools for the isolation and characterization of polyubiquitylated proteins from cells or tissues. Another important development over the recent past has been the availability of linkage-specific DUBs and the realization that these enzymes can be useful tools for studying the ubiquitin modifications. In particular, treating ubiquitinated proteins with a panel of complementary linkage-specific DUBs in parallel reactions (a.k.a. Ubiquitin Chain Restriction (UbiCRest), followed by gel-based analysis can provide useful insights into the ubiquitin linkage types attached to proteins and their architecture (Hospenthal, Mevissen, & Komander, 2015; Mevissen et al., 2013; Panda & Gekara, 2018). In this chapter we describe how to identify the types of polyubiquitin chains present in proteins and to define their architectural organization. This protocol combining TUBE-based isolation of polyubiquitinated proteins, UbiCRest, SDS-PAGE and immunoblot analysis with specific antibodies has successfully applied to elucidate polyubiquitin dynamics in NOD2 signaling pathway (Panda & Gekara, 2018).

2. Isolation of ubiquitin conjugates by TUBE pulldown

2.1 Materials

Reagents

- Anti-ub Agarose- TUBE 2 (Lifesensors, cat. no. UM402)
- Trizma base (Sigma-Aldrich, cat. no. T1503)
- Sodium chloride solution (Sigma-Aldrich, cat. no. 71386)
- Ethylenediaminetetraacetic acid (Sigma-Aldrich, cat. no. 431788)
- Nonidet P 40 (Sigma-Aldrich, cat. no. 74385)
- Glycerol (Sigma-Aldrich, cat. no. G5516)

- N-ethylmaleimide (Sigma-Aldrich, cat. no. 04259)
  CAUTION N-ethylmaleimide is very toxic; avoid exposure.
- Iodoacetamide (Sigma-Aldrich, cat. no. I1149)
  CAUTION Iodoacetamide is toxic; avoid exposure
- PVDF transfer membrane (Millipore, cat. no. IPVH00010)
- NaF (Sigma-Aldrich, cat. no. 201154)
  CAUTION NaF is toxic; avoid exposure.
- Complete protease inhibitors, EDTA-free (Roche Diagnostics, cat. no. 04693132001)
- Primary bone marrow derived macrophages (BMDMs)
- NuPAGE Novex LDS Sample Buffer (Life Technologies, cat. no. NP0008)
- NuPAGE Novex MES SDS Running Buffer (20 ×) (Life Technologies, cat. no. NP0002)
- NuPAGE Novex MOPS SDS Running Buffer (20 ×) (Life Technologies, cat. no. NP0001)
- NuPAGE Novex 4–12% Bis-Tris Protein Gels, 1.0 mm (Life Technologies, cat. no. NP0323BOX (15 well), cat. no. WG1402BOX (20 well))
- MilliQ water

Preparation of buffers

TUBE lysis buffer. Prepare TUBE lysis buffer containing 20 mM Tris (pH7.4), 150 mM NaCl, 1 mM EDTA (ph.8), 1% (v/v) NP-40,10% glycerol,1 × EDTA free Complete protease inhibitor, 100 mM NaF, 5 mM *N*-ethylmaleimide (NEM) and 10 mM iodoacetamide. This buffer should be freshly prepared.

TUBE wash buffer. Prepare TUBE wash buffer containing 20 mM Tris (pH7.4), 150 mM NaCl, 1 mM EDTA, 1% (v/v) NP-40, 10% glycerol and 1 × EDTA free Complete protease inhibitor. This buffer should be freshly prepared.

20× TBST. Prepare a 20× stock of TBS by combining appropriate amounts of 19 mM Tris base, 137 mM NaCl and 2.7 mN KCl. Autoclave and store at room temperature. Prepare a working solution of TBS (1 ×, pH7.4) by diluting the 20 × stock in Milli-Q water. For TBST, add 1 mL of Tween 20 per liter of TBS.

2.2 Procedure

2.2.1 Equilibration of Agarose TUBE

- Allow TUBE 2 Agarose (UM402) to thaw at room temperature for 10–15 min.
- Gently mix TUBE 2 Agarose by inverting the vial several times to ensure a homogeneous suspension.

- Transfer a desired volume of resin (as 50% slurry) with a wide bore (cut) pipet tip to a 1.5 mL micro centrifuge tube. Larger volumes (e.g., 100 μL) of resin can be equilibrated and stored for up to 1 week at 4 °C.
- Collect resin by low speed centrifugation (1000–5000 × *g*) for 5 min at room temperature (RT).
- Remove and discard the supernatant.

  Critical step: When working with agarose matrices, use gel-loading tips to collect supernatants to avoid losing beads.
- Suspend the resin in 5–10 volumes of 1 × TBST and incubate for 5 min on a rocker platform at RT.
- Repeat collection/wash at least two times prior to pull down.
- Remove and discard the final wash supernatant making sure to leave 50% slurry.

2.2.2 Pulldown of poly-ubiquitinated proteins

- Pre-chill inhibitor containing cell lysis buffer and micro centrifuge tubes to 4 °C.
- Treat and wash BMDM cells (2 × with 1 × ice-cold PBS) appropriately as an initial starting point, add 500 μL lysis buffer to 10 × 10^6 cells (BMDMs).
- Collect the cells by scraping and transfer the lysate to 1.5 mL tube, incubate on ice for 30 min.
- Clarify the lysate by high speed centrifugation (~14,000 × *g*) for 10 min at 4 °C.
- Remove an "Input" sample for analysis by western blotting.
- Add the amount of cell lysate to the amount of equilibrated Agarose TUBEs and incubate for O/N on a rocker platform at 4 °C.
- Collect beads by low speed centrifugation (1000–5000 × *g*) at 4 °C for 5 min. Save supernatant as the "Unbound Fraction".
- Wash beads with 1 mL of 1 × TBST, collect by low speed centrifugation and aspirate the supernatant leaving a small volume cushion. So as to avoid disturbing the beads.
- Repeat twice (1000–5000 × *g*, 4 °C) for 5 min.
- For western blot analysis, re-suspend the resin in LDS reducing sample buffer, boiled at 100 °C for 5 min.
- Clarify the purified protein by centrifugation (1000–5000 × *g*) at 4 °C for 5 min.
- Analyze the eluted samples by SDS-PAGE/Western blotting in parallel with input or sample can be stored at −20 °C for later use.

3. Deubiquitinase based assay to determine ubiquitin linkages and their architecture

3.1 Materials

Reagents

- UbiCREST Deubiquitinase Enzyme Kit (Boston biochem, cat. no. K-400)
- Trizma base (Sigma-Aldrich, cat. no. T1503)
- Sodium chloride solution (Sigma-Aldrich, cat. no. 71386)
- Ethylenediaminetetraacetic acid (Sigma-Aldrich, cat. no. 431788)
- Nonidet P 40 (Sigma-Aldrich, cat. no. 74385)
- Glycerol (Sigma-Aldrich, cat. no. G5516)
- PR-619 (Lifesensors, cat.no SI9619)
- N-ethylmaleimide (Sigma-Aldrich, cat. no. 04259)
  CAUTION N-ethylmaleimide is very toxic; avoid exposure
- Iodoacetamide (Sigma-Aldrich, cat. no. I1149)
  CAUTION Iodoacetamide is toxic; avoid exposure
- PVDF transfer membrane (GE Healthcare, cat. no. 10600023)
- NaF (Sigma-Aldrich, cat. no. 201154)
  CAUTION NaF is toxic; avoid exposure
- Complete protease inhibitors, EDTA-free (Roche Diagnostics, cat. no. 04693132001)
- Primary bone marrow derived macrophages (BMDMs)
- NuPAGE Novex LDS Sample Buffer (Life Technologies, cat. no. NP0008)
- NuPAGE Novex MES SDS Running Buffer (20 ×) (Life Technologies, cat. no. NP0002)
- NuPAGE Novex MOPS SDS Running Buffer (20 ×) (Life Technologies, cat. no. NP0001)
- NuPAGE Novex Tris Acetate SDS Running Buffer (20 ×) (Life Technologies, cat. no. LA0041)
- NuPAGE Novex 4–12% Bis-Tris Protein Gels, 1.0 mm (Life Technologies, cat. no. NP0323BOX (15 well), cat. no. WG1402BOX (20 well))
- NuPAGE Novex 3–8% Tris-Acetate Midi Protein Gels, 1.0 mm (Life Technologies, cat. no. WG1602BOX (20 well))
- MilliQ water

Preparation of buffers

UbiCRest lysis buffer. Prepare UbiCRest lysis buffer containing 20 mM Tris (pH 7.4), 150 mM NaCl, 1 mM EDTA (pH.8), 1% (v/v) NP-40,

10% glycerol, 1 × EDTA free Complete protease inhibitor, 100 mM NaF, 20 μM PR-619, 5 mM *N*-ethylmaleimide (NEM) and 10 mM iodoacetamide. This buffer should be freshly prepared.

UbiCRest wash buffer. Prepare UbiCRest wash buffer containing 20 mM Tris (pH 7.4), 150 mM NaCl, 1 mM EDTA, 1% (v/v) NP-40, 10% glycerol and 1 × EDTA free Complete protease inhibitor. This buffer should be freshly prepared.

10 × DUB reaction buffer. Prepare 10 × DUB reaction buffer containing 500 mM Tris (pH 7.5), 500 mM Nacl and 50 mM DTT. This buffer can be stored at −20 °C for at least 6 months to 1 year.

DUB dilution/assay buffer. Prepare DUB dilution/assay buffer containing 500 mM Tris (pH 7.5), 500 mM NaCl and 10 mM DTT. This buffer can also be stored at −20 °C for at least 6 months to 1 year.

3.1.1 Linkage specificity of DUBs

The DUBs that will be used in the UbiCRest analysis have the following linkage specificity.

DUBs	Conc	Type of linkage	References
USP2	0.5 μM	Cleave all eight linkages Positive control	Ye et al. (2011)
OTUD3	3 μM	K6 and K11 polyubiquitin chains	Hospenthal, Freund, and Komander (2013) and Mevissen et al. (2013)
CEZZANE	2 μM	K11 polyubiquitin chains	Bonacci et al. (2018) and Bremm, Freund, and Komander (2010)
UCHL3	3 μM	K27 polyubiquitin chains	Pan et al. (2018) and Panda and Gekara (2018)
TRABID	3 μM	K29, K33 and K63 polyubiquitin chains	Evans et al. (2001), Kristariyanto et al. (2015), and Licchesi et al. (2011)
OTUB1	10 μM	K48 polyubiquitin chains	Mevissen et al. (2013) and Wang et al. (2009)
AMSH	3 μM	K63 polyubiquitin chains	Komander et al. (2009) and Sato et al. (2008)
OTULIN	1 μM	Linear (M1) polyubiquitin chains	Fiil et al. (2013) and Keusekotten et al. (2013)

3.2 Procedure

3.2.1 Purification of ubiquitinated substrates for UbicRest analysis

Here we describe isolation of the NOD2-RIP2 receptor complex as an example for UbiCRest analysis.

1. Seed WT BMDMs at a density of 5×10^6 cells per dish into two 10 cm dishes and culture them in the presence of 10 mL of complete cell culture medium (IMDM).
2. Next day morning, stimulate cells with 200 ng/mL L18-MDP (NOD2 ligand) for 60 min at 37 °C.
3. After 1 h, scrape cells off the 10 cm dishes and transfer them to prechilled 50 mL falcon tubes.

   Critical step: Don't trypsinize the cells, as surface receptors would get destroyed.
4. Centrifuge the cells at 300 × *g* for 2 min at 4 °C.
5. Remove the supernatant and wash cells once with 2 mL of ice-cold PBS.
6. Remove PBS, add ice-cold UbiCRest lysis buffer and lyse the cells on ice for 30 min.
7. Centrifuge the lysate at 16,000 × *g* for 30 min at 4 °C.
8. To the clarified lysate, add 125 μL/ of Agarose TUBE 2, pre-equilibrated by washing in cold TUBE-w buffer (without DUB inhibitor).
9. Then incubate the mixture on a rotating wheel at 4 °C O/N.
10. The following day, wash the beads five times. Each wash done by centrifugation the beads 1000 × *g* for 3 min, removing the supernatant and adding 1 mL of UbiCRest wash buffer.

3.2.2 Ubiquitin restriction analysis

11. Resuspend the beads in 1 × *DUB reaction buffer.*
12. Transfer the beads into nine separate tubes for subsequent DUB incubations, 50 μL (~45 μL) of beads were aliquoted per reaction.
13. Prepare 5 × concentrated DUBs stock on ice by diluting enzymes using *DUB dilution buffer* and pre-incubate them at RT (22 °C) for 10–15 min. The following final enzymes were used (USP2 [0.5 μM], OTUD3 [3 μM], Cezanne [2 μM], UCHL3 [3 μM], Trabid [3 μM], OTUB1 [10 μM], AMSH [3 μM], OTULIN [1 μM]).

    Critical step: Pre-incubation of the enzymes with DUB dilution buffer containing DTT is required to achieve the full activity.

14. Then add 5 μL of 5 × concentrated DUB stock to their respective beads, which finally give 1 × concentrated DUBs in a volume of 50 μL.
15. Place the samples on a dry heat block at 37 °C for 1–3 h
16. Stop the reaction after 1–3 h by adding 5 μL of LDS sample buffer.
17. Once the samples are mixed with LDS sample buffer, they can be stored at −20 °C and analyze by SDS-PAGE at a later time.

4. Sample analysis

By using these protocols, the visualization of a polyubiquitin chains or a polyubiqutinated substrate can be achieved by using the following technique.

4.1 Western blotting using specific antibodies

Linkage-specific antibodies (K63, K48, K27 and M1) and tagged ubiquitins (such as FLAG-ubiquitins and HA-ubiquitins) are available and can be used to evaluate the levels of specific polyubiquitin chains in samples (Fig. 1). Alternatively, following ubiquitin restriction (UbiCRest) with different DUBs samples can be separated by SDS-PAGE and immunoblotted with

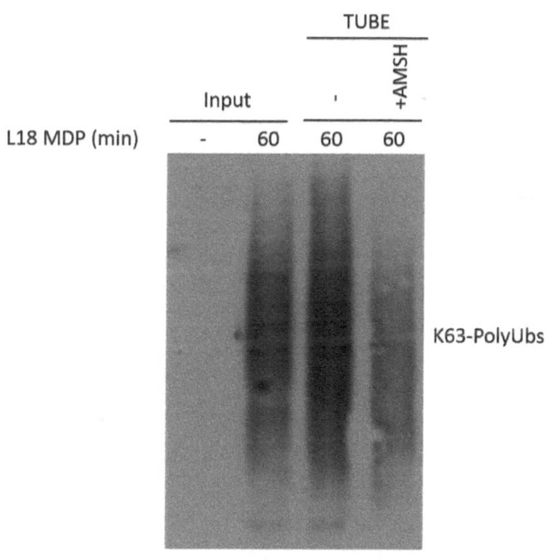

Fig. 1 UbiCRest assay to elucidate NOD2-mediated K63-linked polyubiquitination. TUBE-pulldowns from BMDMs stimulated with L18 MDP (for 60 min) were treated with AMSH (a K63-specific DUB) then analyzed by immunoblotting with an anti-K63 antibody. Input: total lysates from control cells (0) or stimulated cells (60).

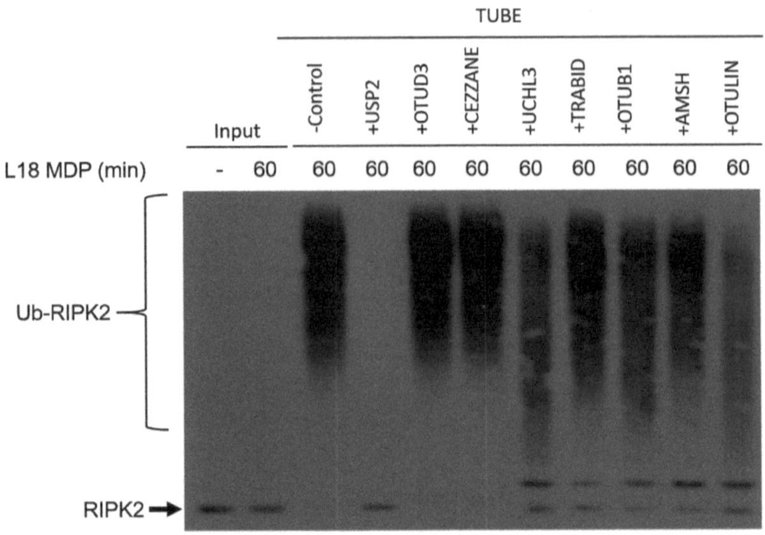

Fig. 2 UbiCRest analysis to determine type of ubiquitin linkages on a specific substrate (RIPK2). TUBE-pulldowns from BMDMs stimulated with L18 MDP (for 60 min) were treated with indicated DUBS then analyzed by immunoblotting with an anti-RIPK2. The sensitivity of RIPK2 ubiquitin conjugates to UCHL3 (K27), OTUB1 (K48), AMSH (K63), OTULIN (M1) is indicative of the presence indicated ubiquitin linkages. Input: total lysates from control cells (0) or stimulated cells (60).

antibodies against specific proteins or linkages of interest. Polyubiquitinated proteins migrate at high molecular weight. Therefore, estimating the molecular weight of specific proteins and their sensitivity to different linkage-specific DUBs, can provide insights into the type of polyubiquitin chains attached to the proteins of interest (Fig. 2).

5. Strengths and limitations

UbiCRest is a quick and easy method to detect substrate ubiquitination, the type of linkages present on the substrate and their architecture. It does not require any specialized instruments and hardly takes 1–3 h depending upon the reaction set up and concentration of enzymes used. Most linkage types can be distinguished using linkage specific DUBs. This assay is particularly useful for the detecting K6, K11, K27, K48, K63 and Linear ubiquitin linkages since highly specific DUBs for these linkage types are commercially available.

UbiCRest has a number of limitations that need to take into consideration. Some DUBs may be unspecific at higher concentrations (Hospenthal et al., 2013; Keusekotten et al., 2013). Therefore careful titration of the DUBs, for example using linkage specific Di-ubiquitins is essential.

Another issue is that some DUBs commoly used in UbiCRest, for example TRABID and OTUD3 targets more than one ubiquitin linkages in homotypic as well as heterotypic chains. Moreover, complex branched/mixed structure are likely less amenable to DUBs. Hence where necessary, validation by other approaches such as mass spectrometry or the use of ubiquitin mutants should be considered on a case by case basis.

Acknowledgments

This work was supported by the Laboratory for Molecular Infection Medicine Sweden (MIMS), Umeå University, Stockholm University, the Swedish Research Council (grants 2015-02857 and 2016-00890 to N.O.G.) and the Swedish Cancer Foundation (grant, CAN 2017/421 to N.O.G.).

References

Aillet, F., Lopitz-Otsoa, F., Hjerpe, R., Torres-Ramos, M., Lang, V., & Rodriguez, M. S. (2012). Isolation of ubiquitylated proteins using tandem ubiquitin-binding entities. *Methods in Molecular Biology*, *832*, 173–183.

Bonacci, T., Suzuki, A., Grant, G. D., Stanley, N., Cook, J. G., Brown, N. G., et al. (2018). Cezanne/OTUD7B is a cell cycle-regulated deubiquitinase that antagonizes the degradation of APC/C substrates. *The EMBO Journal*, *37*, e98701.

Bremm, A., Freund, S. M., & Komander, D. (2010). Lys11-linked ubiquitin chains adopt compact conformations and are preferentially hydrolyzed by the deubiquitinase Cezanne. *Nature Structural & Molecular Biology*, *17*, 939–947.

Evans, P. C., Taylor, E. R., Coadwell, J., Heyninck, K., Beyaert, R., & Kilshaw, P. J. (2001). Isolation and characterization of two novel A20-like proteins. *The Biochemical Journal*, *357*, 617–623.

Fiil, B. K., Damgaard, R. B., Wagner, S. A., Keusekotten, K., Fritsch, M., Bekker-Jensen, S., et al. (2013). OTULIN restricts Met1-linked ubiquitination to control innate immune signaling. *Molecular Cell*, *50*, 818–830.

Hershko, A., & Ciechanover, A. (1998). The ubiquitin system. *Annual Review of Biochemistry*, *67*, 425–479.

Hospenthal, M. K., Freund, S. M., & Komander, D. (2013). Assembly, analysis and architecture of atypical ubiquitin chains. *Nature Structural & Molecular Biology*, *20*, 555–565.

Hospenthal, M. K., Mevissen, T. E. T., & Komander, D. (2015). Deubiquitinase-based analysis of ubiquitin chain architecture using ubiquitin chain restriction (UbiCRest). *Nature Protocols*, *10*, 349–361.

Hrdinka, M., Fiil, B. K., Zucca, M., Leske, D., Bagola, K., Yabal, M., et al. (2016). CYLD limits Lys63- and Met1-linked ubiquitin at receptor complexes to regulate innate immune signaling. *Cell Reports*, *14*, 2846–2858.

Keusekotten, K., Elliott, P. R., Glockner, L., Fiil, B. K., Damgaard, R. B., Kulathu, Y., et al. (2013). OTULIN antagonizes LUBAC signaling by specifically hydrolyzing Met1-linked polyubiquitin. *Cell, 153*, 1312–1326.

Komander, D., & Rape, M. (2012). The ubiquitin code. *Annual Review of Biochemistry, 81*, 203–229.

Komander, D., Reyes-Turcu, F., Licchesi, J. D., Odenwaelder, P., Wilkinson, K. D., & Barford, D. (2009). Molecular discrimination of structurally equivalent Lys 63-linked and linear polyubiquitin chains. *EMBO Reports, 10*, 466–473.

Kristariyanto, Y. A., Abdul Rehman, S. A., Campbell, D. G., Morrice, N. A., Johnson, C., Toth, R., et al. (2015). K29-selective ubiquitin binding domain reveals structural basis of specificity and heterotypic nature of k29 polyubiquitin. *Molecular Cell, 58*, 83–94.

Licchesi, J. D., Mieszczanek, J., Mevissen, T. E., Rutherford, T. J., Akutsu, M., Virdee, S., et al. (2011). An ankyrin-repeat ubiquitin-binding domain determines TRABID's specificity for atypical ubiquitin chains. *Nature Structural & Molecular Biology, 19*, 62–71.

Matsumoto, M. L., Wickliffe, K. E., Dong, K. C., Yu, C., Bosanac, I., Bustos, D., et al. (2010). K11-linked polyubiquitination in cell cycle control revealed by a K11 linkage-specific antibody. *Molecular Cell, 39*, 477–484.

Mevissen, T. E., Hospenthal, M. K., Geurink, P. P., Elliott, P. R., Akutsu, M., Arnaudo, N., et al. (2013). OTU deubiquitinases reveal mechanisms of linkage specificity and enable ubiquitin chain restriction analysis. *Cell, 154*, 169–184.

Newton, K., Matsumoto, M. L., Wertz, I. E., Kirkpatrick, D. S., Lill, J. R., Tan, J., et al. (2008). Ubiquitin chain editing revealed by polyubiquitin linkage-specific antibodies. *Cell, 134*, 668–678.

Pan, M., Zheng, Q., Ding, S., Zhang, L., Qu, Q., Wang, T., et al. (2018). Chemical protein synthesis-enabled mechanistic studies on the molecular recognition of K27-linked ubiquitin chains. *Angewandte Chemie (International Ed. in English), 58*, 2627–2631.

Panda, S., & Gekara, N. O. (2018). The deubiquitinase MYSM1 dampens NOD2-mediated inflammation and tissue damage by inactivating the RIP2 complex. *Nature Communications, 9*, 4654.

Sato, Y., Yoshikawa, A., Yamagata, A., Mimura, H., Yamashita, M., Ookata, K., et al. (2008). Structural basis for specific cleavage of Lys 63-linked polyubiquitin chains. *Nature, 455*, 358–362.

Swatek, K. N., & Komander, D. (2016). Ubiquitin modifications. *Cell Research, 26*, 399–422.

Wang, T., Yin, L., Cooper, E. M., Lai, M. Y., Dickey, S., Pickart, C. M., et al. (2009). Evidence for bidentate substrate binding as the basis for the K48 linkage specificity of otubain 1. *Journal of Molecular Biology, 386*, 1011–1023.

Ye, Y., Akutsu, M., Reyes-Turcu, F., Enchev, R. I., Wilkinson, K. D., & Komander, D. (2011). Polyubiquitin binding and cross-reactivity in the USP domain deubiquitinase USP21. *EMBO Reports, 12*, 350–357.

FSC
www.fsc.org
MIX
Papier | Fördert gute Waldnutzung
FSC® C083411

Kolibri 360 GmbH
Ferdinand-Jühlke-Straße 7
99095 Erfurt, Deutschland
produktsicherheit@kolibri360.de